Archimedes

NEW STUDIES IN THE HISTORY AND PHILOSOPHY OF SCIENCE AND TECHNOLOGY

VOLUME 56

Archimedes has three fundamental goals; to further the integration of the histories of science and technology with one another: to investigate the technical, social and practical histories of specific developments in science and technology; and finally, where possible and desirable, to bring the histories of science and technology into closer contact with the philosophy of science. To these ends, each volume will have its own theme and title and will be planned by one or more members of the Advisory Board in consultation with the editor. Although the volumes have specific themes, the series itself will not be limited to one or even to a few particular areas. Its subjects include any of the sciences, ranging from biology through physics, all aspects of technology, broadly construed, as well as historically-engaged philosophy of science or technology. Taken as a whole, *Archimedes* will be of interest to historians, philosophers, and scientists, as well as to those in business and industry who seek to understand how science and industry have come to be so strongly linked.

More information about this series at http://www.springer.com/series/5644

Martin Jähnert

Practicing the Correspondence Principle in the Old Quantum Theory

A Transformation through Implementation

Springer

Martin Jähnert
Institut für Philosophie, Literatur-,
Wissenschafts- und Technikgeschichte
Fachbereich Wissenschaftsgeschichte,
Technische Universität Berlin
Berlin, Germany

Max Planck Institute
for the History of Science
Berlin, Germany

Zugl. Berlin, Technische Universität, Diss., 2016

ISSN 1385-0180 ISSN 2215-0064 (electronic)
Archimedes
ISBN 978-3-030-13299-6 ISBN 978-3-030-13300-9 (eBook)
https://doi.org/10.1007/978-3-030-13300-9

Library of Congress Control Number: 2019933897

This Springer imprint is published by the registered company Springer Nature Switzerland AG.
The registered company address is: Gewerbestrasse 11, 6330 Cham, Switzerland

Acknowledgments

It is a pleasure to have the opportunity to acknowledge and to thank those who generously have supported and who have accompanied me in my research for this book. I thank Friedrich Steinle at the Technische Universität Berlin and Jürgen Renn at the Max Planck Institute (MPI) for the History of Science, Berlin, for their advice and support in all stages of my dissertation. Their constant intellectual as well as institutional support, their personal encouragement and their patience made this book possible. I thank Alexander S. Blum with whom I discussed the book in its various stages during countless reading sessions and lunch breaks. His detailed comments on the entire manuscript have been much enjoyed and equally appreciated.

Likewise, conversations and heated debates in the project "History and Foundations of Quantum Physics" have given me invaluable food for thought. Especially, I want to thank Christoph Lehner and Christian Joas. Furthermore, I am grateful for the countless discussions with my colleagues in the project "History and Foundations of Quantum Physics" and at the Technische Universität Berlin. I want to thank Finn Aaserud, Guido Bacciagaluppi, Massimiliano Badino, Arianna Borrelli, Elise Crull, Anthony H. Duncan, Michael Eckert, Markus Ehberger, Clayton A. Gearhart, Dieter Hoffmann, Jeremiah James, Michel Janssen, Marta Jordi Taltavull, Shaul Katzir, Roberto Lalli, Daniela Monaldi, Jaume Navarro, Robert Rynasiewicz, Matthias Schemmel, Suman Seth, and Adrian Wüthrich. I also want to thank John Heilbron, Súli Sigurdsson, Anke te Heesen, and everyone else who has attended one of my talks or read parts of my manuscript. I thank my editor Jed Z. Buchwald for including me in the series Archimedes. He and Tilman Sauer provided thorough and helpful feedback on my manuscript, for which I am deeply grateful.

Financial support from several organizations allowed me to write my dissertation and then turn it into this book. I want to thank the Max Planck Society and Jürgen Renn's Department I at the MPI for the History of Science for a predoctoral scholarship and for a wrap-up position, the German-Israeli Foundation for a predoctoral scholarship, the Berlin Center for the History of Science for a postdoctoral fellowship, as well as the American Institute of Physics, the History of Science Society, and the German Academic Exchange Service (DAAD) for travel grants.

I am grateful for the authorization to consult archives and to publish quotations from holdings of American Philosophical Society (Philadelphia), the Niels Bohr Archive (Copenhagen), University of Chicago Library, Albert-Ludwigs-Universität Freiburg, Niedersächsische Staats-und Universitätsbibliothek Göttingen and archive of the Deutsches Museum (Munich). I thank Urs Schoepflin, Esther Chen, Ellen Garske, Ruth Kessentini, Matthias Schwerdt, and all other members or former members of staff at the MPI's library for their invaluable work. Similarly, I am grateful to the terrific administrative, linguistic and diplomatic support from Lindy Divarci, Petra Schröter, Shadiye Leather-Barrow, and Nina Krampitz.

Last but not least, I wish to thank my parents, family and friends for their unconditional support. For a while, Alrun and I anticipated that our son's first word might be "Korrespondenzprinzip" – it wasn't; however, 2-year-old Anton found his own clever comment on the topic, which might have amused one or the other protagonist in this story, too: "Das mag ich nicht. Da ist Prinzip drin." Thank you for this and everything else.

Contents

List of Figures

List of Tables

List of Tables

Chapter 1
Introduction

In the summer of 1923, Niels Bohr gave a lecture at Harvard University on the quantum theory of the atom. He had done so many times during the 1920s in Germany, Britain, the Netherlands, and now also in the U.S. For almost 300 listeners in the audience, Bohr's presentation was anything but an exposition of well-known ideas. At least the professor for theoretical physics at Harvard, Edwin Kemble, thought so, as he would give another talk "to restate Bohr a little more slowly and simply" shortly after Bohr had left. In the very first sentence of his talking notes, Kemble formulated the central theme of his talk and explained: "the correspondence principle is the important tool."[1]

This book presents a history of this tool. Looking at detailed case studies on Arnold Sommerfeld and his work on multiplet intensities in Munich, on James Franck and Friedrich Hund and their work on the Ramsauer effect in Göttingen, and on Fritz Reiche and his work on the quantum theory of radiation in Breslau, it studies what physicists called the "applications of the correspondence principle." Examining these applications to different phenomena, I study the implementation of a research tool as an interaction between a theoretical tool and the scientific problems addressed by it. I argue that the use of this theoretical tool changed the tool itself. This process which I call *transformation through implementation*

[1]Bohr's lecture was announced in *The Harvard Crimson* ("Spectra and Atoms to be Bohr's Topic," *The Harvard Crimson*, 25 October 1923) and reported to have 300 listeners the next day ("Professor Bohr tells how he invented the model atom," *The Harvard Crimson*, 26 October 1923). For Bohr's one-page note see BSM 11.3; for Kemble's talking points see AHQP 55.2. Kemble's identification of the correspondence principle as a tool is exceptionally clear and explicit. This is not surprising, given that Kemble was a student of Percy Bridgman and closely followed his pragmatist position. See Schweber (1986) and Assmus (1990, 1992a,b). The pragmatic understanding of the principle and its function, which Kemble made explicit, was not limited to the U.S. Physicists working in Europe—mostly Germany and the Netherlands—largely refrained from speaking about the correspondence principle as a "Werkzeug," "werktuig," or "tool." Nonetheless they used it as such rather than focusing on its development as a fundamental principle.

M. Jähnert, *Practicing the Correspondence Principle in the Old Quantum Theory*, Archimedes 56, https://doi.org/10.1007/978-3-030-13300-9_1

was responsible for the conceptual development of the correspondence principle in highly local contexts from 1922 onwards and played a central role for the development of quantum mechanics by Werner Heisenberg in 1925.

1.1 Quantum Physics in the 1920s: The *Patchwork of Problems*, Research Tools and the State-Transition Model

The thesis of *transformation through implementation* and the approach to the history of the correspondence principle rely on a set of historiographical presuppositions. Most importantly, it builds on the assumption that the old quantum theory or rather quantum physics prior to 1925 is best understood as a social as well as an epistemic network and can be described as a *patchwork of problems*.[2]

Within this network, physicists' everyday practice focused on specific physical problems, which were strongly tied to particular actors and their academic institutions: Multiplet spectra were central for physicists like Sommerfeld and his students in Munich, optical dispersion was the problem occupying Rudolph Ladenburg and Fritz Reiche in Breslau, collisions of electrons with gas atoms were studied by James Franck in Göttingen. As such, they formed a *patchwork of problems* or a network of research fields.

To tackle these problems, physicists gradually came to adopt a shared description of quantum systems, which emerged with Bohr's trilogy in 1913 and which was codified as the core of his quantum theory of multiply periodic systems in 1918. This description rested on two fundamental assumptions of the Bohr model: The first is that a physical system has a set of stationary states. The second is that changes in the system take place in transitions from one such state to another.[3]

As Bohr pointed out as early as 1918, these basic assumptions were not sufficient to tackle a wider range of physical phenomena.[4] To do so, additional assumptions

---

[2]See Renn (2013). These assumptions emerged in the context of the project *History and Foundations of Quantum Physics*, which was undertaken jointly at the *Max Planck Institute for the History of Science* and the *Fritz Haber Institute* and aimed to understand the conceptual development of quantum physics in a new way.

[3]This is not to say that this description was universally accepted, nor that it became the predominant one after Bohr's seminal papers from 1913. Especially between 1913 and 1916, as Olivier Darrigol, Alexi Assmus, and Marta Jordi Taltavull have pointed out, this was not the case. Bohr himself thought in 1914 that the Zeeman effect could not be explained on the basis of transitions between stationary states and only adopted the universality of the frequency condition in 1915. Similar developments can be observed in the case of molecular spectroscopy, dispersion or in the case of collisions between electrons and atoms. For these phenomena, a description in terms of states and transitions did not appear as a natural description in the 1910s. The conviction that quantum systems should be described in terms of states and transitions emerged only gradually throughout the 1910s.

[4]Bohr (1918a, 5–6). See also Bohr's extensive discussion of the relation between his fundamental assumptions and classical mechanics and electrodynamics in Bohr (1923b).

were needed and physicists took up a wide range of theoretical resources from different domains when working on the *patchwork of problems*. In the quantum theory of multiply periodic systems, for example, Bohr used classical mechanics and quantum conditions. Sommerfeld took up the classificatory term schemes of spectroscopy in his work on multiplet spectroscopy, while Reiche developed his arguments on the basis of Einstein's quantum theory of radiation, and Franck and Hund used a qualitative classical description of scattering in their work on the Ramsauer effect.[5]

To describe this situation, I believe that both the term "Bohr model" and the term "quantum theory" are insufficient. The former is too narrow since it implies that Bohr, Sommerfeld or Reiche operated with the same spatiotemporal model rather than following very different approaches. The latter suggests a coherent and general theoretical structure, which the historical actors did not have at their disposal. To avoid both pitfalls, I use the analytic term *state-transition model* to describe the general conception of quantum systems. This term highlights that there was a shared basis, on which various phenomena in the *patchwork of problems* became interconnected. At the same time, it is possible to discuss the instantiations of this model, which depended on the specific approach to the respective phenomenon and thus remained separated operationally.

From this perspective, the solution of a particular problem in the patchwork of quantum physics depended on the operationalization of the state-transition model on the basis of different theoretical resources. The question thus becomes how these different resources were transferred between different empirical and theoretical domains, how new techniques were integrated into existing representations of the state-transition model and interacted with other theoretical resources. This leads to an analysis of the way in which the implementation of a theoretical tool reshaped the tool, the conception of the respective research problem, and the state-transition model.

[5]For example, the approach first taken in Bohr's 1913 trilogy and then prolonged in its extensions up to 1918 utilized the techniques of classical mechanics like Hamilton-Jacobi theory or perturbation theory, used quantum conditions to select quantum states from classical solutions, and applied the frequency condition to account for spectral lines. Most of these techniques, by contrast, appeared in a quite different capacity or played no constructive role at all in Albert Einstein's work on the quantum theory of black-body radiation, or in the work on multiplet spectroscopy by Sommerfeld or Alfred Landé. Both cases left assumptions on the mechanical description of stationary states and their quantization aside and instead built their arguments on different specifications of the state-transition model: Einstein, on the one hand, assumed that atoms were distributed over their stationary states according to statistical mechanics and that transitions between these states obeyed the frequency condition. Introducing probabilities for the transitions into the state-transition model, he derived Planck's radiation law. Sommerfeld, on the other hand, interpreted spectroscopic terms as energy levels of the stationary states and their combinations in terms of the frequency condition. Without using Bohr's or Einstein's approaches, he introduced a new quantum number as a label for the different states and selection rules to describe the possible transitions.

1.2 The Correspondence Principle in the Historiography of the Old Quantum Theory

The present characterization of quantum physics in terms of the *patchwork of problems*, the *state-transition model* and the implementation of theoretical tools has important implications for the historiographic choices underlying this book. It means to break with the standard narrative of the old quantum theory, which dates back at least to the 1950s and the 1960s when the generation of physicists around Werner Heisenberg, Max Born, and Wolfgang Pauli actively engaged in writing their own history. They established a narrative of the history of quantum physics that has persisted to this day in historical essays in physics textbooks and philosophical reflections on theory change, and also structures the lines along which histories of quantum physics have been written. In this standard account, the history of quantum physics was the history of a theoretical development, whose main hallmarks were the emergence of the Bohr model, its extension into the old quantum theory, its crisis, and ultimately the revolutionary creation of quantum mechanics.[6]

This book moves away from this narrative of a linear theoretical development and looks at diverse phenomena within the *patchwork of problems*, considering the approaches taken to them as relatively autonomous from each other. It therefore follows multiple developments and studies the implementations of the correspondence principle as a research tool in each of them.

Moreover, the focus of this book rests thoroughly on the analysis of theoretical practices and seeks to understand conceptual development as a result of it. This perspective, focusing on physical problems and tools for their solution, is not new. Already in 1966, Max Jammer noted that:

> Every single quantum-theoretic problem had to be solved first in terms of classical physics; its classical solution had then to pass through the mysterious sieve of the quantum conditions or, as it happened in the majority of cases, the classical solution had to be translated into the language of quanta in conformance with the correspondence principle. Usually, the process of finding "the correct translation" was a matter of skillful guessing and intuition rather than of deductive and systematic reasoning. In fact, quantum theory became the subject of a special craftsmanship or even artistic technique which was cultivated to the highest possible degree of perfection in Göttingen and in Copenhagen.[7]

For Jammer, craftsmanship and artistic technique had nothing to do with "deductive and systematic reasoning" and, as it seems, they had nothing to do with sound physics at all. Consequently, the old quantum theory appeared to be a hopelessly deficient theoretical framework and presented an incoherent mix of classical and quantum concepts or a "halfway house"—to borrow a term used by Anthony

[6]For a brief summary of the standard account, see Kragh (2002, 155–163) or Darrigol (2010). For book-length expositions of the standard account, see Jammer (1966), Mehra and Rechenberg (1982a,b) and Darrigol (1992). For a more nuanced differentiation of research strands in quantum physics followed in Munich, Copenhagen, and Göttingen, see Seth (2007, 2013).

[7]Jammer (1966, 196).

Duncan and Michel Janssen—between classical physics and the new quantum mechanics.[8]

This book accepts that physicists used a set of tools to solve problems in an eclectic way. Rather than following Jammer's diagnosis, however, it seeks to analyze the "skillful guessing." The lack of internal consistency is reduced to the question as to whether and how historical actors explicitly or implicitly hit upon limitations of their approaches and how they reacted to them. The main objectives are to spell out how quantum theoretical arguments were made, what technical resources went into it and to what ends they were put to use.

Studying the *practice of theory* of quantum physics prior to 1925, this book connects to the vast literature on theoretical practices of the last decades—to mention just a few examples—by Andrew Warwick, Ursula Klein, David Kaiser and by Suman Seth, Christian Joas and Jeremiah James, who dealt specifically with the history of quantum physics.[9] These studies built on central themes of science studies and the historiography of experimentation from the 1980s, which focused on scientific practices. They have analyzed theoretical practices as based on tacit knowledge, skills and calculational techniques that often needed to be acquired in face-to-face training.

What makes these studies relevant for the present discussion is their focus on theoretical tools and their circulation, application and manipulation. In this respect, Ursula Klein's analysis of "paper tools" and David Kaiser's extension of it are particularly important.[10] On the one hand, Kaiser's work on Feynman diagrams led me to think about how the correspondence principle traveled from Copenhagen to other places in the network of quantum theory in the 1920s, and which kind of introduction to it was necessary to make correspondence arguments.[11] On the other hand, Ursula Klein's analysis stresses the importance of visualization and materialization on paper for reasoning with paper tools. This raised the question of what exactly was put on paper in the case of the correspondence principle and what this meant for its development.[12] Answers to these questions have to be empirical ones and will be given throughout this book in relation to studies on *practice of theory*.

Writing the history of theoretical tools—as Kaiser has put it—"cuts orthogonally across conceptual histories of theoretical physics."[13] At the same time, tools are embedded in certain conceptual frameworks and have to be applied to some kind of

---

[8]Duncan and Janssen (2014). Duncan and Janssen used the term in a different manner to describe the Wentzel-Kramers-Brilluion approximation that connects wave mechanics to the old quantum theory.

[9]Warwick (1989, 2003), Klein (1998, 1999, 2001, 2003), Kaiser (2005), Seth (2010) and James and Joas (2015).

[10]See Klein (2003) for her most elaborate description of paper tools as well as Kaiser (2005) for his elaboration on paper tools in quantum field theory.

[11]Kaiser (2005).

[12]Klein (2003).

[13]Kaiser (2005, 9).

problem. Histories of theoretical tools thus overlap with historical studies of larger theoretical developments as well as case studies on individual scientific problems. As such, they benefit from them and open up new questions that do not arise when looking primarily at certain physical phenomena or analyzing the development of a theoretical framework. In the following, I will discuss the relation of the present history of the correspondence principle and these historiographical approaches.

With respect to historical studies of individual research problems, there are a number of studies on both the development of multiplet spectroscopy and the development of dispersion theory within the old quantum theory.[14] The case studies on Sommerfeld and Reiche are indebted to these works. It is not by chance, however, that Sommerfeld's work on multiplet intensities and Reiche's work on the f-sum rule have been absent from the literature. Not always rightfully, they have been regarded as peripheral for the development of their respective research fields and therefore have not received thorough treatment. The case studies in this book analyze the applications of the correspondence principle to these specific problems and make genuine contributions to the understanding of the historical development of these research fields. More importantly, however, they are relevant for the present history of the correspondence principle as they give detailed insights into the use of the correspondence principle and provide the basis for a comparative study of its different applications.

With respect to the historical scholarship engaging directly with the correspondence principle, Olivier Darrigol's *From C-numbers to Q-numbers* presents the standard reference to this day.[15] In his work, Darrigol focused on the role of formal analogies in the history of quantum theory and studied the correspondence principle as part of the foundational core of Bohr's quantum theory. Within this analysis, Darrigol clarified the content of Bohr's formulation of the correspondence principle and revised an interpretation that was widely accepted by physicists and discussed by philosophers of science. According to this "textbook interpretation", the correspondence principle demands that quantum theory ought to reproduce the results of classical physics or even to reduce to it in the socalled classical limit or the limit of high quantum numbers.[16] Darrigol's work clearly distinguished this notion of a classical limiting case from the historical formulation of the correspondence

[14]For multiplet spectroscopy, see Forman (1970), Cassidy (1976, 1979), Serwer (1977) and, more recently, Seth (2010). For dispersion theory and BKS theory, see Hendry (1981, 1984), Darrigol (1992, 224–234), Konno (1993), Duncan and Janssen (2007a,b) and Jordi Taltavull (2013). By contrast, theoretical approaches to scattering in the old quantum theory are underrepresented in the historiography. Among the few noteworthy exceptions are works by Roger Stuewer, Bruce Wheaton, and recently Michael Eckert on the history of X-ray physics, see Stuewer (1975), Wheaton (1983), and Eckert (2015).

[15]Darrigol (1992).

[16]The interpretation of the correspondence principle as a metatheoretical principle was held, for example, by Jammer (1966, 110), Konno (1993, 120) and Kragh (2002, 156) and was widely accepted in more general philosophical discussions on the "generalized correspondence principle." For a summary of this discussion, see Radder (1988) and Radder (1991, 198–203). I will not go further into this discussion as it does not elucidate the actual historical version of the correspondence principle.

principle and showed that Bohr's main idea was a different one. For Bohr, the correspondence principle was a physical law of quantum theory. It stated that there was a relation between the motion of an electron in a stationary state and its radiation.[17]

Moreover, Darrigol analyzed the role of the correspondence principle in the transition from the old quantum theory to quantum mechanics. Following the line of the standard view on the history of quantum physics outlined above, he discussed the inception of the Bohr model, the emergence of the old quantum theory, its crisis, and the eventual emergence of quantum mechanics. Along this line, he discussed the work of Bohr, Kramers, Born and Heisenberg as the main protagonists of the theoretical development and focused on physical problems like the helium spectrum or dispersion theory as the key empirical problems leading to the emergence of quantum mechanics. Within this analysis, Darrigol diagnosed that the interpretation of the correspondence principle changed dramatically after 1923. As he saw it, the correspondence principle was stripped of its visualizable content and became a prescription for a "symbolic translation" of the formal laws of classical into quantum mechanics. In other words, the connection between radiation and motion no longer served as the principle's core idea. It was reinterpreted as a "vertical analogy," which postulated a formal translation of the structures of classical physics into quantum theory in the sense of modern day quantum mechanics.[18]

While this book agrees with Darrigol's interpretation of the correspondence principle in Bohr's early writings, its perspective on the development of the principle in the 1920s and its applications is different. I will argue that a shift of interpretation did not occur. The idea of a connection between radiation and motion remained the stable core of the correspondence principle for physicists in Copenhagen as well as in other places in the quantum network. Heisenberg's pivotal paper "Über die quantentheoretische Umdeutung kinematischer und mechanischer Beziehungen", I will show, emerged from an attempt to formulate a new quantum kinematics and thereby a coherent mathematical representation of the connection between radiation and motion.

In connection to this specific claim, this book takes a different approach to the transition from the old quantum theory to quantum mechanics. Building on the idea of a *patchwork of problems*, it will not follow a linear theoretical development, which underlies the standard account of the old quantum theory. Instead, I analyze the trajectories of many different research fields and study the applications of the correspondence principle by Bohr, Kramers and Heisenberg as well as of Sommerfeld, Franck and Hund, or Reiche. This is not meant as a more comprehensive addendum to the history of the correspondence principle. As outlined above, this book analyzes the applications of the correspondence principle as a historical phenomenon in its own right. It shows how these applications shaped the principle's conceptual development and how they fueled the development of quantum theory as a larger theoretical framework.

[17] Darrigol (1992, 147).

[18] Darrigol (1992, xviii).

1.3 Sources

The present analysis is based on published research papers, scientific correspondence between the historical actors, and research notebooks. In principle, these sources have long been available and some of them have been discussed in previous studies.[19] The bulk of the material analyzed in the case studies on Sommerfeld, Franck and Hund, and Reiche (Chaps. 4, 5, and 6), however, has not received a thorough discussion, nor have these materials been analyzed with respect to the use of the correspondence principle as a research tool. This book brings these sources together and uses them to study the applications of the correspondence principle within the day-to-day work on the old quantum theory.

Research papers have often provided the most detailed formulation of the arguments under investigation. As results of the research process, they often present constructive arguments that can be interpreted as descriptions of their production. At the same time, they were created for communicating within a community and thus always aimed at convincing others, sometimes contorting the pathway that led to them.

As usual, the analysis of arguments in research papers thus has to consider other materials like scientific correspondence, research notebooks, lecture notes or previous works, wherever they are available. With respect to the availability of such material, the historiography on quantum physics is in a privileged position: In addition to the published correspondence and papers of key actors like Werner Heisenberg, Max Born, Albert Einstein, Wolfgang Pauli, Niels Bohr, Arnold Sommerfeld, and Erwin Schrödinger,[20] the *Archive for the History of Quantum Physics* (AHQP), put together by Thomas S. Kuhn, John Heilbron, Paul Forman and others, brings together a plethora of sources held in various archives in the U.S. and Western Europe. In addition, there is still material which was not incorporated into the AHQP for various reasons. On the one hand, archives situated behind the Iron Curtain remained out of reach for Kuhn in the 1960s and even today we do not know what kind of material might be found in places like Wroclaw. On the other hand, the materials of physicists like Friedrich Hund or Helmut Hönl were not incorporated. Both were still active in physics and the history of physics in the 1960s and their papers were only given to the university archives in Göttingen and Freiburg at the end of their (working) lives.

Searching for such materials, I found among others the scientific diary of Friedrich Hund and letters by Arnold Sommerfeld and Helmut Hönl, which had

---

[19]Especially the development of Bohr's formulation of the correspondence principle (Chap. 2) and the developments in Copenhagen leading up to matrix mechanics (Chap. 7) have been analyzed in large parts within the standard account. Likewise, David Cassidy's work on Heisenberg's first rump model analyzed the letters of Heisenberg to Landé in the AHQP and mentioned the correspondence arguments in this context. Due to different research questions, however, his analysis of the same material is different from the one presented in Chap. 4.

[20]Kramers (1956), Born et al. (1969), Bohr (1976, 1977), Pauli (1979), Bohr (1981), Sommerfeld (1968a, 2000, 2004) and Schrödinger (2011a,b).

not been used in previous studies, and which turned out to be central for the analyses in Chaps. 4 and 6. By contrast, recollections on the history of quantum physics by historical actors will not be used extensively for methodological reasons. Like every retrospective account, such recollections are subject to shifts in interpretation and thus have to be analyzed with special care. In the case of the history of quantum physics, there are additional problems that impede their use as a source. As mentioned above, these accounts emerged in the 1960s, as the creators of quantum mechanics were making active attempts to reassure themselves of their own history. In this process, they revisited their own letters and papers and produced commemorative articles in *Festschrift* publications and obituaries, produced the influential anthology *Sources of Quantum Mechanics*, and gave oral history interviews conducted in the context of Kuhn's *Sources for the History of Quantum Physics* project.[21] These recollections codified the crucial contributions to the development of matrix mechanics. Rather than being considered as a source, they should be seen as historical narratives in their own right.

Even without the oral history interviews, the sources in the AHQP present ample material for studying the work and the interactions of not just one or two actors, but of a large group of quantum physicists situated in different places. This group comprises physicists like Wolfgang Pauli, Werner Heisenberg, Niels Bohr, and Arnold Sommerfeld, who are famous beyond the history of quantum physics, as well as those only known to historians of quantum physics like Hendrik Antoon Kramers, Alfred Landé, Fritz Reiche, Edwin Kemble, John H. Van Vleck, Ralph Kronig, and Helmut Hönl. Through archival research alone, the work of the latter physicists, especially, would hardly be accessible at the level of detail presented here.

1.4 The Argument

This book conceptualizes the history of the correspondence principle as a research tool. It describes and analyzes the principle's applications as *transformations through implementation*. This term describes the pattern I see underlying the development of correspondence arguments. The historical case studies flesh out

[21]For accounts of the pathway to quantum mechanics by the historical actors, see for example Heisenberg (1960), Kronig (1960), and van der Waerden (1968). The oral history interviews for the AHQP were prepared by sending lists of apparently crucial papers and by asking for preliminary accounts from the former actors to refresh their memories. Most of the interviews were conducted by Kuhn, who was—as Michel Janssen put it—"leading the witness." For similar observations on Kuhn's interviews, see Duncan and Janssen (2007a, 600) and Seth (2007, 29). In her forthcoming book, Anke te Heesen analyzes Kuhn's approach to the oral history interviews in the AHQP and provides a deeper understanding of his underlying conception of historical sources. She shows that Kuhn saw the interview as a means to "climb into the heads" of the historical actors. Using sociological and, more importantly, psychoanalytic techniques, he sought to unearth and record memories and thereby to produce sources, whose epistemic value differed not in essence but in degree from the letters, papers and notebooks from the 1920s. See te Heesen (forthcoming).

the characteristics of this process and show how it fueled the development of the correspondence principle. In order to give structure to the individual chapters, I will now sketch the main idea and show how it is analyzed in the following chapters.

The applications of the correspondence principle took off with its identification as a problem-solving tool, its integration into ongoing research, and its actual implementation. The implementation of the principle was followed by a second step. In it, physicists recognized conceptual tensions between the correspondence principle and their theoretical notions, techniques, models, and larger research agendas. Moreover, they encountered discrepancies between results obtained from the correspondence principle, observed experimental results, or conceptualized physical intuitions. Such tensions could translate into problems of theoretical consistency or empirical adequacy. In the third step, solutions to these problems emerged in the form of *adaptive reformulations*, in which the correspondence principle was adjusted to the structure of the problems and techniques at hand, leading to a wide range of different correspondence arguments.

On the whole, the correspondence principle was reshaped through the process of *transformation through implementation*. This transformation took place both on an operational level and on a conceptual level, turning the principle from a general, qualitative idea into a quantitative tool for calculation and eventually into the conceptual backbone of matrix mechanics. Naturally, this reshaping did not just affect the tool. Its integration into a new framework went hand in hand with a change in the conception of the respective research problem and the framework it was approached in.

Studying this *transformation through implementation* of the correspondence principle, this book begins in Chap. 2 with the formulation of the correspondence principle by Bohr and Kramers. This chapter reconsiders the emergence of the principle and puts a particular focus on the role of the physical problems driving its development. Its main function is to introduce the correspondence principle in its original formulation, and thereby to establish a baseline for the ensuing historical analysis of its dissemination, reception, and application.

This analysis is undertaken on two different levels in the main part of this book. Chapter 3 surveys and analyzes the dissemination and the reception of the correspondence principle. It presents the context in which the correspondence principle was adapted outside of Copenhagen. Building on the work of Peter Robertson and Alexi Kojevnikov, I discuss the role of the positioning of Bohr's institute in neutral Denmark after World War I, Bohr's lecture tours and the importance of postdoc fellowships for young physicists in the 1920s.[22] This chapter looks at the various paraphrases of Bohr's correspondence argument given by physicists working outside of Copenhagen between 1922 and 1926. These paraphrases in research papers, overview articles, and textbooks show how the principle was first received outside of Copenhagen and to what extent the interpretations of Bohr's writings differed from each other. Taking these paraphrases and the large historical

[22]Robertson (1979) and Kojevnikov (forthcoming).

context together, I discuss some of the preconditions for the dissemination of Bohr's correspondence arguments within Europe and the U.S.

The overarching description in Chap. 3 is complemented in the following Chaps. 4 through 6 by three case studies of the applications of the correspondence principle. These case studies analyze the work of Arnold Sommerfeld, Werner Heisenberg and Helmut Hönl, James Franck and Friedrich Hund, and Fritz Reiche, Rudolf Ladenburg and Willy Thomas. They form the core of the discussion of the process of *transformation through implementation* and provide the basis for analyzing how the correspondence principle was integrated into differing approaches; what kind of tensions arose from its implementation; under which circumstances they translated into problems, and how *adaptive reformulations* emerged from them.

These studies hardly present an all-encompassing survey of the applications of the correspondence principle.[23] The individual actors and their applications are not chosen as central or typical for a particular research field—though some of them may arguably be considered as such. Rather, the individual case studies represent a spectrum of different social and institutional settings, research agendas, theoretical approaches, and research problems.[24] This diversity and multidimensionality of contexts—and not the typical or representative character of a particular approach—is what is necessary for my argument. It makes it possible to characterize the pattern of *transformation through implementation*; to study the relation between research tools, problem solving and conceptual development; and to distinguish conditions that were essential for making correspondence arguments from those conditions that were important in a particular case but appear historically contingent in comparison with others.

[23]Necessarily, they do not cover all applications of the correspondence principle in a particular research field. At the level of detail presented here, this would have gone well beyond the scope of this book. In addition to the material covered in the case studies, such a more comprehensive study would have to include works on atomic and molecular spectroscopy, dispersion theory and the study of collisions. For an overview, see Appendix A. This extensive, but not necessarily complete material cannot be analyzed on the same level of detail as the case studies presented in this book, as the available archival material does not suffice for a detailed analysis of these papers and the developments connecting them.

[24]This spectrum could have been extended by including other cases. Perhaps the most important one would have been Edwin Kemble's. Kemble, with whom I began the introduction, applied the correspondence principle to the intensity of band spectra at Harvard from 1923 to 1925. His work with the correspondence principle belonged to the research field of molecular spectroscopy and paralleled the case of Sommerfeld and Heisenberg in important ways. See Kemble (1924, 1925a,b). Paradoxically, Kemble's proximity to the work of Sommerfeld and Heisenberg was essentially the reason not to include it: The possibility—and in fact the need—for a comparison between the two cases would have shifted the focus of this book. It would have meant to focus the discussion on applications that effectively dealt with the same problem, tackled using the same models in different institutional and experimental contexts. In this way, the book would have been a comparison of theoretical practices in Europe and the U.S. and an analysis of the interaction between the two communities. Instead, I made the historiographic choice to keep the focus on applications that featured both different institutional and social settings as well as different physical problems and conceptual approaches. For a discussion of Kemble's work, see Assmus (1990).

Chapter 7 concludes the discussion. It looks at Bohr's, Kramers', and Heisenberg's responses to the correspondence arguments of Sommerfeld, Hönl, and Kronig in the context of multiplet spectroscopy and thus contrasts the applications of the correspondence principle in Munich, Breslau, and Göttingen with the developments within the Copenhagen community. Thereby it shows how *transformation through implementation* became significant beyond the context of multiplet spectroscopy and fueled the search for a new quantum mechanics. In other words, it shows how the adaptation of the principle as a research tool played into the conceptual development of quantum theory.

Chapter 2
The Correspondence Principle in Copenhagen 1913–1923: Origin, Formulation and Consolidation

This chapter discusses the work of Niels Bohr and Hendrik Antoon Kramers on the quantum theory of the atom in relation to the correspondence principle. Within the main argument of this book, this chapter plays a pivotal role. It presents a historical analysis of the genesis of the principle and provides the baseline for the historical analysis in the following chapters. Following unpublished as well as published papers, it analyzes the origin, the formulation, and the consolidation of the correspondence principle from 1913 until 1923 and puts a special emphasis on the role of physical problems for the development of the principle.[1]

The core idea of the correspondence principle, as pointed out by Olivier Darrigol, was the connection between the motion of an electron and its radiation. Analyzing the transformation of this idea, Sect. 2.1 traces the origins of the correspondence idea within Bohr's work on his model of the atom from 1913 up to 1915.[2] In Sects. 2.2 and 2.3, I then discuss the transformation of this idea in Bohr's quantum theory of (multiply) periodic systems from 1916–1918. Showing how this transformation emerged from Bohr's attempts to extend his initial model to different periodic and multiply periodic systems, I analyze how Bohr formulated and used his correspondence idea as a statement on the possibility for the emission of radiation. I thereby show that the correspondence argument published in his paper "On the Quantum Theory of Line Spectra"[3] does not represent Bohr's original train of thought. In addition to the search for a rational foundation of quantum theory

[1]This analysis builds on the extensive scholarship on the history of Bohr's quantum theory of the atom in general given by Jammer (1966), Heilbron (1964), and Heilbron and Kuhn (1969) as well as the correspondence principle, which was discussed by Darrigol (1992), Assmus (1990, 1992b), Tanona (2002) and Bokulich (2009).

[2]This section follows the work of John Heilbron and Thomas S. Kuhn, Rud Nielsen and Olivier Darrigol (see Footnote 1). A previous version has been published as Jähnert (2013).

[3]Bohr (1918a,b).

M. Jähnert, *Practicing the Correspondence Principle in the Old Quantum Theory*, Archimedes 56, https://doi.org/10.1007/978-3-030-13300-9_2

presented in the paper, a more comprehensive account needs to include the concrete problems that played a crucial role in the development of Bohr's arguments.[4]

Section 2.4 concludes the discussion of the formulation of the correspondence principle in Copenhagen and introduces the formulation of the principle, which was received by physicists in the 1920s. On the one hand, this section looks at the consolidation of Bohr's correspondence argument in his writings from 1920 until 1923 and briefly analyzes how Bohr turned his correspondence argument into a principle. On the other hand, it discusses the limitations and the first extension of the correspondence principle, developed by Bohr's assistant Hendrik Antoon Kramers.

2.1 Formulating the Problem, Part I: The State-Transition Model and the Relation Between Radiation and Motion

Introducing the state-transition model for the first time, Bohr's 1913 trilogy "On the Constitution of Atoms and Molecules"[5] inaugurated an approach to atomic spectra that would come to be the prevalent one of the 1910s and 1920s. At the same time, it is widely recognized as the first of many expositions in Bohr's writings of the central tensions between the state-transition model and the description of a quantum system in terms of classical mechanics and electrodynamics. This first published exposition entails—as Jammer, Heilbron and Kuhn as well as Darrigol have emphasized—the "germ of the Correspondence Principle," and, as such, marks the point of departure for the present reconstruction.[6]

This reconstruction, however, has to take into account that Bohr's presentation of the state-transition model in the 1913 trilogy does not follow the lines along which Bohr had arrived at his model. Rather than presenting a genetic argument, as Heilbron and Kuhn have shown, the paper should be read as "a *post hoc* rationalization."[7] In this rationalization, the ideas of stationary states and transitions which had emerged in several, temporally distinct steps were introduced as postulates of Bohr's new theory of the atom. More importantly, these ideas were justified in a manner that was strikingly different from the way Bohr had arrived at them.[8] While

[4]This aspect was pointed out by Darrigol (1992) and also by Alexi Assmus in her work on the molecular tradition in quantum theory. See Assmus (1990, 1992b).

[5]Bohr (1913a).

[6]For the discussion of the 1913 trilogy see Heilbron (1964, 268–294), Jammer (1966, 70–88), Heilbron and Kuhn (1969, especially 266–283) and Darrigol (1992, 85–89). The quote given in the main text is taken from Heilbron and Kuhn (1969, footnote 98 on page 251); for other identifications of the correspondence idea in the trilogy see Jammer (1966, 80) and Darrigol (1992, 89).

[7]Heilbron and Kuhn (1969, 251).

[8]Heilbron and Kuhn (1969). In general, Heilbron and Kuhn have reconstructed the following pathway that led Bohr to his model: After working on his dissertation on the electron theory of metals, Bohr focused on the problem of mechanical stability in the Rutherford atom and developed

the argument in the trilogy can thus not be read as a genetic argument, one can interpret it in a more constructive manner: what Heilbron and Kuhn identified as Bohr's *post hoc* justification was at the same time a reflection, which formulates the new model, presents a way to apply it, and discusses its central conceptual problems.

Within this reflection, the aforementioned relation between radiation and motion plays a central role in two interconnected ways. On the one hand, Bohr recognizes that such a relation is in conflict with the main assumption he uses to implement the state-transition model. As such, his paper marks the formulation of the central problem which his thinking on the correspondence principle would address. On the other hand, Bohr's thinking about a relation between radiation and motion in the state-transition model is more than a consideration of consistency. As the main assumptions about the state-transition model do not suffice to make the argument, the relation between radiation and motion is crucial for its explication. In other words, Bohr's reflection serves a double purpose. On the one hand, it aims to clarify the foundations of the new model and the extent to which it can and cannot be reconciled with classical mechanics and electrodynamics. The other hand, it establishes and justifies a procedure for the quantization of the atom's stationary states. Both—the clarification of foundations and the development of a quantization procedure—are not only *post hoc* rationalizations, marking the end of the genesis of the Bohr model, they also point towards its extension and consolidation.

Bohr's 1913 Trilogy: The "Germs of the Correspondence Principle"

The main argument presented in Bohr's 1913 trilogy is a derivation of the Balmer formula from the newly proposed atomic model, which Bohr goes through three times, focusing on different aspects each time. As Heilbron and Kuhn have argued, these three derivations represent a shift in Bohr's conception of his new model: While the first derivation assumes the emission of τ quanta of frequency $\omega/2$, the latter two assume the emission of a single quantum $\tau\omega_\tau/2$. This shift in interpretation of the same formula, they argued, distinguishes the first derivation conceptually from the second and the third. Only the latter two derivations drop the

---

a first quantum theoretical description of the atom. This description, as Heilbron and Kuhn have emphasized, focused on the idea that there are certain states of the atom which are assumed to be mechanically stable by definition. They are selected by a quantum condition which does not involve assumptions about radiation or even transitions between states. Instead Bohr assumed a constant ratio of the kinetic energy of the atom and its mechanical frequency. Bohr only introduced the idea of transitions between stationary states in reaction to the work of J. W. Nicolson, which led him to incorporate the Balmer formula into his considerations. Working his way backwards from the empirical Balmer formula, Heilbron and Kuhn have suggested, Bohr introduced the idea that radiation was emitted during transitions between stationary states, which he now reinterpreted as stable because they were radiation-free.

close analogy to Planck's theory in which the radiation is emitted by an oscillator in a certain state, and embrace the frequency condition as the new description of the radiation process.[9]

Regardless of his shifting interpretation, Bohr's argument follows the same steps in all three cases. After the introduction of the stationary states and the frequency condition, each derivation consists of the calculation of the energy E of the atom in a stationary state from classical mechanics, the quantization of the energy to construct a set of discrete energy levels, and finally the calculation of the radiation frequencies ν of the spectrum via the frequency condition:

$$\Delta E = h\nu.$$

Given the simplicity of the system considered and the assumed validity of classical mechanics for the description of the stationary states, the first step did not call for an extended mathematical argument. Bohr simply gave the equations for the energy, frequency and radius of the stationary state without any calculation. Likewise, he obtained the hydrogen spectrum from the frequency condition once the discreteness of the energy levels was established.

While these moves were already straightforward within the newly proposed model, Bohr's main argument deals with the quantization of the energy of the atom, i.e. the construction of a discrete set of energy levels from the continuous range of energies of the classical planetary model. This quantization could not be achieved on the basis of the two main assumptions and is based on a relation between the frequency of the emitted radiation and the mechanical frequency of the electron. Regardless of the conceptual shift identified by Heilbron and Kuhn, it is this relation which is of major importance for Bohr in his early works on his new model of the atom. As such, it recurs beyond the successful derivation of the Balmer formula in Bohr's subsequent papers on the spectrum of the Stark and Zeeman effect, and becomes the touchstone of Bohr's thinking on the correspondence idea.

Bohr introduced the relation in his first derivation of the Balmer formula. Considering the binding of a free electron from infinity into a stationary state inside the atom, Bohr argued that the mechanical frequency of the free electron with respect to the atom equals zero and that—after the binding and the associated emission of radiation—this frequency is ω. The arithmetic mean value $\frac{\omega}{2}$ of these frequencies, he assumed, could be equated with the radiation frequency ν emitted in the binding process so that

> [...] a homogeneous radiation is emitted of a frequency ν, equal to half the frequency of revolution of the electron in its final orbit.[10]

This assumption has two parts: one is the homogeneity of the radiation frequency, the other is the numerical identity of the radiation frequency and the mechanical

[9] Heilbron and Kuhn (1969, 251).

[10] Bohr (1913a, 4–5).

frequency. This second part connects the radiation and the motion of the electron and allows Bohr to quantize the energy of the stationary states: using the equation $\nu = \omega/2$, he could rewrite the Planck relation:

$$W = \tau h\nu = \tau h\frac{\omega}{2}.$$

Still interpreting this equation as a statement about the number of quanta emitted during the binding process, as Heilbron and Kuhn have pointed out, Bohr solved this equation for the mechanical frequency ω and used it to eliminate the mechanical frequency from the classical equation for the energy of the system in a stationary state. Thereby, he expressed the atom's energy in terms of the quantum number τ:

$$W = \frac{2\pi me^4}{h^2\tau^2}.$$

Arriving at these quantized energy levels, Bohr obtained the Balmer formula for the hydrogen spectrum by simply applying the frequency condition:

$$\nu = R\left(\frac{1}{\tau_2^2} - \frac{1}{\tau_1^2}\right).$$

Through this formal manipulation, Bohr reinterpreted the Planck relation in a substantial way. Planck and others regarded this relation as a description of the energy of a resonator oscillating and radiating with the same frequency ν. For Bohr it was a relation restricting which frequencies were allowed for the radiating system and which were not. As such, it became a quantum condition which allowed him to express the energy of the stationary states in terms of the quantum number τ.

The derivation of the Balmer formula resulting from this reinterpretation was not the final result of Bohr's paper. The basic assumption used for quantization—the relation between the frequency of the radiation and the mechanical frequency of the electron—was highly problematic with regard to its consistency with the definition of the stationary states and the frequency condition. In the stationary states electrons were supposed to move without emitting radiation. Expressed in terms of frequencies, this meant that the radiation frequency ν is 0, not half the mechanical frequency. Likewise, the frequency condition stated that the radiation frequency depends on the initial and the final state of the atom, i.e. on two mechanical frequencies.

In light of these contradictions, Bohr's quantization procedure was in conflict with the main physical assumptions of his model. Its roots came from classical electrodynamics, specifically from electron theory and its classical radiation mechanism. In classical radiation theory, the accelerated motion of a charge causes varying electric and magnetic fields and thereby becomes the source of radiation. A direct consequence of this radiation mechanism is that the oscillatory properties of light like frequency and intensity are determined by the frequency and amplitude of the electron's motion. This classical radiation mechanism of electron theory was contradicted entirely within Bohr's model and, more importantly, was replaced by it, as radiation is produced in the entirely different process of a transition between stationary states.

Bohr explicitly acknowledged that the major assumptions of his model introduced a radical break with classical electrodynamics. While he knew that the classical radiation mechanism would not find a place in his model, the formal relation between radiation and motion still played a major role in his thinking. This is emphasized by the fact that Bohr knew that he could quantize the orbits of his atom by assuming that the angular momentum of the atom is equal to $\frac{h}{2\pi}$;[11] nonetheless, he based his quantization procedure on the purely formal relation between radiation and motion. Seeing how this connection could be reestablished was a vital concern for Bohr, first, because he needed a justification for it in light of the obvious contradictions with the basic assumptions of his model; second, because he hoped for insights on how to develop the theory by investigating which parts of classical radiation theory could still be used within the new model.

The investigation of this question loomed large in Bohr's 1913 paper and continued to play an important role in this later research. For the preceding argument Bohr justified the radiation-motion relation by means of a comparison between the radiation process and the motion of the electron before and after the radiation process. This justification itself was not very convincing: it rested solely on its empirical success and the questionable mean-value argument, so that Bohr decided to give a second derivation of the Balmer formula to add further plausibility to his initial assumption.

The first argument, however, already points to some of the central aspects of his attempt to connect radiation and electronic motion: In light of the missing radiation mechanism, Bohr tried to recover a relation between radiation and motion within quantum theory that was formally analogous to classical radiation theory. This attempt was based on experimental, mathematical and intertheoretical arguments and aimed to establish insights into the properties of a radiation mechanism that had yet to be found.

This mixture of empirical, mathematical and intertheoretic arguments is also apparent in Bohr's second derivation of the Balmer formula. In this derivation Bohr reversed the direction of the argument: the aim was no longer to derive the Balmer formula from the proposed atomic model. Rather, he intended to demonstrate the consistency of his atomic model, the relation between radiation and motion and the empirical spectrum. He argued this point by making the slightly more general claim that the energy of the quantized atom is $f(\tau)h\omega$. This approach still replaces the radiation frequency by a multiple of the orbiting frequency and thus tacitly assumes a connection between radiation and motion. Determining $f(\tau)$ as a function of the quantum number τ, Bohr followed the same quantization procedure as in the first derivation and obtained the spectrum of the hydrogen atom with the "generalized"

[11] Bohr (1913a, 15).

assumption as:

$$\nu = \frac{\pi m e^2 E^2}{2h^3}\left(\frac{1}{f(\tau_2)^2} - \frac{1}{f(\tau_1)^2}\right).$$

This shows that the function $f(\tau)$ has to be $f(\tau) = c\tau$, in order to be equivalent to the empirical Balmer formula. In the determination of c Bohr compared the frequency of a quantum transition and the orbiting frequencies of the stationary states in the case of high quantum numbers, in which classical electrodynamics and quantum theory predict the same numerical value for the radiation frequency[12]:

> If N is great the ratio between the frequency before and after the emission will be very near equal to 1; according to the ordinary electrodynamics we should therefore expect the ratio between the frequency of radiation and the frequency of revolution also is very nearly equal to 1. This condition will only be satisfied if $c = \frac{1}{2}$.[13]

By tuning his model in this way, Bohr showed the consistency of his major assumptions, the radiation-motion connection and the Balmer formula, which entered as an empirical constraint. In this proof of consistency, the high quantum number limit played an important role. It rendered precise the condition for the consistency. Moreover, it showed that the assumption of a relation between radiation and motion within quantum theory led to the agreement of quantum theory and classical theory in this limit.

In the last part of the argument, Bohr went one step further. Extending his proof of consistency to transitions between two states with energy E_N and E_{N-n}, he discussed the relation between radiation and motion in general. Showing that in this case the relation between the radiation frequency and the mechanical frequency was $\nu = n\omega$, he argued that there was some kind of analogy between the classical and the quantum theory of radiation:

> The possibility of an emission of a radiation of such a frequency may also be interpreted from analogy with ordinary electrodynamics, as an electron rotating round a nucleus in an elliptical orbit will emit a radiation which according to Fourier's theorem can be resolved into homogeneous components, the frequencies of which are $n\omega$.[14]

This recapturing of the relation between orbital frequency and the radiation frequency was a central part of Bohr's conclusion: While classical electrodynamics in general was in contradiction with the basic assumptions of the Bohr model, the classical relation between the frequency of radiation and the mechanical frequency could be reestablished as a formal relation. Demonstrating this was as important to Bohr as showing that his new model predicted the correct Balmer formula. As such, the classical relation between motion and radiation was indispensable for Bohr's

---

[12] The frequency of radiation is then: $\nu = \frac{\pi^2 m e^2 E^2}{2c^2 h^3}\frac{2N-1}{N^2(N-1)^2}$. For the orbiting frequency before and after emission, Bohr writes: $\omega_N = \frac{\pi^2 m e^2 E^2}{2c^2 h^3 N^3}$ and $\omega_{N-1} = \frac{\pi^2 m e^2 E^2}{2c^2 h^3 (N-1)^3}$.

[13] Bohr (1913a, 13).

[14] Bohr (1913a, 14).

early work as a silent touchstone of his derivations, and of vital concern in the attempt to explore the break and the continuity with classical radiation theory.

The Stark Effect 1913

Given its centrality in the 1913 trilogy, it is not surprising that the radiation-motion relation continued to play an important role in Bohr's subsequent work. After he had finished the trilogy, Bohr adapted his model to account for the effects of external electric and magnetic fields on the atomic spectrum.[15] The two effects, which had been observed by Pieter Zeeman for the magnetic field in 1897, and by Johannes Stark for the electric field in 1913, showed that an external field "splits up" a spectral line of an element into different components in a such a way that the frequencies of the "split-up" lines differ only slightly from the original line, and that they are symmetrical with respect to the original line.[16]

Based on the state-transition model, Bohr saw two possible explanations for the splitting of the spectral lines: the external field either affected the energy levels of the atom, or it changed the frequency condition. Following the first alternative for the Stark effect,[17] Bohr considered a simplified model for the atom in a weak, constant electric field, described by the orbital frequency of the electron ω and its total energy A as a function of the major axis of the orbit $2a$ and a constant C:

$$\omega^2 = \frac{e^2}{4\pi^2 ma^2}\left(1 \mp 3E\frac{a^2}{e}\right) \text{ and } A = C - \frac{e^2}{2a} \mp 3aeE.$$

As in the 1913 trilogy, the central problem addressed in Bohr's argument was the quantization of this energy, for which—as Bohr wrote to his teacher Ernest Rutherford—he followed a procedure that was "exactly analogeous [sic!]"[18] to the one used in his original paper. Returning to the relation between radiation frequency and orbital frequency for the quantization, Bohr reconsidered his argument on the limit of high quantum numbers and adapted it to his new problem: In this limit, he

[15] See Bohr (1913b) as well as the discussion by Darrigol (1992), on which the following is based.

[16] See Zeeman (1897a,b) and Stark (1914), Stark and Wendt (1914) and Stark and Kirschbaum (1914a,b). For discussions of the Zeeman and Stark effect, see Heilbron and Kuhn (1969), Arabatzis (1992) and Kox (1997).

[17] The second one, Bohr thought, needed to be invoked for the Zeeman effect. As Darrigol (1992, 92) has pointed out, Bohr assumed that the frequencies of the Zeeman effect did not obey Ritz's combination principle and that the magnetic field did not change the energy levels of the atom. Following these assumptions, Bohr thought, the only possibility to account for the Zeeman effect was to change the frequency condition and to assume that the energy of the transition was associated with different frequencies. These arguments did not involve the connection between radiation and motion and will therefore not be discussed in the following.

[18] Bohr to Rutherford, 31 December 1913 in Bohr (1981, 591).

argued, one could rewrite the frequency condition as a differential equation. As the radiation frequency equals the mechanical frequency in this case, the Planck relation can again be rewritten in terms of a mechanical frequency of the electron ω_n:

$$A_{n+1} - A_n = h\nu \longrightarrow \frac{dA_n}{dn} = h\omega_n.$$

This equation, which Bohr found was satisfied by the energy and frequency describing the atom in his original trilogy, provided the quantization condition for his model of the Stark effect. Formally manipulating the equation,[19] Bohr used it to quantize the major axis a and consequently found the quantized energy levels of the linear Stark effect:

$$A_n = C - \frac{2\pi^2 e^4 m}{h^2 n^2}\left(1 \pm E\frac{3h^4}{16\pi^4 e^5 m^2}n^4\right).$$

The first term in this expression was identical to one for the hydrogen atom without an electric field, and thus returned the terms of the Balmer formula for vanishing electric fields. In the presence of an electric field, each energy level of the atom was subdivided into two different energy levels, $E_{Balmer} + E_{Stark}$ and $E_{Balmer} - E_{Stark}$. According to the frequency condition, there are four possible transitions between these twofold states, leading to four split-up lines spaced symmetrically around the initial line:

$$\nu = \nu_{Balmer} \pm \nu_{Stark,1} \pm \nu_{Stark,2}.$$

[19] He took the reciprocal $\frac{dn}{dA_n}$ and multiplied it by $\frac{dA_n}{da}$, yielding:

$$\frac{dn}{dA_n}\frac{dA_n}{da} = \frac{dn}{da} = \frac{1}{h\omega_n}\frac{dA_n}{da}.$$

Calculating the derivative and introducing the expression for ω the right-hand side of the equation can be evaluated easily as:

$$\frac{dn}{da} = \frac{\pi e\sqrt{m}}{h\sqrt{a}}\frac{\left(1 \mp 4E\frac{a^2}{e}\right)}{\sqrt{\left(1 \mp 3E\frac{a^2}{e}\right)}}.$$

A Taylor expansion of the square root term then yields the given formula, which Bohr used for the quantization of his model. In the limiting case of a weak electric field, Bohr obtained his quantization condition by neglecting higher powers of a beyond the quadratic term:

$$\frac{dn}{da} = \frac{\pi e\sqrt{m}}{h\sqrt{a}}\left(1 \mp \frac{5}{2}E\frac{a^2}{e}\right)$$

Integrating this equation leads to an expression of the major axis a in terms of n, so that Bohr could reexpress the energy as a function of the quantum number n.

Comparing this prediction with Stark's observation, Bohr thought that his treatment was not yet satisfactory. While the number of lines predicted matched the number of lines observed, Stark had found that the inner two lines were polarized parallel to the electric field, while the outer two lines were polarized perpendicular to the field. As his simplified model for the Stark effect only considered the effect of the electric field parallel to the field, Bohr assumed that his model should only account for the two lines polarized parallel to the field. This assumption in itself shows the extent to which Bohr was thinking in terms of the classical radiation-motion connection. In classical electron theory, the relation between the polarization of the spectral lines and the orientation of the electron's orbit with respect to the electric field is a necessary consequence of the radiation mechanism; such a relation does not follow from the two major assumptions of the state-transition model.

Assuming that his model should only account for the lines parallel to the field, Bohr assumed that there were essentially two series of energy levels between which no transitions occurred. Energy levels involving a plus-sign did not form a transition with energy levels with a minus-sign and vice versa, so that the quadruple splitting was reduced to a double splitting.[20] With this assumption, Bohr thought he could predict the frequency of the lines in the Stark effect polarized parallel to the field:

$$\nu = \frac{2\pi^2 e^4 m}{h^3}\left(\frac{1}{n_2^2} - \frac{1}{n_1^2}\right)\left(1 \pm E\frac{3h^4}{16\pi^4 e^5 m^2}\right)(n_2^2 n_1^2) = \nu_{Balmer} \pm \nu_{Stark}.$$

As Darrigol (1992, 90–91) has pointed out, this is the first argument in Bohr's writings that involves a selection rule. While the non-occurrence of certain transitions, as we will see in the next section, was crucial for the formulation of the correspondence principle, Bohr's later correspondence arguments were not developed as extensions of the present argument.

Interim Conclusion

As we have seen in this section, the relation between radiation and motion played a twofold role in Bohr's initial work on the planetary model of the atom. On the one hand, his arguments tried to pin down the extent to which this relation could be reestablished within the state-transition model in light of the conceptual break with classical radiation theory. On the other hand, Bohr's early work with the state-

[20] Bohr (1913b, 515–516). To justify this constraint, Bohr argued that the continuous classical electrodynamics and quantum theory should yield the same results in the limit of large quantum numbers. As it was implausible in classical radiation theory that an electron with the energy $E = E_{Balmer} + E_{Stark}$ turn into an energy $E = E_{Balmer} - E_{Stark}$, Bohr argued, this should also be impossible in quantum theory.

transition model relied heavily on the relation between radiation and motion, which provided the quantum condition necessary to make his arguments.

This second aspect ceased to play a role after 1915/16, when Bohr adopted a different quantum condition. This quantum condition did not involve assumptions about the radiation process or the interaction between radiation and matter, but quantized the energy or the action of a mechanical system.[21] Due to the fruitfulness of the new quantization condition, Bohr dropped his early quantization procedure and took the frequency condition as the sole relation pertaining to the radiation process.[22]

2.2 Formulating the Problem, Part II: The State-Transition Model and the Radiation Process in Quantum Theory

While the new quantization procedure made Bohr's original use of the relation between the orbiting frequency and the frequency of radiation obsolete, the relation and the asymptotic limit between quantum theory and classical radiation theory continued to play a central role in Bohr's reasoning on the state-transition model. These central elements of Bohr's early work came to be used very differently from 1916 onwards and thereby took on new meaning. As will be discussed in Sect. 2.3, Bohr returned to these elements when he ran into difficulties with the description of the radiation process within the state-transition model, which was based on the frequency condition alone.

In order to understand the transformation in Bohr's thinking about the relation between radiation and motion, which eventually led to the formulation of the correspondence principle, we need to understand the specific challenge Bohr was attempting to resolve. As Darrigol and Alexi Assmus have argued, Bohr's correspondence argument took shape in his attempts to explain selection rules, i.e. the necessity to restrict energetically possible transitions. Following Assmus' work rather than Darrigol's on this point, this section discusses how the challenge emerged from attempts to describe simple harmonic systems like the harmonic oscillator and the rotator in terms of the state-transition model.[23]

---

[21] Bohr used the relation $\overline{T}/\omega = \oint T dt = 1/2nh$, where $\overline{T}$, ω and n are the mean kinetic energy, the mechanical frequency and the associated quantum number, respectively. Later he adopted the Sommerfeld quantum condition: $J = \oint p dq = nh$, where J is the action determined from the phase integral of the generalized momentum p, conjugated to a particular generalized coordinate q.

[22] Heilbron and Kuhn (1969, 279–280).

[23] Darrigol (1992, 123). Identifying the issue of selection rules as the starting point for Bohr's formulation of the correspondence principle, Darrigol has argued that Bohr formulated "the first systematic generalization of the correspondence idea" in his discussion of the Stark and Zeeman effect. Although Bohr's earlier arguments certainly qualify as less systematic and are not as clearly directed towards the formulation of selection rules, Assmus shows that Bohr's arguments on the

Reformulating the Problem: Harmonic Motions and the State-Transition Model

These models were situated in what Alexi Assmus has called the “molecular tradition in the early quantum theory” and identified as the mainstream of quantum physics in the 1910s. Leading to a first quantum theory of specific heats and Bjerrum’s theory of band spectra, as she sees it, this tradition provided the missing link between Planck’s theory of black-body radiation and the quantum theory of atomic and molecular structure developed in the 1910s and the 1920s.[24]

With respect to the description of radiation, there was a central conceptual difference between the molecular tradition and Bohr’s atomic theory. The former assumed that molecules, represented by harmonic oscillators and rotators, radiated according to classical radiation theory. Electrons in atoms, by contrast, had to be described within the state-transition model and hence radiated according to the frequency condition. From this perspective, consider the quantized energies E and frequencies ω for the oscillator and the rotator given by:

$$E_{osc} = nh\omega \quad E_{rot} = m^2 \frac{h^2}{8\pi^2 I}$$
$$\omega = const \quad \omega_{rot} = m\frac{h}{4\pi^2 I},$$

where I is the moment of inertia of the rotator and n and m are the vibrational and rotational quantum numbers, respectively: According to classical radiation theory, both systems radiate with their proper frequencies. Especially for the rotator, this prediction was in marked agreement with experimental data for band spectra. In the infrared region, the spectrum of diatomic molecules showed an approximately linear progression of spectral frequencies, just as expected. In this situation, Assmus has argued, the classical radiation mechanism was still in play at least until 1916 and molecules were not treated within the state-transition model.

The move towards a description of molecules in terms of the state-transition model, she argued further, came from Bohr in an attempt to incorporate his atomic model with its new radiation mechanism into the mainstream of the quantum physics of the 1910s. Based on the frequency condition alone, however, this attempt was quite problematic. In this case, the state-transition model leads to radiation frequencies:

$$\nu_{osc} = (n' - n'')\omega \quad \nu_{rot} = (m'^2 - m''^2)\frac{h^2}{8\pi^2 I}.$$

harmonic oscillator and rotator not only predated these considerations, but also presented Bohr’s first explanation of selection rules from the harmonic character of the motion in a stationary state.

[24] Assmus (1992b).

As the quantum numbers could take all possible values $0, 1, 2, 3 \ldots \infty$, the state-transition model predicts far more lines than are actually observed. The oscillator radiates with its proper frequency ω and its overtones $2\omega, 3\omega \ldots$. The radiation frequencies of the rotator do not follow a linear progression, but instead depend quadratically on the quantum numbers of the initial and final states. To be in agreement with experimental evidence, however, the mechanical frequencies of the oscillator and the rotator needed to be directly proportional to the radiation frequencies, as predicted by classical radiation theory.[25]

Assmus' assessment is an important one for reconstructing the formulation of the correspondence principle. As will be discussed in detail below, Bohr's solution to the above mentioned problem foreshadowed central aspects of his first correspondence argument. This shows that the correspondence principle was not only rooted in general metatheoretical considerations on the relation between classical and quantum theory. Rather, Bohr's work was driven by a specific challenge emerging from the attempt to extend the state-transition model to harmonic motions.

To assess the role of this problem and its solution, however, I believe one has to reevaluate an essential part of Assmus' argument: As she portrays it, physicists did not attempt to incorporate the state-transition model in their work on molecules and kept the classical radiation mechanism as it led to predictions in agreement with experiment. They adopted the state-transition model only when Bohr, Sommerfeld and others had sufficiently extended the quantum theory of the atom into a form in which a unification of both descriptions was desirable. Following this line of argument, Bohr, who had a strong motivation to extend his atomic model to other domains, first needed to adapt the state-transition model and to develop essential aspects of the correspondence principle before the state-transition model could be accepted as a general description of atoms and molecules.

This certainly gives a possible and plausible way in which the development could have taken place. However, it does not fit well with the available evidence. Consider the case of Karl Schwarzschild and his treatment of the rotator in 1916 in his important paper "Zur Quantenhypothese," on which Assmus' claims rest: Obtaining the expressions for the energy of the rotator given above, Schwarzschild explicitly applied the frequency condition to account for the ultraviolet lines in molecular spectra. Doing so, he stated that

> [t]he same rotations of the molecule, whose frequencies are expressed directly in the equidistant absorption bands in the infrared, according to the BOHRian approach, produce bands in the ultraviolet, in keeping with the quadratic DESLANDRESian law.[26]

Assmus interprets this comment in the following way: The radiation of molecules in the infrared was still governed by the classical radiation mechanism, while the

---

[25] Assmus (1992b, 229).

[26] Schwarzschild (1916, 568). "Es wird also hier die Anschauung nahegelegt, daß dieselben Rotationen des Moleküls, deren Frequenzen im Ultrarot unmittelbar in äquidistanten Absorptionsstreifen zum Ausdruck kommen, im Ultraviolett nach dem BOHRschen Ansatz wirkend Banden gemäß dem quadratischen DESLANDRESschen Gesetz erzeugen."

same molecules radiate in the ultraviolet region of the spectrum according to the state-transition model.[27] This conceptual distinction between the classical and the quantum theoretical radiation mechanism could be made, Assmus argued, because the historical actors believed that the frequency condition applied only to radiation emitted by electrons, as in the photo effect or during transitions in the Bohr atom, while molecular motions were governed by classical radiation theory.

Assmus' reading of the above quotation, I think, is an overinterpretation, which becomes unlikely after a closer inspection of Schwarzschild's work. Prior to submitting his paper, Schwarzschild communicated a previous version of his work to Sommerfeld in the following way:

> An essay on band spectra: Electrons orbit a rotating molecule of moment of inertia J. Energy of electronic motion A of the rotation following Planck $\frac{h^2}{8\pi^2 J}n^2 (n =, 1, 2, 3\ldots)$. From this, according to Bohr, the frequency series [follows]:
>
> $$\nu = \frac{A}{h} + \frac{h^2}{8\pi^2 J}n^2$$
>
> This is Deslandres' formula.[28]

Arguing that the quadratic progression of Deslandres' law followed from Bohr, i.e. from the state-transition model, Schwarzschild initially associated the quadratic expression of the energy in a particular state directly with the radiation frequency in the spectrum. This approach was inconsistent with both Bohr's radiation mechanism and classical radiation theory. In the former, one would have to apply the frequency condition and thus consider the energy difference of two stationary states. In the latter, one would have to consider the mechanical frequency of the rotator, which depended linearly on the rotational quantum number.

This inconsistency became apparent through Schwarzschild's correspondence with Sommerfeld. Initially Sommerfeld responded to Schwarzschild that he was "curious about the development of your band spectrum" and did not yet point

---

[27] Assmus (1992b, 228). Though this interpretation appears to be plausible, it is based on a tentative and incorrect translation of a small comment at the end of Schwarzschild's paper. Assmus translates the German key phrases "unmittelbar [...] zum Ausdruck kommen" as "revealed directly," creating the impression that Schwarzschild explicitly saw the rotations as producing radiation according to the classical radiation mechanism. Moreover, she mistranslates "nach dem Bohrschen Ansatz wirkend" as "according to the Bohr condition, effective bands." As "wirkend" is not an adjective for bands but an adverb describing the Bohrian approach, it should be translated as "acting according to Bohr's approach."

[28] Schwarzschild to Sommerfeld, 21 March 1916 in Sommerfeld (2000, 543–544). "Ein Versuch zu den Bandenspektren: Elektronen umkreisen ein rotierendes Molekül vom Trägheitsmoment J. Energie der Elektronenbewegung A, der Rotation nach Planck $\frac{h^2}{8\pi^2 J}n^2 (n =, 1, 2, 3\ldots)$. Daraus [folgt] nach Bohr die Frequenzserie:

$$\nu = \frac{A}{h} + \frac{h^2}{8\pi^2 J}n^2$$

Das ist die Deslandres'-Formel."

towards a discrepancy in Schwarzschild's work.[29] In the next letter, however, he remarked that:

> Your band formula is indeed very peculiar. Here you take, as far as I see, the total energy for the determination of ν, not the energy transition from one orbit to another. Isn't this an inconsequence from the Bohrian standpoint?[30]

Following Sommerfeld's comment, it appears, Schwarzschild included the frequency condition in his consideration and made the comment quoted above.[31] It is noteworthy that Sommerfeld did not mention a discrepancy with the classical radiation theory when he pointed to the oddness of Schwarzschild's explanation of Deslandres' law. Rather, he cautiously commented that Schwarzschild's account was not in keeping with Bohr's approach, while at the same time indicating that he had opted for the state-transition model.

Likewise, Schwarzschild immediately introduced the frequency condition into his work without discussing other possibilities. This makes it highly doubtful that the clear conceptual demarcation between molecular radiation governed by classical radiation theory and atomic radiation subject to quantum theory played a role in his work. Instead, it appears that Schwarzschild had not considered the radiation process and its intricacies in detail when writing his paper, which instead focused almost entirely on the quantization of the mechanical system. The brief comment on the difference between ultraviolet and infrared bands, which Assmus took as a formulation of a clear conceptual boundary, might thus better be understood as an initial attempt to describe a puzzling situation.

This tentative description tellingly stresses that the "same rotations of molecules" are responsible for both the infrared and the ultraviolet part of the molecular spectrum. If anything, Schwarzschild thus underlined the unity of the underlying radiation mechanism. Consistently, he argued that radiation was "produced [...] according to the BOHRian approach" in the ultraviolet, whereas mechanical frequencies were only "revealed directly" in the infrared but not explicitly produced according to classical radiation theory. If one wants to lend more weight to Schwarzschild's tentative comment, it formulates an inclination towards Bohr's theory rather than distinguishing two theories valid within their own proper domain.

While Schwarzschild's case is somewhat unclear, and at first glance allows for different interpretations, a more definite position on the subject was taken by Johann M. Burgers in 1916 and 1917. When he addressed the problem in an

[29]Sommerfeld to Schwarzschild, 24 March 1916 in Sommerfeld (2000, 545). Without going into details, Sommerfeld only made a short remark on Schwarzschild's band spectrum: "Auf die Entwicklung ihres Banden-Spektr[ums] bin ich gespannt."

[30]Sommerfeld to Schwarzschild, 29 March 1916 Sommerfeld (2000, 546). "Ihre Bandenformel ist ja sehr merkwürdig. Hier nehmen Sie, soviel ich sehe, die ganze Energie zur Bestimmung des ν, nicht den Energie-Übergang aus einer Bahn in eine andere. Ist das nicht vom Bohr'schen Standpunkt eine Inconsequenz?"

[31]Schwarzschild probably did so while proofreading his paper, which he sent one day after Sommerfeld's letter.

extension of Schwarzschild's treatment of the rotator,[32] Burgers explicitly discarded Bjerrum's theories based on classical radiation theory and based his work on the state-transition model, or as he put it, "the principles of the theory of quanta":

> One will assume that a given spectral line is not emitted by a vibrating electron, but that it is emitted when the electron passes *discontinuously* from a certain definite state of motion to another definite state.[33]

Burgers thought that a quantum theory of molecular spectra based on the state-transition model had to cover not only the visual and ultraviolet, but also the infrared spectrum. Abandoning Bjerrum's theory, the question thus became, how "the rotation of the molecule exerts an influence of the same kind on the frequency of the light emitted, as it does in Bjerrum's theory?"[34]

Burgers' explanation of infrared spectra was not widely received within the community of molecular spectroscopy[35]; however, his work shows very clearly that for him Bohr's and Bjerrum's theory could not coexist in their own proper empirical and theoretical domains. Not only for Bohr, but also for a growing number of other physicists, this meant abandoning Bjerrum's theory and choosing the state-transition model as the basis of the new "theory of quanta."

This choice was made, however, before the state-transition model was able to cope with the rotator, the harmonic oscillator or for that matter the simple harmonic motions which, as we will see below, emerged in the treatment of the Stark and Zeeman effect. This challenge was recognized simultaneously by Bohr and other physicists like Burgers and Sommerfeld. Like Bohr, they, too, proposed solutions to this challenge, albeit in a different, ultimately forgotten form. These solutions show that Sommerfeld and Burgers saw the necessity of restricting the possibilities of transitions to cope with the challenge of extending the state-transition model.[36]

---

[32]Burgers' major contribution to the developing quantum theory of multiply periodic systems was to show that action-angle variables of a non-degenerate system were adiabatic invariants just like the energy of a system.

[33]Burgers (1917, 170–171).

[34]Burgers (1917, 171).

[35]Burgers' work published in the Proceedings of the Royal Academy of Amsterdam during the war was cited only by Torsten Heurlinger (Heurlinger 1920).

[36]Burgers' solution to the problem was to include the interaction between rotation and electronic motion in Schwarzschild's quantization procedure. In this way, he found an additional term in the energy expression, depending linearly on the quantum number for the rotation n_4. With this new term, the frequency condition gave an additional term in the series formula and offered the possibility of recovering the equidistance of Bjerrum's theory on the condition, which he did not interpret physically, that the quadratic term vanishes for transitions between certain initial and final states. Sommerfeld restricted the possibilities of transitions by imposing what he called "quantum inequalities." According to these additional relations, each quantum number could only decrease during a transition.

2.3 Bohr's Adaptive (Re-)Formulation: The Emergence of the Correspondence Principle (1916–1918)

After this discussion of the molecular tradition and its relation to the state-transition model up to 1916, we can return to the reconstruction of Bohr's arguments leading to the formulation of the correspondence principle. Bohr developed these arguments from 1916 to 1918 within his attempt to establish a quantum theory of (multiply) periodic systems in a systematic form. In this context, his main goal was to show that the state-transition model was internally consistent and that it provided the basis for discussing the existing "applications" of quantum theory from a "uniform point of view."[37]

In 1916, the first of these attempts culminated in an article entitled "On the Application of the Quantum Theory to Periodic Systems."[38] As is well known, Bohr retracted this paper from publication when he received the extensions of his atomic model by Sommerfeld, Epstein, and Schwarzschild. In the next 2 years, he reworked his article and extended his considerations beyond simple periodic to multiply periodic systems. The resulting article, "On the Quantum Theory of Line Spectra" published in 1918, was his first grand treatise on the quantum theory of multiply periodic systems. The treatise, which would eventually be identified as an exposition of the old quantum theory, was intended as Part One and Two of a four-part, famously incomplete exposition on quantum theory. It became the basis for Bohr's talks and lectures on quantum theory, which he would present in the following years and publish subsequently in *Zeitschrift für Physik*.[39]

Building on Darrigol's and Assmus' work, I follow Bohr's arguments on the relation between radiation and motion in these published and unpublished papers. In this way, I show how his understanding of the relation between radiation and motion changed in response to the specific challenge arising from the treatment of harmonic motions within the state-transition model. This reaction to a specific problem marked an adaptive reformulation of Bohr's earlier considerations about the relation between radiation and motion. It led to the formulation of the first published correspondence argument in 1918 and became the nucleus for Bohr's formulation of the correspondence principle.

---

[37]Bohr (1981, 433). With these presentations, Darrigol has argued, Bohr hoped to grasp the conceptual challenges for the further development of quantum theory.

[38]Bohr (1981, 433–461).

[39]See Nielsen (1976) for an overview of the publishing history and development of Bohr's grand treatises.

Restricting Transitions for Harmonic Motions

Bohr's unpublished article "On the Application of the Quantum Theory to Periodic Systems" of April 1916 presents one of his first attempts to establish a quantum theory of periodic systems based on the state-transition model. Bohr's main aim was to show that the state-transition model was internally consistent and thus presented a unification of the existing "applications" of quantum theory.[40] In his discussion, Bohr focused on three separate issues: the possibility of describing the stationary states by means of classical mechanics and a quantum condition; the description of the radiation process according to the frequency condition; and considerations about the statistics of the atom in different states.

For the present discussion, the second issue is relevant. Discussing the radiation process based on the frequency condition, Bohr attempted to show that it is possible "to construct a consistent theory of line spectra by means of a formal generalization of Planck's original theory of an harmonic vibrator."[41] Such a formal generalization, Bohr thought, had to show that the frequency condition was able to account for the radiation spectrum of the harmonic oscillator and of the spectrum of hydrogen.

The argument in which he developed this formal generalization was the following: Bohr first considered Planck's theory of the harmonic oscillator, which was not based on the frequency condition, and "derived" the frequency condition:

> On simple kinematic considerations it is natural to assume that the frequency of any radiation emitted or absorbed by an harmonic vibrator of constant frequency is equal to the frequency of vibration of the particles. Denoting the energy in the nth state by E_n, we therefore get from (1) [$E = nh\omega$] the following relation between the frequency of the radiation ν and the amount of energy emitted during the passing of the system between two successive stationary states:
>
> $$E_{n+1} - E_n = h\nu.$$
>
> This equation expresses Planck's original assumption of emission and absorption in Quanta $h\nu$.[42]

Using the equality of the radiation frequency and the mechanical frequency for the harmonic oscillator, Bohr thus considered the energy difference between two adjacent stationary states of the oscillator and arrived at an expression that is identical to the frequency condition.

In his next step, Bohr demanded that the frequency condition was valid for adjacent states of periodic systems in general. As he observed this implied that "we cannot on this assumption expect a simple relation between the frequency of the

[40] Bohr (1981, 433).

[41] Bohr (1981, 443).

[42] Bohr (1981, 443).

radiation and the frequency of vibration in the stationary states."[43] Consequently Bohr dropped the former derivation of the frequency condition in the case of the harmonic oscillator and instead considered the frequency condition to be a postulate of the quantum theory of periodic systems.

Next, Bohr extended this frequency condition for transitions between adjacent states into his general frequency condition for transitions between all possible states. Bohr was fully aware that "[t]his generalization may at first sight appear inconsistent with the assumption made above as to the frequency of the radiation emitted by an harmonic vibrator."[44] To justify it, he argued:

> In the first place, there is no essential difference between the transition between successive and distant states, if the frequencies in the various states are not the same, and if the radiation nevertheless is assumed to be monochromatic. Next in the limit where the ratio ω_{n1}/ω_{n2} differs very little from unity, the application of [the frequency condition, MJ] gives a frequency which, instead of converging to ω, converges to $(n_2 - n_1)\,\omega$; but this is just what we would expect from analogy with the ordinary theory of radiation, since the motion of any periodic system which is not an harmonic vibrator can be resolved in harmonic terms corresponding with frequencies which are entire multiples of the frequency of revolution ω.[45]

Extending his argument to the hydrogen spectrum and Ritz's combination principle more generally, Bohr concluded that "it thus seems possible to construct a theory [...] which covers the spectrum of an harmonic vibrator as well as that of the hydrogen atom."[46]

Bohr's admittedly formal argument rested on one key assumption, which he introduced as an aside and did not discuss in detail at this point. As he had come to understand, the treatment of the harmonic oscillator in terms of the state-transition model required that the oscillator could only make transitions to adjacent states. It was this restriction of possible transitions that resolved the challenge discussed in the previous section. It ensured that the energy radiated according to the frequency condition was equal to the energy lost by the harmonic oscillator posited in Planck's theory and its classical radiation mechanism. Incorporating the harmonic oscillator into the state-transition model in this way, Bohr obviously did not deduce restrictions for the transition of an harmonic system. Rather, this requirement was a further specification of the state-transition model which had yet to be understood.

While Bohr tried to justify his generalization on the basis of the asymptotic relation between classical and quantum theory in the limit of high quantum numbers, he did not yet attempt to explain why the harmonic oscillator could only make transitions to adjacent states. Yet, this new specification of the state-transition model

[43] Bohr (1981, 443–444). Bohr justified this move by invoking the asymptotic relation between classical and quantum theory in the limit of high quantum numbers, which was discussed in Sect. 2.1.

[44] Ibid.

[45] Bohr (1981, 445).

[46] Ibid.

already played an essential role in his thinking about the relation between radiation and motion.

This can be seen most clearly in Bohr's treatment of the rotator in the state-transition model.[47] In this case, he encountered the challenge discussed in Sect. 2.2. Using the expression for the rotator's energy:

$$E_{rot} = m^2 \frac{h^2}{8\pi^2 I},$$

he applied the frequency condition and obtained the radiation spectrum:

$$\nu_{rot} = \frac{h^2}{8\pi^2 I}(m''^2 - m'^2).^{48}$$

This quadratic expression for the radiation frequency, Bohr knew, was in flat contradiction with the linear progression shown in the infrared spectrum of diatomic molecules. He resolved this problem by applying the main insight he obtained from his consideration of the harmonic oscillator:

> From analogy with the theory of the hydrogen spectrum we should expect that the lines corresponding to transitions between successive states are by far the strongest. In fact the considerations on p. 269 [concerning the harmonic oscillator, MJ] would even suggest that these lines are the only ones possible, *since the motion of the molecules in the stationary states are of a simple harmonic character.*[49]

Considering transitions between the initial state $m + 1$ and the adjacent final state m, the frequency condition gave the frequency emitted by the rotator as:

$$\nu_{rot} = \frac{h^2}{4\pi^2 I}((m+1)^2 - m^2) = \frac{h^2}{4\pi^2 I}\frac{2m+1}{2}.$$

In other words, Bohr was able to recover the arithmetic progression of the spectral lines predicted by the classical radiation mechanism in Bjerrum's theory.

Bohr's argument for the rotator is of central importance for the present reconstruction. For Bohr, the rotator could only make transitions between adjacent states because its motion was of a "simple harmonic character." This was the main insight he took away from his consideration of harmonic systems: The harmonic character of the motion determined which transitions were possible and which were not.

Emerging from Bohr's attempt to integrate the oscillator into the consolidated state-transition model, this interpretation changed the meaning of the relation between radiation and motion. The relation now implied that a particular frequency

---

[47] Bohr (1981, 447).

[48] Note that Bohr tacitly assumed that the moment of inertia I was the same in both the initial and the final state.

[49] Bohr (1981, 448, my emphasis).

in the motion of the radiating system was associated with a particular transition. As Bohr put it, there were "kinematic relations between the phenomenon of transition and the ordinary theories of radiation."[50]

Extension to the Stark and Zeeman Effect

As mentioned above, Bohr retracted his paper from publication when he was confronted with the considerable extension of his model of the atom through Sommerfeld's theory of the fine structure, Schwarzschild's and Epstein's treatment of the Stark effect, and Debye's and Sommerfeld's explanation of the normal Zeeman effect. With the help of Hamilton-Jacobi theory and action-angle variables, Sommerfeld, Schwarzschild, and Epstein developed a quantum theory of multiply periodic systems.[51]

In contrast to Bohr's theory of periodic systems, which quantized the energy of a system as a whole, the extended quantum theory quantized the momentum conjugated to a canonical variable of the physical system separately. Thereby, it became possible to understand the motion of the system as a superposition of different motions. In Sommerfeld's theory of the fine structure, for example, one could distinguish between the original elliptical orbit and the rotation of this orbit in its plane. While the former was an anharmonic motion, the latter was a purely harmonic one.

This possibility fit right into Bohr's new interpretation of the relation between radiation and motion. Reacting to the extensions of his model, Bohr revised his theory of periodic systems and extended it to include multiply periodic systems. This extended version provided the basis for a series of papers entitled "On the Quantum Theory of Line Spectra." In one draft for the first of these papers,[52] Bohr extended his previous argument on selection rules to the line splitting in the Stark and Zeeman effects and the respective accounts by Schwarzschild, Epstein, and Sommerfeld.

Applying this argument to the Stark and Zeeman effects, he explained the splitting of the lines in analogy to his reasoning on the harmonic oscillator. As Sommerfeld's work on the normal Zeeman effect and Epstein's treatment of the Stark effect indicated, the new terms in the energy expressions were associated with additional harmonic motions of the electronic orbit: a precession of the atom around the axis of the magnetic field in the case of the Zeeman effect, and an oscillation parallel to the electric field in the linear Stark effect.

As these new motions were simply harmonic, Bohr argued, the respective electric and magnetic quantum numbers should change only by ± 1 or 0 during a transition.

---

[50] Bohr (1981, 448).

[51] Sommerfeld (1916a,b), Epstein (1916a,b,c), Schwarzschild (1916), and Debye (1916a,b).

[52] Bohr (1976, 48–52).

The absence of additional lines in the Stark and Zeeman effect was thus due to "a *systematic non-appearance* of lines which might be expected by an unrestricted application" of the frequency condition.[53]

From Possible Transitions to Transition Probabilities

In addition to his extension to harmonic motions other than the rotator and the harmonic oscillator, Bohr revisited his previous argument on the restriction of transitions and its connection with the kinematic relation between radiation frequency and mechanical frequency. In this process Bohr considered the harmonic and the anharmonic oscillator in classical radiation theory and quantum theory as he had done in the retracted 1916 paper. This time he used this example to develop the main line of the argument for the restriction of transitions. In classical radiation theory, he argued, the harmonic oscillator can only emit and absorb radiation equal to the mechanical frequency, while the anharmonic oscillator with the same fundamental frequency emits the fundamental frequency ω and its overtones $2\omega, 3\omega \ldots m\omega$. Assuming an analogy between quantum theory and classical radiation theory, Bohr argued:

> [i]n order to retain the analogy to this theory we may therefore expect that in case of harmonic motions the system can pass in one step only between [adjacent] states [...]. In case, however, that the motion is only approximately harmonic we [?] shall expect that there will be <a> small tendency of the system to pass between states corresponding to more distant values of n.[54]

In analogy with classical radiation theory, Bohr continued to argue, the quantum harmonic oscillator should only be capable of emitting radiation with the fundamental frequency ω, corresponding to a transition to an adjacent state. An anharmonic oscillator, on the other hand, which oscillates with the fundamental frequency and its overtones, should give rise to transitions to other states as well.[55]

[53]Bohr (1976, 50–51, emphasis MJ). See Darrigol (1992, 123–125) for his explanation of the selection rule argument. Darrigol identified this argument as the first systematic exposition of a selection rule in Bohr's work and hinted at the possibility that Bohr might have remembered his earlier restrictions of possible transitions from 1915. Bohr's explanation of selection rules was clearly different from his early arguments. In his admittedly preliminary account, Bohr had limited transitions for the Stark effect on the assumption that there were different series of stationary states between which no transitions occurred. The conception that the orbit of an atom in an electric field performed a precession within its plane, and by extension the harmonic character of such a motion, did not play a role. Bohr's argument on the restriction of transitions became possible only with the new approach, which allowed him to quantize different motions separately and thereby to impose restrictions on each one independently.

[54]Bohr (1976, 49, annotations in the edition).

[55]Ibid.

With this argument, Bohr still restricted the possibilities of transitions in terms of frequencies; however, he also began to consider the intensity of spectral lines. The "approximately harmonic oscillator," with its overtones in addition to the fundamental frequency, had a "small tendency" to undergo transitions in which the quantum number changed by more than one unit. This "small tendency," Bohr suggested, became apparent in the faintness of the intensity of the additional lines.

> [These transitions, in which the quantum number changes by more than one] giv[e] rise to the emission or absorption of frequencies equal to entire multiples of the fundamental frequency ω_0. This is in conformity with the recent interesting observation <by E.C.Kemble, Phys. Rev. 8 (1916) 701> of the presence in the ultrared spectra of some diatomic gases of faint lines [...] probably due to an approximately harmonic vibration of the two atoms in the molecule relative to each other.[56]

Bohr thus began to consider that a spectral line was characterized not only by a frequency but also by its intensity. The quantum theory of the atom and its minimalistic account of radiation by the frequency condition, however, was unable to make statements about this property. Bohr's thinking about the relation between radiation and motion led him to consider it. A more detailed formal description, however, would come not from the quantum theory of the atom, which continued to develop as a mechanical theory, but from the quantum theory of black-body radiation and the arguments based on statistical mechanics that were made in this context.

In 1916, Einstein developed an idea which became one of the cornerstones of Bohr's correspondence arguments. In his derivation of Planck's radiation formula, Einstein introduced a probability for the transitions between two stationary states to describe the thermal equilibrium between radiation and an ensemble of quantum theoretical resonators.[57] The transition probabilities Einstein introduced to the state-transition model were constants associated with the respective transitions between two states. Receiving Einstein's work, Bohr adopted the transition probabilities from Einstein's statistical argument and interpreted them in the light of his thinking about the selection rules for harmonic motions: The transition probabilities allowed him to express the "tendency" of a system to pass from one state to another state in a formal way.

As Darrigol has pointed out, however, as along as Bohr tried to account only for selection rules, transition probabilities were not necessary to make the argument. All that mattered was the possibility of a transition and thereby the occurrence or absence of a certain harmonic component.[58] As Bohr's pathway towards the formulation of the correspondence principle shows, this argument had emerged from considerations on the frequency correspondence and had not involved assumptions on the "tendency" of the system for a certain transition, let alone its transition probabilities. As we will see below, however, Einstein's transition probabilities

---

[56]Ibid.

[57]Einstein (1916).

[58]Darrigol (1992, 126–127).

became central in the published version of Bohr's correspondence argument. In it, Bohr expressed his explanation of selection rules in terms of intensities rather than frequencies. He would come to argue that the absence of a certain transition was due to a zero transition probability, in other words, that an energetically possible line simply remained dark.

Putting the Pieces Together: Bohr's Correspondence Argument of 1918

As we have seen in the previous discussion, Bohr's thinking on these issues followed a particular historical trajectory. Bohr first tackled the explanation of selection rules and connected this problem with his considerations on the relation between radiation and motion. These considerations resulted in a new understanding of this relation as a statement about the possibility of transitions. This reinterpretation initially focused on the relation between radiation frequency and mechanical frequency, and was then extended to intensities and Einstein's transition probabilities. These ideas—the description of the intensity through transition probabilities and the explanation of selection rules through the harmonic character of the electron's motion—became the cornerstones of Bohr's reinterpretation of the connection between radiation and electronic motion, which he had pondered in his early papers.

The above reconstruction differs considerably from Bohr's first correspondence argument in the published version of "On the Theory of Line Spectra."[59] In it, Bohr wove the different threads together in a different way, which he summarized in the beginning of his paper:

> In order to obtain the necessary relation to the ordinary theory of radiation in the limit of slow vibrations, we are therefore led directly to certain conclusions about the probability of transition between two stationary states in this limit. This leads again to certain general considerations about the connection between the probability of a transition between any two stationary states and the motion of the system in these states, which will be shown to throw light on the question of the polarization and intensity of the different lines of the spectrum of a given system.[60]

Bohr's argument thus starts off with the general analogy between classical and quantum theory in the limit of high quantum numbers and establishes a connection between radiation and motion to make a statement about the transition probabilities in this limit. This connection is then extrapolated to all quantum numbers in general and finally makes it possible to describe the intensities and polarization of spectral lines. Following the previous reconstruction, this first published correspondence argument is an inversion of Bohr's pathway, or, as Bohr himself put it in a letter

[59] Bohr (1918a).

[60] Bohr (1918a, 8).

to Sommerfeld, an expression of his "unfortunate tendency to let all results appear in systematic order."[61]

This interpretation has important consequences for the overall interpretation of the correspondence principle in Bohr's writings. Beginning in 1918, Bohr counted the principle among the fundamental assumptions of his quantum theory and applied it to different phenomena. As we have seen, however, his formulation of the principle was driven by two interlocking motivations: on the one hand, his attempts to clarify the meaning of the relation between radiation and motion within the state-transition model in general, and on the other, his response to a specific challenge arising from the extension of the state-transition model.

The correspondence argument from his 1918 paper thus does not follow the lines along which Bohr arrived at it. Like his 1913 trilogy, we have to interpret it as a reflection on the relation between radiation and motion within the developing quantum theory of multiply periodic systems. Taking this historical qualification into account, Darrigol's analysis remains largely unchanged. In his explicit formulation, Bohr followed the train of thought summarized in the quote given above.

As in 1913, his starting point was the limiting case, in which the different frequencies of two stationary states and the frequency of radiation coincide with each other. In 1913, as discussed in Sect. 2.1, Bohr had rewritten the frequency condition as a differential equation in the limit of large quantum numbers. Assuming that the radiation frequency was equal to the mechanical frequency in this case, he arrived at the expression:

$$\frac{dA_n}{dn} = h\omega_n.$$

In 1918, Bohr derived this expression again, this time without making questionable assumptions about the radiation process and its connection to the mechanical description of the atom. Rather, his argument relied solely on the mechanical description of a multiply periodic system, leading to the purely mechanical relation from his derivation of Ehrenfest's adiabatic hypothesis:

$$dE = \omega dI.$$

This equation, which Darrigol has called Bohr's "Golden Rule," describes the relation between the energy E, the action I and the frequency ω of the electron

[61] Bohr to Sommerfeld, 27 July 1919 in Bohr (1976, 689). "Ich leide aber so sehr von Schwierigkeiten Abhandlungen in befriedender Form zu bringen und von einem unglücklichen Hand alle Resultate in systematischer Reihenfolge erscheinen zu lassen."

in one stationary state.[62] It is identical to the formulation of 1914, taking $A_n = E_n$ and considering the Sommerfeld quantum condition $I = hn$.

In the high quantum number limit, Bohr then again replaced the differentials dE and dI in this formula by differences:

$$dE = \omega\, dI \longrightarrow E' - E'' = \omega\,(I' - I'').$$

While this move seems to be a mathematical triviality, it marks a key shift in the argument. The former equation $dE = \omega dI$ is a statement about one stationary state that does not radiate. The difference equation, by contrast, can be reinterpreted as a statement about two separate states between which transitions are possible. This is exactly what Bohr did: using the frequency condition and the quantum condition $I = nh$, he rewrote his formula:

$$\begin{aligned} E' - E'' &= (I' - I'')\omega \\ h\nu &= h(n' - n'')\omega \\ \nu &= (n' - n'')\,\omega. \end{aligned}$$

As in the 1913 trilogy, this result reconnects the frequency of a quantum transition with the orbiting frequency of the electron in the classical limit. As the radiation frequency is an entire multiple of the harmonic frequency of the electron's motion, this is nothing other than the classical Fourier relation, which decomposes the radiation frequency into the fundamental frequency ω and its overtones $2\omega, 3\omega \ldots n\omega$. While he had already achieved this result in 1913, Bohr's new argument was not merely an elegant restatement. In contrast to the 1913 version, Bohr no longer assumed the mechanical frequency and the radiation frequency to be equal in the classical limit; instead, this relation resulted from his argument without recourse to classical radiation theory.

Bohr comments that this recapturing of the Fourier relation implies "the close relation between the ordinary theory of radiation and the theory of spectra." It is important for Bohr and the nature of the correspondence argument that this relation is not based on the classical radiation mechanism:

> while on the first theory [classical electrodynamics, M.J.] radiations of different frequencies $\tau\omega$ [...] are emitted and absorbed at the same time, these frequencies will on the present theory [...] be connected with entirely different processes [different transitions, MJ].[63]

---

[62]Whereas this equation can be obtained from Hamilton-Jacobi theory in action-angle-variables, Bohr derived it from Ehrenfest's theorem in Paragraph 2 of his paper. Bohr immediately marked the importance of the "equation [...] which will be often used in the following." Bohr (1918a, 12). As Darrigol has pointed out, the rule and its generalization to systems of s degrees of freedom establishes the unambiguity of the quantum condition and other important properties of multiply periodic systems. See Darrigol (1992, 115).

[63]Bohr (1918a, 15).

This was what Bohr had envisioned from the start: to explore the relation between quantum theory and classical electrodynamics and to see which parts also appeared within quantum theory, while assuming an entirely different radiation process.

In his next step, Bohr extended his correspondence argument from radiation frequencies to intensities, which are also determined from the motion of an electron in classical radiation theory. To this end, he first introduced the Fourier representation of the electron's trajectory $x(t)$ in a stationary state:

$$x(t) = \sum_{\tau} C_\tau \cos(\tau\omega t + c_\tau),$$

where C_τ is the Fourier coefficient and ω is the frequency of the respective overtone τ. Given this representation in classical radiation theory, it is possible to simply read off the radiation frequency. Likewise, the energy E emitted by a multiply periodic system is given by the expression:

$$\frac{dE}{dt} = \frac{2e^2}{3c^3}\ddot{x}(t)^2.$$

The energy is therefore determined by the Fourier coefficients C_τ associated with the respective frequency. By extension, the intensity of the radiation I associated with a particular frequency becomes proportional to:

$$I \propto C_\tau^2 \omega_\tau^4.$$

Extending his correspondence argument from frequencies to intensities, Bohr postulated a relation between the Fourier coefficients of the electron's motion and the intensity of radiation in analogy to classical theory:

> [w]e must further claim that a relation, as that just proved for frequencies, will, in the limit of large n hold also for the intensities of the different lines in the spectrum. Since now on ordinary electrodynamics the intensities of the radiations [sic!] [...] are directly determined from the coefficients C_τ [...], we must therefore determine the *probability of spontanuous transition*[sic!] from a given stationary state [...] to a neighbouring state [sic!].[64]

The argument advanced here is that within the classical limit the intensity of a spectral line can be described both by the transition probabilities in quantum theory and by the Fourier coefficients of the electron's motion in classical radiation theory. Hence—according Bohr—the Fourier coefficients somehow determine the transition probabilities.

So far Bohr's argument focused on the limiting case, in which both the frequency and intensity of a spectral line could be determined from classical radiation theory and from the statistics of discrete transitions. Bohr extended his considerations, arguing that the parallelism in the classical limit implied a general relation within

[64]Bohr (1918a, 15–16. emphasis in the original).

quantum theory. This relation would link the mechanical frequencies and Fourier coefficients of the motion in a stationary state with the frequency and probability of a certain transition:

> Now this connection between the amplitudes of the different harmonic vibrations into which the motion can be resolved [...] and the probabilities of transition [...] may clearly be expected to be of a general nature.[65]

This relation between the Fourier coefficients of the electron's motion and the transition probabilities establishes a second correspondence. In place since 1913, the first correspondence relation connects radiation and mechanical frequencies. The second, newly established correspondence relation connects the transition probabilities with the Fourier coefficients of the motion. Together, these two correspondence relations are formally analogous to the relations produced on the basis of the classical radiation mechanism. However, except for the classical limit, where this correspondence recovers the identity of classical radiation theory, Bohr's correspondence argument does nothing more than postulate a general connection between electronic motion and the radiation process within quantum theory. Bohr could not formulate this general connection between radiation and motion explicitly in the form of a mathematical equation.

The correspondence argument, based solely on the intertheoretic analogy between quantum theory and classical radiation theory, was of little use. To see how the principle could play a fruitful role in quantum theory, it is necessary to consider the continuation of Bohr's argument. While Bohr presents this part as a mere application of a more general idea, it was in this context that he had reconceptualized the relation between radiation and motion. Moreover, it was here that his bold claim of an analogy of quantum theory and classical radiation theory unfolded its potential.

Making no further assumption that would allow him to mathematize his idea, Bohr used his general correspondence idea to explain selection rules and more, generally, to interpret atomic spectra.[66] In light of his correspondence argument, Bohr was able to restate his explanation of the selection rules:

> Thus in general there will be a certain probability of an atomic system in a stationary state to pass spontaneously to any other state of a smaller energy, but if for all motions of a given system the coefficients C [...] are zero for certain values of τ, we are led to expect that no transition will be possible.[67]

While he had accounted for the selection rules in terms of frequencies in his 1916 draft, his new explanation was built on the intensity of the respective harmonic components. No matter what the explicit relation between the transition probability

[65] Bohr (1918a, 15–16).

[66] Bokulich (2009, 1). Bokulich has identified this argument as the core of the principle, but did not comment on the context in which Bohr developed his correspondence idea. For a short discussion of the problem of selection rules before Bohr's correspondence principle and the relation to the principle, see Darrigol (1992, 123).

[67] Bohr (1918a, 16).

and the Fourier coefficients actually is, he argued, if certain Fourier coefficients in the motion of the electron are zero there will be no transition probability. Hence no transition will be observed even though it would be possible according to the frequency condition.

The simplest example with which Bohr illustrated the power of such an explanation was the harmonic oscillator with one degree of freedom. The Fourier "series" of the oscillator is simply $x = C \cos \omega t$. According to the correspondence relations, the single frequency and amplitude of the oscillator are associated with one frequency and one transition probability. Without resorting to the radiation process of classical radiation theory, this explains why the harmonic oscillator only makes transitions to adjacent states and why other transitions do not occur.[68]

Extending this kind of explanation to the more complicated multiply periodic systems like Sommerfeld's, Epstein's, and Schwarzschild's models for the Stark and Zeeman effect showed the full potential of the early correspondence idea. In this case the Fourier series is more complex, as more fundamental frequencies and their respective overtones occur for each degree of freedom in the motion:

$$x(t) = \sum_{u=1}^{u} C_{\tau_1 \ldots \tau_u} \cos 2\pi ([\tau_1 \omega_1 + \tau_2 \omega_2 + \ldots + \tau_u \omega_u] + \gamma_{\tau_1 \ldots \tau_u}).$$

With the more complicated Fourier series, new aspects entered the non-occurrence argument, as in this case not only the coefficients could be zero, but also the cosine terms in the case of degenerate systems. With his correspondence argument Bohr could explain the selection rules for the hydrogen atom in an external electric or magnetic field and the general effect that the field has on the spectrum. As in the 1916 draft, he argued that external fields introduce new motions with new mechanical frequencies, causing the electron's orbit to precess around the nucleus. The frequency of this motion ω_{ext} is set by the external field and is orientated either clockwise or counterclockwise. Hence there are two fundamental frequencies $+\omega_{ext}$ or $-\omega_{ext}$ associated with the external field, while the Fourier coefficients of the overtones of these frequencies are zero. According to the correspondence idea, this implies that all transition probabilities are zero, except those corresponding to a transition in which the quantum number associated with the external field changes by ± 1. This explanation was equivalent to the one given in the 1916 draft, except that Bohr had come to understand that the Fourier coefficients of the electron's motion were the crucial parameter expressing whether or not certain harmonic components occurred.

This understanding made it possible to think about the difference between perturbed and unperturbed systems in general. If in the unperturbed case some Fourier coefficients were zero, an external perturbation could change these coefficients to a value different from zero and therefore lead to the occurrence of other lines in the

[68] Bohr (1918a, 16).

spectrum.[69] Perturbations thus manipulated the probability of a transition, as they affected the dynamics of the mechanical system.

Through this new explanation Bohr had come to understand selection rules as an intensity phenomenon: Although certain transitions were energetically possible, they were not observed because their intensity was zero. The line remained dark. This offered a new possibility of interpreting atomic spectra. In Bohr's early works and in most works focusing on the quantization of the atom's stationary states, the intensity of a spectral line was of little or no importance. Spectra were understood as a series of frequencies spaced according to the series formula. The brightness or darkness was not essential for understanding this spacing and was treated as a secondary property, or externalized as a problem for a yet to be developed quantum theory of radiation. With the intensity explanation of the selection rules, this perspective was contrasted with a different view on spectra, according to which there was a continuous range of possible frequencies with regions of non-zero intensity (spectral lines) in between regions of zero intensity. These latter regions form the dark parts of the spectrum, but there could be radiation within this part of the spectrum in principle. Consequently, the intensity of a spectral line was not a negligible, but a primary property for the description of the spectrum.

2.4 Consolidation and Extension of the Correspondence Principle (1919–1923)

With the correspondence argument published in "On the Quantum Theory of Line Spectra" and the subsequent formulation of the correspondence principle, Bohr reestablished a general relation between the motion of a radiating system and the radiating system. In this capacity, the correspondence principle, as we have seen, provided an answer first of all for the problem of selection rules. At the same time, Bohr's correspondence argument marked a transitory stage in the formulation of the correspondence principle in two respects. On the one hand, the correspondence idea stood at the end of an argument that involved several questionable steps, like the asymptotic connection to classical electrodynamics in the high quantum number limit, and the generalization that rested solely on its success in explaining selection rules. On the other hand, Bohr had not provided a clear strategy for the mathematical operationalization of the correspondence principle: while the correspondence argument implied that the transition probabilities should be determined from the

[69] Bohr (1918a, 32–35). For a discussion, see Darrigol (1992, 128–132), who argues that this last conclusion was based on Bohr's and Kramers' new way of dealing with perturbation theory. It allowed them to circumvent the problems of the perturbation techniques Sommerfeld, Schwarzschild, and Epstein had used, most of which were due to the fact that a perturbed system was no longer separable and hence not solvable in Hamilton-Jacobi theory.

Fourier representation of the electronic motion, the details of such a determination remained obscure.

Concluding the discussion of the formulation of the correspondence principle in Copenhagen, this section discusses how Bohr and his assistant Hendrik Antoon Kramers reacted to this situation. On the one hand, it shows how Bohr finally coined the term "correspondence principle" and counted it among the foundations of his quantum theory of multiply periodic systems. On the other hand, I analyze Kramers' extension of Bohr's correspondence argument, which presented a first tentative operationalization of the correspondence relations based on a mathematical relation and actually allowed the intensity of spectral lines to be determined.

The Consolidation of the Correspondence Principle

The consolidation of the correspondence principle in Bohr's writings was closely connected to Bohr's attempts to establish his quantum theory of multiply periodic systems as the unifying framework for quantum physical research. This consolidation of the correspondence argument into a principle of quantum theory took shape through several lectures, which Bohr was invited to give at almost every major physics center in the quantum network. The context of these lectures and their impact on the dissemination of the correspondence principle throughout the quantum network will be discussed in the next chapter. Here I discuss Bohr's conception of the correspondence principle without tying it to the way it was received by his peers.

In his lecture in Berlin, in which he introduced the term "correspondence principle," Bohr explained its core idea as follows:

> Moreover, although the process of radiation connected to a transition between two stationary states cannot be described in detail on the basis of the ordinary theory of electrodynamics, according to which the nature of the radiation emitted by an atom is directly determined by the motion of the system and its resolution into harmonic components, there is found, nevertheless, to exist a far-reaching *correspondence* between the various types of possible transitions between the stationary states on the one hand and the various harmonic components of the motion on the other hand.[70]

[70]Bohr (1920, 427, emphasis in the original). "Ferner, obgleich es unmöglich ist, den Strahlungsvorgang, mit welchem ein Übergang zwischen zwei stationären Zuständen verbunden ist, in Einzelheiten zu verfolgen mit Hilfe der gewöhnlichen elektromagnetischen Vorstellungen, nach welchen die Beschaffenheit einer von einem Atom ausgesandten Strahlung direkt von der Bewegung des Systems und von ihrer Auflösung in harmonische Komponenten bedingt ist, hat es sich nichtsdestoweniger gezeigt, dass zwischen den verschiedenen Typen der möglichen Übergänge zwischen diesen Zuständen einerseits und den verschiedenen harmonischen Komponenten, in welche die Bewegung des Systems zerlegbar ist, andererseits eine weitgehende *Korrespondenz* stattfindet."

Bohr expressed this central idea again and again in his lectures. Though he introduced it through a mixture of historio-critical analysis and physical derivation,[71] Bohr now considered the correspondence principle as a starting point for a "rational generalization" of electrodynamics, and not as a result of a more or less sound argument. Accordingly, the correspondence principle was no longer an idea that was argued for. Rather one's reasoning had to be in accordance with it. The most concise formulation was given in Bohr's last grand survey of the old quantum theory:

> [T]he possibility of the occurrence of a transition, accompanied by radiation, between two stationary states of a multiply periodic system, whose quantum numbers are respectively $n'_1 \ldots n'_u$ and $n''_1 \ldots n''_u$, [is] considered as conditioned by the presence of certain harmonic components [in the Fourier series] for the electric moment of the atom, for which the frequencies $\tau_1\,\omega_1 + \ldots + \tau_u\,\omega_u$ are given by the following equation:
>
> $$\tau_1 = n'_1 - n''_1, \ldots, \tau_u = n'_u - n''_u$$
>
> We, therefore, call these the "corresponding" harmonic components in the motion, and the substance of the above statement we designate as the "Correspondence Principle" for multiply periodic systems.[72]

Following this statement Bohr reacted to the label "Analogieprinzip" which had been attached to his argument. Distancing himself from his previous use of the term "formal analogy," he emphasized that the correspondence principle was not primarily a metatheoretical analogy between classical radiation theory to quantum theory. Instead, he considered the correspondence principle to be a "purely quantum theoretical law [...], which is in no way capable of diminishing the contrast between the postulates [of quantum theory, MJ] and electrodynamical theory."[73] This interpretation of the correspondence principle as a law of quantum theory,

[71]The historio-critical analysis would usually be part of Bohr's introduction, in which he reviewed quantum theory and emphasized the results on which the correspondence principle could be based. See for example Bohr (1920, 424–427). The physical derivation would involve a shortened form of the argument made in "The Quantum Theory of Line Spectra"; see Bohr (1920, 430–432).

[72]Bohr (1923b, 142). "[Die] Möglichkeit des Auftretens eines von Strahlung begleiteten Übergangs zwischen zwei stationären Zuständen eines mehrfach periodischen Systems, deren Quantenzahlen bzw. gleich $n'_1 \ldots n'_u$ und $n''_1 \ldots n''_u$ sind, als bedingt ansehen von der Gegenwart derjenigen harmonischen Schwingungskomponente in dem durch (2) gegeben Ausdruck für das elektrische Moment des Atoms, für deren Schwingungszahl $\tau_1\,\omega_1 + \ldots + \tau_u\,\omega_u$ die Gleichungen gelten:

$$\tau_1 = n'_1 - n''_1, \ldots, \tau_u = n'_u - n''_u.$$

Diese nennen wir deshalb die 'korrespondierende' Schwingungskomponente in der Bewegung, und den Inhalt der obigen Aussage bezeichnen wir als das 'Korrespondenzprinzip' für mehrfach periodische Systeme."

[73]Bohr (1923b, footnote on page 142–143). "In Q.d.L. [Quantentheorie der Linienspektren] wird diese Bezeichung noch nicht benutzt, sondern der Inhalt des Prinzips ist dort als eine formale Analogie zwischen Quantentheorie und klassischer Theorie bezeichnet. Eine solche Ausdrucksweise könnte jedoch Missverständnisse veranlassen, da ja [...] das Korrespondenzprinzip als ein rein quantentheoretisches Gesetz betrachtet werden muss, das in keiner Weise den Kontrast zwischen den Postulaten und der elektrodynamischen Theorie zu vermindern vermag."

Darrigol has pointed out, is essential for understanding Bohr's perspective on the principle:

> Most important, the "correspondence" was not between classical and quantum theory—a common misinterpretation—but between quantum-theoretical concepts of motion and radiation.[74]

Counting the principle among the foundational postulates and principles, Bohr clarified its conceptual place in the quantum theory of multiply periodic systems. He did so most clearly in his 1923 paper "Über die Anwendung der Quantentheorie auf den Atombau, I. Die Grundpostulate der Quantentheorie." In it, Bohr bisected his quantum theory into two conceptually interconnected but operationally largely independent parts. The first part, discussed in the chapter "Die stationären Zustände," dealt with the mechanical constitution of the atom, the possibility to describe it in terms of classical mechanics, and the appropriate quantum conditions. The second part, discussed in the chapter "Die Strahlungsprozesse," was concerned with the transitions of such a quantum system and dealt with the radiation emitted according to the frequency condition.[75]

For Bohr, the correspondence principle clearly belonged to the second part. It finally came to govern the possibility of transitions in the state-transition model and determined the transition probabilities from the motion of the system in a stationary state. In this capacity the relation between radiation and motion was a statement about the kinematics of the atom in general, independent of questions concerning the specific dynamics governing the stationary states.

Fixing its status, as Darrigol has observed, Bohr's hope was to develop a "rational generalization of classical electrodynamics" on the basis of the principle, which would be formally analogous to classical radiation theory, despite the conceptual divide between the two.[76] In the meantime he used the principle in a deductive and in an inductive way. On the one hand, he intended to show that various spectral phenomena could all be deduced consistently on the basis of the principle and the state-transition model. On the other hand, he aimed to obtain hints, *Fingerzeige*, on the mechanical constitution of the stationary states from the interpretation of empirically given spectra.[77]

---

[74]Darrigol (1992, 138).

[75]Bohr (1923b).

[76]See Darrigol (1992, 138) for a discussion of Bohr's idea of a "rational generalization."

[77]Darrigol (1992, 151). The most frequent example for this use of the principle as an interpretational device was the explanation of selection rules and the effect of external fields on the spectrum discussed above. Bohr came to regard selection rules and the effect of external fields as different aspects of the same phenomenon. On the one hand, the appearance of new harmonic components due to the perturbation of the external field explained why new lines occurred; on the other hand, the fact that only two new fundamental frequencies without overtones entered into the Fourier series explained why this splitting was limited to two additional lines. See Bohr (1920, 444–452), Bohr (1921), and Bohr (1923b, 146). In addition to these paradigmatic examples, Bohr began to incorporate the research of other physicists into his approach by giving explanations of their results in terms of the correspondence principle. For example, Bohr incorporated Stern and Voelmer's

Limitations and Extensions: The Initial-Final-State Problem and the Zwischenbahn

With the correspondence argument published in "On the Quantum Theory of Line Spectra" and the subsequent formulation of the correspondence principle, Bohr reestablished a general relation between the motion of a radiating system and the radiating system. In this capacity, the correspondence principle, as we have seen, provided an answer first of all for the problem of selection rules. Yet, the principle did not provide a clear strategy for its mathematical operationalization.

In the end, Bohr had stipulated that the correspondence between Fourier coefficients of the electron's motion and the transition probability applied to all quantum numbers and argued for its plausibility through the explanation of selection rules and other spectroscopic phenomena. These plausibility arguments were fruitful and stabilized Bohr's bold claim, but they did not imply a procedure for the actual determination of intensities. This—so Bohr felt—required assumptions about the mechanism of radiation itself. As long as such a mechanism was not available, he postponed further extensions of the correspondence idea.[78]

To complete the analysis of the consolidation of the correspondence principle, this section discusses the attempts to go beyond this position. These attempts emerged from the first attempt to implement the correspondence principle in the actual determination of the intensity of spectral lines, and to resolve the central ambiguity in Bohr's original correspondence argument.

One of the first physicists to comment on this issue was Peter Debye. In a letter of 6 June 1918, he wrote:

> In particular your *ansatz* for the calculation of the intensities is evidently of major importance! Personally, I feel a little dissatisfied when I see that you are relating the intensity to the Fourier coefficient of a single orbit. It seems to me that if a system goes over from one orbit $n_1, n_2, n_3 \ldots$, to another orbit $n_1', n_2', n_3' \ldots$ the Fourier coefficients will be different for the first orbit, say C, and for the second, say C'. Wouldn't it be more in the sense of the consideration to measure the probability of the transition by $(C + C')/2$ or perhaps $\sqrt{CC'}$? Or is it the case, that I did not understand you correctly at all?[79]

work on the broadening of spectral lines (Stern and Volmer 1919), and Ehrenfest and Breit's work on weak quantization (Ehrenfest and Breit 1922).

[78]Bohr (1918a, 16). Bohr stated that "we cannot without a detailed theory of the mechanism of transition obtain an exact calculation" of the transition probabilities.

[79]Debye to Bohr, 6 June 1918 in Bohr (1976, 607). "Insbesondere ist Ihr Ansatz zur Berechnung der Intensitäten offenbar von grösster Wichtigkeit! Ein kleines unbefriedigendes Gefühl bleibt mir noch, wenn ich sehe, dass Sie die Intensität in Beziehung setzen zu den Fourier Coefficienten einer einzigen Bahn. Es scheint mir doch so zu liegen, dass wenn ein System von einer Bahn $n_1, n_2, n_3 \ldots$ auf eine andere Bahn $n_1', n_2', n_3' \ldots$ geht, die fraglichen Fourier Coefficienten für die erste Bahn etwa C und für die zweite Bahn davon verschieden etwa C' sein werden. Würde es nicht dem Sinne der Ueberlegungen besser entsprechen, wenn die Wahrscheinlichkeit des Ueberganges durch $(C + C')/2$ oder vielleicht $\sqrt{CC'}$ gemessen würde? Oder liegt die Sache so, dass ich Sie nur nicht richtig verstanden habe?"

With this comment, Debye attempted to formulate Bohr's relation between transition probabilities and Fourier coefficients in a mathematical way. Thereby, he pinpointed a specific ambiguity in Bohr's correspondence argument. As Debye found, Bohr had implicitly taken the initial orbit to be significant. For him this was inconsistent with the state-transition model: In analogy to the frequency condition, the mathematical formulation of the correspondence idea had to involve the Fourier series of both the initial and the final states. While these are approximately identical in the limit of high quantum numbers, the Fourier series of two stationary states differ from one another considerably for small quantum numbers. Consequently, it is relevant whether the initial or the final state is used or whether—as proposed by Debye—some kind of average over the initial and the final state is needed.

With this assessment, Debye formulated what I call the initial-final-state problem. This problem, as we will see throughout this book, recurred time and again when physicists implemented the correspondence principle. It had also been present in Bohr's original correspondence arguments and Bohr had been fully aware of it, following a twofold strategy for its resolution. First, he established the connection between radiation and motion in the limit of high quantum numbers, where the problem did not play a decisive role. Second, he had given up an explicit mathematical relationship between the radiation spectrum and the motion of the radiating system when extrapolating his relation to all quantum numbers.

So why had this approach been successful in the case of selection rules? If the transition probabilities depended on both the initial and the final state, why did Bohr's assumption that the initial state was responsible suffice to make the argument? The answer to this question is that Bohr's argument about the selection rules did not require a decision. The motion of a system associated with a particular selection rule was a simple harmonic one in both the initial and the final state. The argument about the "systematic non-occurrence" of particular transitions hinged on the harmonic character of the motion in general, not on the precise connection of transition probabilities and Fourier coefficients in particular.

Debye's line of thought did not remain limited to the private correspondence between himself and Bohr; at about the same time the young Dutch physicist Hendrik A. Kramers was working on it in more detail in his dissertation.[80] In his work, which had been assigned to the new assistant by Bohr, Kramers worked out an estimate for the intensities of arbitrary transitions.

[80]For this particular period in Kramers' biography, see Dresden (1987, 97–110). The young Dutch physicist had become Bohr's assistant in 1916. He did most of the calculations for Bohr, especially on helium, which was one of the most important research topics in the old quantum theory. For his dissertation in Copenhagen, Bohr had assigned Kramers to work out an estimate of the intensities of spectral lines. Kramers' dissertation was concerned mostly with the hydrogen atom. In his dissertation he used the by then standard derivations for the hydrogen atom with and without external fields by means of celestial mechanics and dealt with the problem of intensities for spectral lines (Dresden 1987, 101). Dresden indicates that even the technicalities of the original "Bohr" argument were already worked out by Kramers, as he was the expert on Hamilton-Jacobi theory upon which Bohr relied.

To do so, Kramers followed Bohr's exposition very closely: like Bohr, he went back to Einstein's argument on the transition probabilities and pointed to the "intimate formal connection" between classical radiation theory and quantum theory shown by Bohr's correspondence argument.[81] Revisiting Bohr's original argument, Kramers changed it in order to extend the correspondence relations beyond the classical limit and selection rules. He dropped the classical limit to establish the frequency and the intensity correspondence. Instead, he compared a transition in the Bohr model with what he called "the multitude of mechanically possible states of the system lying 'between' the initial state and the final state."[82]

To describe a state within this multitude, Kramers reformulated the quantum condition as:

$$I_k = [n_k'' + \lambda(n_k' - n_k'')]h,$$

where the auxiliary parameter λ varies from the initial state $\lambda = 0$ to the final state $\lambda = 1$. Integrating over this parameter Kramers could "easily prove" that:

> the frequency ν of the radiation emitted during the transition under consideration is equal to the mean value, taken over all states [...] of the frequency $(n_1' - n_1'')\,\omega_1 + \ldots + (n_s' - n_s'')\,\omega_s$, which appears in the motion of the electron when this motion according to [the Fourier expansion] is resolved in its constituent harmonic components.[83]

This "easy" proof was possible by repeating Bohr's correspondence argument discussed above, which led to the expression:

$$\nu = \frac{1}{h}\int_0^1 \delta E = \frac{1}{h}\int_0^1 \delta I_1\omega_1 + \ldots + \delta I_k\omega_k$$

$$= \int_0^1 d\lambda[(n_1' - n_1'')\,\omega_1 + \ldots + (n_s' - n_s'')\,\omega_s].$$

In the classical limit the frequencies ω_k vary only slightly along this integral, so that the relation reduces to Bohr's formulation of the frequency correspondence. Kramers therefore argued that the original correspondence argument could be maintained along with arguments on the selection and polarization rules.

Like Bohr, he could not establish this correspondence relation through a mathematical argument, but introduced the connection between the Fourier coefficients of the electron's motion and the transition probability as a working hypothesis: the transition probability should be determined from a mean value Fourier coefficient

[81] Kramers (1919, 327).

[82] Ibid.

[83] Ibid.

constructed from "the multitude of mechanically possible states of the system lying 'between' the initial state and the final state."[84]

On the whole, Kramers evidently followed Bohr's argument, although his formulation departed from Bohr's considerably. As has been mentioned above, Kramers no longer used the classical limit. Instead, his argument was based on a comparison between the discontinuous radiation of quantum theory and the continuous multitude of mechanical states between two stationary states. The integral over these "states lying in between" results in a frequency and a Fourier coefficient, which were averages over all mechanically possible states between the initial and the final state. This argument was independent of the adequacy of classical electrodynamics in the classical limit and established the relation between radiation and motion within quantum theory.

In addition to this purely quantum theoretical formulation of the correspondence relations, Kramers gave a physical interpretation for them. From the mean-valued frequency and Fourier coefficients he constructed a new Fourier series :

$$\xi = \bar{C} \cos 2\pi \nu t$$

in which $\bar{C}$ is equal to "a suitably chosen mean value of the amplitude C_{λ}."[85]

In classical radiation theory, an electron performing such an oscillation produced radiation with the same frequency as predicted by the frequency condition. From this Kramers stipulated:

> that it might also be possible to obtain an expression for the probability in question by comparing the emitted radiation with the intensity of radiation emitted on ordinary electrodynamics by an electron performing a simple harmonic motion.[86]

Kramers thus proposed that the intensity of a quantum transition was equal to the intensity radiated by an electron in a *state in between* according to classical radiation theory. This opened up new possibilities to go beyond Bohr's and Einstein's description of the transition process, which was based solely on the frequency condition and the idea of transition probabilities. For this description, only the initial and the final states of the system were necessary, whereas the actual dynamics of the transition process was black boxed. By contrast, Kramers' model for the correspondence relation based on a continuous transition from the initial to the final state offered a way to describe the discrete quantum transitions by means of the classical model.

It is characteristic for Kramers' model that it does not change the basic conception of the state-transition model. As it is based on a comparison between a discontinuous transition and the radiation emitted from a *state in between*, Kramers' reasoning did not imply that the electron moves in this state or that it even passed

[84] Kramers (1919, 327).

[85] Kramers (1919, 330).

[86] Ibid.

through it during a transition. As such, it remains non-committal with respect to the actual nature of the transition process or with respect to the nature of radiation.

Allowing the discontinuous transition process to be bypassed by a classical model, Kramers' new formulation of the correspondence relation brought classical reasoning back into quantum theory. Taking the discrete stationary states and the transitions as the core of the Bohr model, Kramers showed that classical models could be constructed which gave the quantum theoretical frequencies and, moreover, allowed for more detailed reasoning within classical radiation theory.

In 1919 Kramers used his *state in between* to estimate the intensities of the hydrogen spectrum in the Stark effect. For this description, Kramers felt, a simple arithmetic average did not suffice. Without a physical argument he took the mean value to be $\bar{C} = e^{\int_0^1 log C_\lambda d\lambda}$, because this form fit perfectly with the Fourier expression of the *state in between*. In other words, Kramers designed the logarithmic mean value so that it would have physical significance. Which mean value was correct remained an unresolved issue throughout the 1920s. Even as late as 1923, after exploring several possible averaging procedures, Franck C. Hoyt came to the conclusion that experimental data did not support a decision for or against any of them.[87]

2.5 Conclusion

In this chapter I have analyzed the origin, formulation, and first extension of the correspondence principle in the writings of Bohr and Kramers from 1913 until 1923. As we have seen, the central problem addressed by the correspondence principle was the conceptual divide between radiation and motion within the state-transition model. Bohr's thinking that led to the formulation of the correspondence principle in 1918 revolved around this problem and centered around the idea to reestablish a connection between radiation and motion within the state-transition model.

Bohr's approach and interpretation of this relation underwent a major transformation. In his initial work from 1913 to 1916, it was used to quantize the atom as a mechanical system. This use disappeared around 1916, when Bohr adopted the Bohr-Sommerfeld quantum conditions. The relation between radiation and motion continued to play a decisive role, however, as Bohr reinterpreted it as a statement about the possibility for the occurrence of a transition. This reinterpretation, as we have seen, was driven by the problem of selection rules on the one hand, and Bohr's continued attempts to clarify the meaning of the relation between radiation and motion, on the other. Taking up Einstein's work on black body radiation, which had

[87]Hoyt (1923a,b, 1925a,b, 1926). Similar to Hoyt, Willy Thomas, as we will see in Chap. 6, and Hans Bartels approached the same problem without success. See Thomas (1924) and Bartels (1925, 1926).

introduced the notion of the transition probabilities into the state-transition model, Bohr found a way to formalize this reinterpretation. Beginning with the published version of his correspondence argument, he focused on the Fourier representation of the electron's motion in order to determine the transition probabilities.

In this manner, his correspondence argument departed considerably from Einstein's original argument. For Einstein, it had only been necessary to assume that the transition probabilities were constants describing the transition of the radiating system. Bohr and Kramers, by contrast, assumed that the transition probability depended on the motion of the radiating system and should, in principle, be calculable from the electron's motion.

Bohr's correspondence arguments did not specify a procedure to determine intensities explicitly and left central conceptual questions unanswered. The most pressing one was the initial-final-state problem, i.e., the question as to whether the transition probability was determined by the motion of the initial or of the final state. While Bohr had circumvented it in his arguments, this problem came to the fore when Debye and Kramers attempted to implement the correspondence principle in the actual determination of spectral intensities. Providing a first tentative solution to the initial-final-state problem, Kramers constructed a physical model for the correspondence relations, which allowed a discrete quantum transition to be described in terms of classical radiation theory.

As has been mentioned, this historical reconstruction not only presents an interpretation of Bohr's and Kramers' work, it also prepares the analysis of the adaptations of the correspondence principle in the following chapters. With respect to implementations of the correspondence principle, Bohr's 1918 argument and, even more so, Kramers' dissertation from 1919 were essential. Later applications of the principle did not use it to quantize the atom as a mechanical system, but rather aimed to determine transition probabilities.

In this approach, the Fourier representation of the electron's motion became the central tool. Apart from a few exceptions, as we will see, physicists adopted this tool to make their own correspondence arguments and turned to Kramers' dissertation rather than Bohr's original papers. The former provided the basis for making correspondence arguments in the form of a physical model, which became known as the *Zwischenbahn* or *intermediate orbit*. As it was considered to be the most elaborate but equally tentative solution to the initial-final-state problem, the model was used to establish explicit mathematical expressions for the intensities.

and selected the notion of the transition probabilities into the state-transition model Bohr found a way to formalize this interpretation. Beginning with the published version of his correspondence argument, he focused on the Fourier components of the electron's motion in order to determine the transition probabilities.

In this manner, the correspondence argument departed considerably from Einstein's original argument. For Einstein, it had only been necessary to assume that the transition probabilities were constants describing the transition of the radiating system. Bohr and Kramers, however, assumed that the transition probabilities depended on the motion of the radiating system and could, in principle, be calculated from the electron's motion.

Bohr's correspondence arguments did not specify a procedure to determine [illegible] explicitly and left central conceptual questions unanswered. The most pressing one was the causal-energetic problem, i.e., the question as to whether the transition probabilities are determined by the motion of the initial or of the final state. While Bohr had hoped to [illegible]

[illegible]

Chapter 3
The Correspondence Principle in the Quantum Network 1918–1926

The discussion in the preceding chapter focused almost entirely on Bohr and his thinking about quantum theory. But how was Bohr's correspondence argument received within the quantum network? When and how did his peers take up the principle in their work and develop their own applications? This chapter surveys the principle's dissemination outside of Copenhagen and discusses the pattern underlying it, arriving at more concrete questions rather than definite answers.

Section 3.1 gives a chronological and general thematic overview of the reception and applications of the correspondence principle from 1918 until 1926 and identifies two, distinct phases. In the first, ranging from 1918 up to 1922, the principle was received and understood by physicists in the quantum network. It was paraphrased and discussed in textbooks, university lectures as well as in review articles. In research papers, Bohr's original correspondence argument was also referenced and reiterated to establish selection rules for atomic and molecular spectra. Characteristically, however, these applications hardly extended the scope of the principle.

The second phase of the principle's reception was different. From 1922 until 1926, physicists did not stop at Bohr's arguments but adopted the principle as a research tool. They extended it to new phenomena and developed their own correspondence arguments to tackle them. Along with this qualitative shift, the number of papers drawing on the correspondence principle increased considerably and correspondence arguments became more elaborate. Prior to 1922, correspondence arguments in research papers remained qualitative for the most part and hardly exceeded a paragraph. After 1922, they became more detailed and technical, filling pages or even entire papers.

Characterizing these clearly distinct phases, this chapter discusses how physicists understood the principle and to which extent they differed in their interpretations. There are two remarkable observations to be made. First, physicists in various places clearly identified the relation between radiation and motion as the core idea of the principle and did not differ widely in their interpretations of it. Second, interpretations developed during the principle's early reception were still

M. Jähnert, *Practicing the Correspondence Principle in the Old Quantum Theory*, Archimedes 56, https://doi.org/10.1007/978-3-030-13300-9_3

in place when physicists first developed their own correspondence arguments. This provides a baseline for discussing the emergence of different applications and the principle's *transformation through implementation* in the following chapters. Physicists adopted one tool rather than many.

The dissemination of the correspondence principle raises additional questions pertaining to the transmission of knowledge within research networks. Such questions have received considerable attention in recent years, most notably in David Kaiser's seminal book *Drawing Theories Apart*, in which he connected the dispersion of Feynman diagrams with the culture of postdoctoral training in a collaborative research center. Section 3.2 discusses whether the dissemination of the correspondence principle and its abrupt shift from reception to adaptation can be understood as a result of the establishment of Bohr's *Universitetets Institut for Teoretisk Fysik* or whether an alternative explanation centering on the growing interest in transition probabilities as a physical quantity is more adequate.

3.1 From Reception to Adaptation: Paraphrases of the Correspondence Principle 1918–1926

As we can gather from private correspondence, published papers and early textbooks on quantum theory, the dissemination of the correspondence idea in the network of quantum theory started as early as 1918. As physicists received Bohr's first published correspondence argument, they referred to it under the label "Korrespondenzprinzip" or "Analogieprinzip."

The existence of these two denominations should not be overestimated. Adolf Kratzer, Alfred Landé, and Paul Peter Ewald, for example, used the two terms interchangeably, even fusing them together into "the important Bohrian *Korrespondenz(Analogie)-prinzip*."[1] Using these different labels, they did not attach different interpretations to the principle or discuss the difference in terminology. The only one to do so was Bohr himself. In his 1918 paper, as we have seen, he had referred to his argument as a "formal analogy between classical and quantum theory." Confronted with the label "Analogieprinzip," he thought it important to clarify that the principle was not a metatheoretical statement about the relation between classical and quantum theory. Rather, it had to be regarded as "a purely quantum theoretical law," which would better be referred to as the "Korrespondenzprinzip."[2] Whether Bohr's prescribed terminology had any impact on the applications of the correspondence principle is highly doubtful. When

[1] Ewald (1920, 305). See also Kratzer (1920, 290 and 294) for Kratzer's use of the two terms. See also Landé (1921a, 231, 233 and 238) and Landé (1921b, 401) for Landé's use of the term "Korrespondenzprinzip" in the first instantiation of his paper "Über den anomalen Zeemaneffekt" and his later adoption of the term "Analogieprinzip" in the second part of the paper.

[2] See Sect. 2.4 in Chap. 2.

his statement appeared in print in 1923, the label "Analogieprinzip" had already disappeared from published papers.

Whether referring to the analogy principle or the correspondence principle, the understanding that physicists formulated in numerous restatements of Bohr's argument was quite homogeneous. This can be seen, for example, in the earliest textbook accounts on Bohr's argument by Fritz Reiche, Arnold Sommerfeld, Léon Brillouin, and Alfred Landé. While they discussed it at different levels of detail, these authors introduced the correspondence principle along the same lines, which can be summarized as follows: As in Bohr's original argument, they first developed the identity between the radiation frequency and the mechanical frequency in the limit of large quantum numbers. Second, this identity was extended into the frequency correspondence for all quantum numbers. Third, this argument was extended to the intensity correspondence. Finally, all textbook accounts pointed to the explanation of selection rules as the most important application of the principle.[3]

Within published research papers, Bohr's correspondence argument, first of all, provided a justification for selection rules in atomic and molecular spectra. In this capacity, the principle was taken for granted and physicists simply referenced Bohr's argument as a well-established part of quantum theory.[4] Within these extensions of Bohr's selection rule argument, there are very few paraphrases of Bohr's argument that reflect the understanding of the principle at the time. Like the textbook accounts, these paraphrases by Fritz Reiche, Adolf Kratzer and Herta Sponer indicate that the core element of Bohr's principle and its use were understood quite homogeneously.[5]

Not by chance, these paraphrases appeared in the context of molecular spectroscopy and focused on the model of the rotator, which was discussed in detail in

---

[3]This summary was given in a qualitative form in Reiche's *Die Quantentheorie: Ihr Ursprung und Ihre Entwicklung* and in the first edition of Sommerfeld's *Atombau und Spektrallinien*. See Reiche (1921, 130–132) and Sommerfeld (1919, 401–403). Sommerfeld's general presentation of the correspondence principle mainly explicated the difference between his and Rubinowicz' approach to the selection rules and the "Analogieprinzip." Even in this characterization, in which he famously referred to the principle as a "magic wand," Sommerfeld summarized Bohr's argument as discussed in the main text. As Eckert and Märker (2004, 18–21) and Seth (2010, 234–235) have pointed out, Sommerfeld extended his presentation considerably in the second and third edition. Including the short paraphrase of the first edition, he gave a more positive interpretation of the principle. More importantly, he added an extensive discussion in an appendix that provided an instructive introduction to the application of the principle on the basis of the Fourier representation of the radiating system. See Sommerfeld (1921, 527–537) and Sommerfeld (1922a, 699–711). This presentation matched the level of detail given in Landé (1922a, 32–36 and 69–73) and in Brillouin (1922, 115–121) and is comparable with the respective expositions in later accounts. See, for example, Buchwald (1923), Born (1925), Pauli (1925), Van Vleck (1926), or Landé (1926).

[4]See Franck and Reiche (1920), Sommerfeld and Kossel (1919), Sommerfeld (1920b), Lenz (1920), and Heurlinger (1920).

[5]In addition, one can consider Alfred Landé's papers on the anomalous Zeeman effect in 1921 and Werner Heisenberg's remarks on the principle in his letters to Landé at the time. Their understanding of the principle is discussed in detail in Chap. 4. As we will see, Heisenberg and Landé understood the general idea of the correspondence principle, but implemented it in different ways.

Chap. 2. Reserving it for the fourth part of his treatise "On the Quantum Theory of Line Spectra," Bohr had not published his argument on the rotator. Reiche, Kratzer and Sponer therefore had to develop the correspondence explanation for selection rules on their own following Bohr's other applications for harmonic systems. For the present survey, the specifics of the problem are not important and I will therefore not describe them in detail again. Instead, I will briefly summarize and compare the main lines of the arguments.

Reiche's correspondence argument followed the line of thought along which Bohr had developed his first explanation of selection rules. Quantizing the energy of the rotator and applying the frequency condition, he found the quadratic dependence of the radiation frequency on the rotational quantum number. Considering the high quantum number limit, Reiche argued that there could only be transitions in which the rotational quantum number changed by 1; otherwise, the required equality between the radiation frequency and mechanical frequency would be lost. He then extrapolated this result to small quantum numbers.[6]

In Munich and Göttingen, by contrast, Adolf Kratzer and Herta Sponer argued on the basis of the Fourier series of the radiating system, thus following the correspondence argument given in Bohr's 1918 paper. Kratzer's correspondence argument, which he developed in his dissertation written with Sommerfeld, was a short one. In it, Kratzer did not restate the principle's content but only presented conclusions drawn from it. Giving the Fourier series of the rotating anharmonic oscillator with the fundamental frequency ν_0 and the frequency of a superimposed rotation ω:

$$x + iy = \frac{1}{2}\sum_{n=0}^{\infty} C_n\left(e^{i(n2\pi\nu_o+\omega)t} + e^{-i(n2\pi\nu_o-\omega)t}\right)$$

$$z = 0,$$

he concluded:

> On the basis of the analogy principle we read from this: the rotational quantum number m only changes its value by ± 1, the oscillational quantum number can change by any value $0, 1, 2, 3\ldots$. It is remarkable that, as soon as rotation is present, a change in the oscillational quantum number is not to be expected without a simultaneous change in the rotational quantum number, while inversely an individual change of the rotational quantum number is possible.[7]

These conclusions show how Kratzer understood the principle and how he put it to work. To establish selection rules, he considered the Fourier representation and

[6] Reiche (1920, 286–287).

[7] Kratzer (1920, 290). "Nach dem Analogieprinzip lesen wir hieraus ab: Die Rotationsquantenzahl m ändert sich nur um den Betrag ± 1, die Oszillationsquantenzahl kann sich um beliebige Werte $0, 1, 2, 3\ldots$ ändern. Bemerkenswert ist, daß eine Änderung der Oszillationsquantenzahl ohne gleichzeitige Änderung der Rotationsquantenzahl nicht zu erwarten ist, sobald Rotation vorliegt, während umgekehrt eine alleinige Änderung der Rotationsquantenzahl möglich ist."

identified which frequency occurred in it and which did not. From this occurrence and absence, he then inferred the possibility and impossibility of transitions, deducing selection rules and a correlation between jumps in the oscillational quantum number and in the rotational quantum number.

Along similar lines and in a more explicit manner, Herta Sponer came to the same conclusions in her unpublished dissertation "Über ultrarote Absorption zweiatomiger Gase" supervised by Peter Debye in Göttingen in 1920. Following a paraphrase of Bohr's correspondence argument, she summarized the steps that had to be taken in the application of the principle:

> Bohr relates these τ's [the overtones of the Fourier series of the motion, MJ] with the quantum theory, as he develops that quantum theory and electrodynamics match each other in the limit of large n, that is, that in this case both expansions:
>
> $$\sum C_{\tau_1,..\tau_s} e^{2\pi i(\tau_1\omega_1+...\tau_s\omega_s)}$$
>
> and
>
> $$\sum C_{\tau_1,..\tau_s} e^{2\pi i((n'_1-n_1)\omega_1+...(n'_s-n_s)\omega_1)}$$
>
> will coincide. Bohr extrapolates this result to the region of small quantum numbers. We will follow his path, and thus have the following task before us: We will write down the χ-coordinate of our dipol (in a x-y-z system), expand it into a Fourier series, infer from the resulting τ's the differences $n' - n$ and insert these into the series formula.[8]

Sponer's—like Kratzer's—correspondence argument was based on the frequency correspondence. In it, she inferred the possibility of the corresponding transitions in quantum theory from the occurrence or absence of frequencies in classical Fourier representation.

In short, this survey of paraphrases of Bohr's correspondence argument in research papers and textbooks shows that the early reception of the correspondence principle focused on the explanation of selection rules. This specific example was central to physicists' understanding of the correspondence principle. According

[8]Sponer (1920, 13–14). "Bohr setzt diese τ's in Beziehung zur Quantentheorie, indem er entwickelt, dass in der Grenze für grosse n sich Quantentheorie und Elektrodynamik decken werden, d. h. dass in diesem Falle die beiden Entwicklungen:

$$\sum C_{\tau_1,..\tau_s} e^{2\pi i(\tau_1\omega_1+...\tau_s\omega_s)}$$

und

$$\sum C_{\tau_1,...\tau_s} e^{2\pi i((n'_1-n_1)\omega_1+...(n'_s-n_s)\omega_1)}$$

übereinstimmen werden. Das Resultat extrapoliert Bohr auf den Bereich kleiner Quantenzahlen. Wir werden seinen Weg beschreiten, haben allso [sic!] folgende Aufgabe vor uns: Wir werden uns die χ-Koordinate (in einem x-y-z-System) unseres Dipols aufschreiben, sie in eine Fourierreihe entwickeln, aus den sich ergebenden $\tau's$ auf die Differenzen $n' - n$ schliessen und diese in die Serienformel einsetzen."

to it, the correspondence principle became a statement about the possibility of transitions of a quantum system. In keeping with Bohr's correspondence argument, the operational core of the principle was the connection between the motion of the radiating system and its radiation properties.

Understanding the principle was one thing, developing further applications was quite another. Apart from the handful of references and adaptations to Bohr's selection rule argument, physicists outside of Copenhagen did not engage in this activity up to circa 1922. This was a disappointment for Bohr. As he saw it, his "attempts to develop the principles of quantum theory [...] were met with very little understanding."[9]

By the time Bohr started complaining about the neglect of his work on quantum theory, the situation was beginning to change. The years 1922 and 1923 became the watershed for the applications of the correspondence principle. From this point onwards, the sheer number of papers featuring correspondence arguments increased considerably. This increase is represented in Tables A.3, A.4, A.5 and A.6 in Appendix A.[10] These tables list papers featuring correspondence arguments for the research fields of atomic and molecular spectroscopy, the quantum theory of absorption and dispersion, and the study of collisions. They are further subdivided with respect to the local communities that produced these papers. In this representation, the chronological order of the papers is maintained for each community. A global chronological order, by contrast, is not given. This is a deliberate choice, informed by the case studies. It reflects that correspondence arguments emerged primarily within specific local settings. Interactions between different communities took place at certain key junctions. They were highly contingent on the personal relations among the actors, which have to be assessed in each individual case and cannot be represented adequately by means of simple tables.

In other words, the tables present a correlation between the centers for the quantum network, the *patchwork of problems* and the applications of the correspondence principle. This correlation opens up two interlocking perspectives on quantum physics during the 1920s. First, it shows how physicists approached different phenomena in various places. In this respect, the applications of the correspondence principle function as a probe into the relation between the quantum network and the *patchwork of problems*. They illustrate a well-known feature of work on quantum physics during the 1920s: the research centers in the quantum network focused on particular problems, while leaving others untouched. For example, the community in Munich worked mainly on multiplet spectra and band spectra. Physicists in Breslau focused on the quantum theory of absorption and dispersion. In Göttingen, the

[9]Bohr to Sommerfeld, 30 April 1922 in Eckert and Märker (2004, 117). "In den letzten Jahren habe ich mich oft wissenschaftlich sehr einsam gefühlt unter dem Eindruck dass meine Bestrebungen, nach besten [sic!] Vermögen die Principien der Quantentheorie systematisch zu entwickeln, mit sehr wenig Verständnis aufgenommen worden ist [sic!]."

[10]Papers written by physicists who developed their correspondence arguments in collaboration with Bohr or Kramers during a stay in Copenhagen are given in a separate Table A.7.

problem of collisions and the quantum theory of dispersion played a major role. In Harvard, band spectra were essential.[11]

Second, we can interpret the correlation with respect to the applications of the correspondence principle itself and look at how the correspondence principle was used from 1922 onwards. As we can see, physicists adopted the principle almost simultaneously. In different places in Europe and the U.S. and in different research fields, these applications went beyond the establishment of selection rules and tackled new phenomena ranging from the intensity of spectral lines to the Ramsauer effect. Despite this wide range, as will be shown in more detail in the next section, these arguments had a common denominator; the correspondence principle was used, first of all, to determine transition probabilities.

To understand how these new applications affected the correspondence principle we need to look at them in detail. Before doing so in the following case studies, however, one overarching observation is crucial: Physicists continued to paraphrase the core idea of the correspondence principle in the same way and departed from Bohr's original formulation in their new correspondence arguments. A prime example for this continuity is Friedrich Hund's formulation of the correspondence principle in his work on the Ramsauer effect in 1922:

> In the application of this principle in atomic physics, the frequencies which are radiated in a quantum jump relate to the frequencies contained in the motion on the initial and on the final orbit; the transition probabilities are determined by the coefficients associated with the respective frequencies in the Fourier series of the motions in the initial and final orbit.[12]

In a similar manner, Fritz Reiche formulated the correspondence principle in his textbook *Die Quantentheorie: Ihr Ursprung und ihre Entwicklung*:

> [T]he "classical" partial vibration $(\tau_1 \ldots \tau_f)$ corresponds to that quantum transition in which the quantum number changes by exactly $(\tau_1 \ldots \tau_f)$. The polarization and intensity of the waves emitted during this quantum transition can thus be read from the form of the oscillation and amplitude of the "respective classical" partial vibration.[13]

Finally, the core idea of the correspondence principle remained unchanged in the work of Sommerfeld and Heisenberg on multiplet intensities from 1922:

---

[11]For a general overview on the quantum community in the 1920s, see Kragh (2002, especially 140–151), Eckert (2001), and Castagnetti and Renn (forthcoming).

[12]Hund (1922, 43). "Bei der Anwendung dieses Prinzips in der Atom-Physik entsprechen die Frequenzen, die bei einem Quantensprung gestrahlt werden, den Frequenzen, die die Bewegung auf der Anfangs- und Endbahn enthalten; die Uebergangswahrscheinlichkeit wird durch die Koeffizienten gegeben, die die betreffende Frequenz in den Fourierreihen der Bewegung auf der Anfangs- und Endbahn hat."

[13]Reiche (1921, 131). "[D]ie 'klassische' Partialschwingung $(\tau_1 \ldots \tau_f)$ entspricht demjenigen Quantenübergang, bei dem die Quantenzahlen sich gerade um $\tau_1 \ldots \tau_f$ ändern. Polarisation und Intensität der bei diesem Quantenübergang emittierten Welle läßt sich also aus der Schwingungsform und Amplitude der 'entsprechenden klassischen' Partialschwingung ablesen."

> The correspondence principle, as is well known, permits one to infer the intensities of spectral lines from the kinematic character of the atomic orbits.[14]
>
> [...]
>
> According to the correspondence principle, the intensities are generally determined as Fourier coefficients.[15]

Whether explaining the principle's content or taking it for granted, Hund, Reiche, and Sommerfeld and Heisenberg followed the formulation of the correspondence principle by Bohr and Kramers. For them, the core idea of the correspondence principle was clear: the motion of the radiating system was associated with the radiation emitted during a quantum transition. To make use of this idea, they set up a Fourier representation of the motion of the system and associated it with the respective transitions to determine the transition probabilities.

The homogeneity with which physicists received the content of the correspondence principle is an important result: for its users in the 1920s, the correspondence principle had a well articulated core. Despite the vagueness, ambiguity, and tentativeness of Bohr's formulation, their applications took off from this core idea. As we will see much more clearly in the case studies, this has important implications for the analysis of the principle's *transformation through implementation*: Its diversification was not rooted in different understandings of the principle. As they applied it to different problems, physicists used one tool rather than many.

3.2 The Dissemination of the Correspondence Principle: Preliminary Considerations

As already mentioned at the beginning of this chapter, the overall chronology outlined above presents a major challenge for understanding the applications and dissemination of the correspondence principle. In light of the previous reconstruction, this challenge appears in the form of a concrete research question: Why did several physicists working independently in different fields and places start to use the correspondence principle around 1922? What motivated them to do so several years after they had understood the principle in terms of both its content and its original range of application?

As we will see in the following case studies, answering these questions is possible in each individual case. Looking at the dissemination of the correspondence principle in general, however, giving such answers is no easy task, given the multitude of actors and the different conceptual situations. In the following, I will

[14]Sommerfeld and Heisenberg (1922a, 131). "Das Korrespondenzprinzip gestattet, wie bekannt, aus dem kinematischen Character der Atombahnen auf die Intensität der Spektrallinen zu schließen."

[15]Sommerfeld and Heisenberg (1922a, 140). "Im allgemeinen sind nach dem Korrespondenzprinzip die Intensitäten als Fourierkoeffizienten bestimmt."

sketch two potentially interlocking candidates for an answer to this question. The first focuses on the research problems addressed in correspondence arguments; the second centers on Bohr's position in the network of quantum theory and his channels of communication.

From Selection Rules to Transition Probabilities

The turn from reception to adaptation in 1922 is marked not only by an immense increase in the number of papers making correspondence arguments but also by a qualitative shift in the way correspondence arguments were made. As indicated in Tables A.3, A.4, A.5 and A.6 in Appendix A, the vast majority of these applications share a common characteristic: Whether concerned with the intensity distribution in series, multiplet, or band spectra, with the number of dispersion electrons, or the explanation of the Ramsauer effect, physicists were concerned with the determination of transition probabilities.

For this task, as Enrico Fermi put it in 1924, the correspondence principle provided the only relevant tool[16]:

> In the present state of atomic theory, the only means that allows one to account for the experimental results on the *intensity of spectral lines* is given by the application of the *correspondence principle*.[17]

In a similar manner, Edwin Kemble spent most of his time explaining the intensity correspondence in his lecture at Harvard in the fall of 1923. The correspondence principle, he pointed out, was "the important tool" of Bohr's quantum theory and provided "a means of calculating" transition probabilities. It thus shed "great light on the question of relative intensities," on which Kemble was working at the same time.[18]

The growing importance of transition probabilities in the applications to the correspondence principle after 1922 is a central feature of the turn from reception to adaptation. From 1918 until 1922, as we have seen, physicists focused on Bohr's

[16]The correspondence principle was the only tool to calculate transition probabilities. As physicists were well aware, however, the intensity of radiation depended not only on the probability for the occurrence of a transition, but also on the number of systems in a certain state given by the statistical weight of that state. These two aspects were conceptually and theoretically separate from each other. The problem of statistical weights was a matter of dividing the state or phase space into equally probable cells. As such, it was an issue for statistical mechanics in quantum theory. The connection between radiation and motion expressed by the correspondence principle, by contrast, was connected to the radiation mechanism of a yet to be developed quantum theory of radiation.

[17]Fermi (1924a, 340, emphasis in the original). "Bij den huidigen stand der atoomtheorie is het eenige middel waardoor men zich theoretisch rekenschap kan geven van de experimenteele resultaten betreffende *de intensiteiten der spectraallinien* gelegen in de toepassing van het *correspondentie beginsel.*"

[18]See page 2 of Kemble's talking points (AHQP 55.2).

selection rule argument. This explanation could be made without setting up a Fourier representation solely on the basis of the frequency correspondence. Even if one considered the Fourier series of the electron's motion, one did so with respect to the occurrence or absence of a certain frequency. The situation changed with the turn to transition probabilities. From 1922 onwards, physicists expressed their correspondence arguments by setting up Fourier series and determined the transition probabilities from the Fourier coefficients of the electron's motion. In this new approach, as we will see in the next chapters, the Fourier coefficients and Kramers' solution to the initial-final-state problem became central. Kramers' "state lying in between the initial and the final state" became known as the "Zwischenbahn" or the "hypothetical intermediate state" and provided *the* physical model used to operationalize the correspondence principle.[19]

How can we characterize the dynamics underlying this shift? At first sight, one might take the following position: initially physicists first of all received the core idea of the correspondence principle and its explanation of selection rules and only gradually came to realize that the principle was far more powerful, providing a means to determine intensities. Intensities, one might further argue, presented a conceptual lacuna within the overall framework of quantum theory and physicists took up the principle in an attempt to establish a more comprehensive description of transition processes.

While presenting a plausible scenario, there are good reasons to reject both the assumption of a gradual recognition of the principle's potential and the assumption of a conceptual lacuna as the central motivation for making correspondence arguments. With respect to the former, it is clear that the intensity correspondence had been part of the reception of the correspondence principle as early as 1918. It was mentioned in the textbook accounts, accepted implicitly in the earliest correspondence arguments, and identified as an important point of departure for further developments.[20] This point is illustrated by Peter Debye's letter to Bohr discussed in Chap. 2. In it, as we have seen, Debye put special emphasis on the intensity correspondence, stating that "your *ansatz* for the calculation of the intensities is evidently of major importance!"[21] Similarly, Sommerfeld pointed to the intensity correspondence as a major feature of Bohr's approach:

> By the way, Bohr has taken another step in abandoning the radiation mechanism by taking over the observed state of oscillation [of radiation, MJ] from classical theory on the basis of

[19]As we will see in Chaps. 6 and 7, the Copenhagen community replaced the *Zwischenbahn* model with the *virtual oscillator* model between 1923 and 1924. This became the source of disagreement with physicists outside of Copenhagen like Reiche or Pauli, for whom the *Zwischenbahn* model presented an integral part of the correspondence approach.

[20]The importance of the intensity correspondence was mentioned, for example, in Reiche (1921, 132) or Landé (1922a, 33–36).

[21]Debye to Bohr, 6 June 1918 in Bohr (1976, 607). See Footnote 79 in Chap. 2.

his correspondence principle. It is well known which benefits he achieves in this way, also in the question of the intensities.[22]

The prospect of determining intensities was thus a well-known part of the correspondence principle. Nonetheless, as we have seen, it remained peripheral in research papers from 1918 until 1922. This is shown most pointedly by the fact that there was only one detailed calculation of an intensity distribution in this period which took up the approach of Kramers' dissertation. Developed by Herta Sponer, this calculation was not deemed important even by the author. In the published version of her doctoral thesis, she merely summarized her results, omitting the details of her calculations.[23]

This brings me to the second issue of the above characterization: the deficient description of intensities in the state-transition model as a lacuna within the framework of quantum theory. Just like the intensity correspondence, this lacuna was well known. Already in the 1910s, physicists had identified the deficient description of intensities in their general descriptions of Bohr's quantum theory of the atom, and had also considered the correspondence principle as a possible point of departure, as is shown exemplarily by the statements of Sommerfeld and Debye. Nonetheless, extensive efforts to develop a general, comprehensive description of intensities within quantum theory were not undertaken. As we will see in the following case studies, physicists turned their attention towards transition probabilities without seeking to improve quantum theory as a conceptual framework.

To understand why this was the case, one has to keep in mind what the correspondence principle was valued for up to 1922. Rather than perceiving it as a fruitful program for a future quantum theory, most physicists saw the correspondence argument as a solution to a specific problem: the explanation of selection rules. As we have seen, this argument could be made solely on the basis of the frequency correspondence and did not require further specifications of the connection between radiation and motion. As such, it neither enforced the conviction that a description of intensities needed to be incorporated into the state-transition model, nor did it suggest how such a description could be developed.

Still, the parallelism of the turn from reception to adaptation and the shift towards the determination of transition probabilities presents a first candidate for interpreting the adoption of the correspondence principle by physicists outside Copenhagen. To see whether this parallelism is more than a correlation or a circular argument, however, requires a more fine grained analysis of the motivations of the historical

---

[22] Sommerfeld (1920d, 418). "Übrigens ist Bohr in dem Verzicht auf den Erregungsmechanismus noch einen Schritt weiter gegangen, indem er den beobachtbaren Schwingungszustand nach seinem Korrespondenzprinzip aus der klassischen Theorie herübernimmt. Es ist bekannt, welche Vorteile er auf diesem Wege, auch in der Frage nach den Intensitäten erzielt."

[23] Sponer (1921). Published in the virtually unknown *Jahrbuch der Philosophischen Fakultät der Georg-August-Universität zu Göttingen*, Sponer's dissertation did not play a role in subsequent discussions.

actors. What turned the determination of transition probabilities into a research subject and did this motivate the development of new correspondence arguments?

Into the Quantum Network: Bohr as the "Director of Atomic Physics" in the 1920s

Another possible candidate for interpreting the dissemination of the correspondence principle may be identified by analyzing the institutional development of the quantum network and the role Copenhagen played within it. Such an interpretation could follow lines similar to David Kaiser's work on the dispersion of Feynman diagrams in the 1940s and the 1950s. In it, Kaiser argued that the dissemination and adaptation of Feynman diagrams has to be understood in connection with their modes of transmission. He identified the *Institute for Advanced Study* in Princeton as the central institution. Their adaptation was mainly due to "some form of informal personal communication" in decidedly pedagogical contexts as young post-doctoral researchers were trained to use them through informal personal communication and then took the diagrams with them to other places and research fields.[24]

Adopting such a perspective in the case of the correspondence principle, we could investigate whether a similar mechanism was in place. We then have to consider whether and to what extent personal discussions with Bohr, exchanges through letters, or lectures were essential for making correspondence arguments. Moreover, we have to ask whether such interactions motivated the adoption of the principle as a tool.

Doing so is not far fetched. The institutional development of Copenhagen, which has recently been analyzed by Alexei Kojevnikov, closely parallels the chronology of the dissemination of the correspondence principle. As Kojevnikov argues, Bohr was not in a position to influence research on quantum physics on a larger scale when he took the chair for theoretical physics at the small university of Copenhagen in 1916. Throughout the 1910s, Copenhagen was at the periphery of the quantum network. This only changed after the end of World War I. As a consequence of their neutrality during the war, Scandinavian countries seized the opportunity to mediate between physicists in Germany and their British, American and French colleagues during the boycott of German science.[25]

Out of this position, Bohr first established Copenhagen as a hub within the growing quantum network. The incipient center was further strengthened in 1921 with the creation of the *Universitetets Institut for Teoretisk Fysik*, Bohr's Nobel prize in 1922 and the extension of the institute in 1924. These developments ran parallel and were fostered in part by Bohr's attempt to build strong connections

[24]Kaiser (2005).

[25]Kojevnikov (forthcoming). See also Robertson (1979) for a history of the institutional development of Bohr's *Universitetets Institut for Teoretisk Fysik*.

with funding agencies in Denmark and the U.S. like the Carlsberg Foundation, the Rask-Ørsted Foundation and, in 1924, the International Education Board of the Rockefeller Foundation (IEB). Becoming the "Director of Atomic Physics,"[26] Bohr drew physicists to Copenhagen by hiring assistants, organizing fellowships granted by different foundations, and by accepting physicists from the U.S. or Japan, who came with their own funding.[27]

This allowed him to work with a close collaborator and to bring physicists to Copenhagen for talks or short visits ranging from a month to a year. As Kojevnikov has argued, this strategy was especially successful in attracting German scientists. Due to the financial instability caused by the hyperinflation in the Weimar Republic, the bonds loosened between young physicists earning their doctorates and their professors. Rather than becoming assistants working toward their *Habilitation*, they became "quasi-free postdocs" able to go abroad.[28]

In contrast to Kojevnikov's main thesis, however, the opportunity to go to Copenhagen opened up for only a few physicists in this younger generation.[29] Despite the new social mobility, only Adam Rubinowicz, Wolfgang Pauli, Werner Heisenberg and one other "postdoc" came from universities in Germany up to 1925. The circumstances under which they did were highly dependent on personal connections to Bohr and were still mediated by German professors. As such, Rubinowicz' visit was arranged by Sommerfeld to support the young researcher from Poland. Pauli and Heisenberg, having been recognized as the most talented physicists of the new generation, caught Bohr's attention during the Wolfskehl lectures in Göttingen in 1922 and became Bohr's assistants to help with publications in German.[30] Though certainly crucial for the individual trajectories of Rubinowicz, Pauli, and Heisenberg, the pathway to Copenhagen was a marginal phenomenon limited to the elite of the new generation in Germany. This was due for the most part to timing. Beginning in 1925, Bohr's connection to the IEB was stable enough to attract and sustain more German "postdocs," like Friedrich Hund, Erich Fues, Lothar Nordheim, Pascual Jordan, Werner Heitler, as well as physicists from Britain like Ralph Fowler and Paul Dirac. Prior to 1925, however, young German physicists

---

[26] Sommerfeld to Bohr, 25 April 1921 in Sommerfeld (2004, 98–99). "Ich fürchte überhaupt, dass Sie sich mit Ihrer Doppelstellung als Institutsdirektor und Direktor der Atomtheorie zu viel übernommen haben."

[27] Kojevnikov (forthcoming).

[28] Kojevnikov (forthcoming).

[29] See Robertson (1979, 156–159) for a list of the visitors in Copenhagen. The new pathway leading to Copenhagen was only beginning to take shape and was restricted to few physicists: Until 1924 only two German professors came to Copenhagen, in part to help get the laboratories up and running. Likewise, a first group of younger physicists from Scandinavia, the Netherlands, Hungary and Germany worked at the institute for a year. Some, like Kramers, even became long-term members.

[30] The episodes leading to Rubinowicz', Pauli's and Heisenberg's pathway to Copenhagen are well known. For a description see Kojevnikov (forthcoming) or Robertson (1979, 62–63). It is unclear how the fourth postdoc, L. Ebert, got invited to Copenhagen.

did not find the path to Copenhagen in large numbers.[31] Unlike the case of Feynman diagrams, there was no cohort of postdocs to take up the principle during its training, and visits to Copenhagen did not play a role in the dissemination of the correspondence principle on a communal scale.

This does not mean that personal communication played only a marginal role. In individual cases, as we will see, private discussions were crucial for the applications of the correspondence principle. Moreover, Bohr reached out to physicists in the quantum network by other means of communication. For example, he gave lectures in Berlin, Göttingen, Leiden, London, Brussels, Liverpool, and at colleges like Yale and Harvard. These lectures, as we have seen in Chap. 2, forced Bohr to clarify his thoughts on the correspondence principle and thereby became essential for the consolidation of the correspondence argument into a principle of quantum theory.[32] At the same time, these lectures brought Bohr into contact with physicists outside of Copenhagen. They can be seen as part of his larger attempt to promote his approach to quantum theory on a communal scale and to establish Copenhagen as a center within the growing network of quantum theory. In this attempt, Bohr also relied on personal correspondence. As 16 microfilm reels in the AHQP testify, letters played an essential role for Bohr even back in the 1910s and continued to do so when Copenhagen became a center of the quantum network in the 1920s. While the strategy of bringing physicists to Copenhagen was thus limited to a small group, these letters allowed Bohr to discuss various approaches to problems in quantum physics with physicists from Germany, Switzerland, Italy, the Netherlands, Great Britain, the U.S., and Japan.

The growing importance of Copenhagen as a center in the quantum network and personal communication with Bohr and the Copenhagen community provided suitable conditions for the dissemination of the correspondence principle. However, in order to see whether these conditions were essential and how the various forms of interaction affected the applications of the correspondence principle, we need to look more closely. For example, we need to examine whether physicists turned

[31] See the list of visitors in Robertson (1979, 156–159). The largest group of visitors came from the U.S. and Japan, where postdoctoral training based on traveling abroad had been established earlier. Sustained by a system of fellowships in their home countries, they made up the largest group of short-term visitors. For the institutionalization of postdoctoral training in the U.S. during the 1920s, see Assmus (1993) and Duncan and Janssen (2007a, 563–566).

[32] See Nielsen (1976) for an editorial account on the relation between Bohr's lectures and his survey articles. Bohr gave lectures in Berlin published as Bohr (1920), in London published as Bohr (1922). Hendrik Antoon Lorentz invited Bohr to the Solvay Conference in 1921. Due to bad health, Bohr had to excuse himself from the conference, so that his talk was elaborated by Paul Ehrenfest in close collaboration with Bohr and Kramers. See Ehrenfest to Kramers, 7 April 1921 (AHQP 8a). The resulting lecture was held at the Solvay Conference and later published as Bohr (1923a). Furthermore Bohr gave a lecture in Copenhagen, published as Bohr (1923b). His Wolfskehl lecture in Göttingen 1922 [AHQP 3 and Bohr (1977, 341–421)] and Leiden (Bohr 1976, 201–217) as well as his lectures at Yale, Harvard, and other universities during his visit to the U.S. in the fall of the 1923, remained unpublished. Munich, the only major center Bohr did not visit, is missing probably because of the particular uneasiness between Sommerfeld and Bohr. See Eckert and Märker (2004).

to the Copenhagen institute to learn the central techniques necessary to make correspondence arguments, or whether their new applications emerged as a result of Bohr's promotion of the correspondence principle.

3.3 Conclusion

As we have seen in this chapter, the correspondence principle was received outside of Copenhagen as early as 1918. Apart from smaller variations, this reception was a uniform one: In places like Munich, Göttingen, Breslau, Leiden, and Harvard, physicists understood that the core idea of the correspondence principle was a connection between the motion of the radiating system and its radiation. This relation, physicists had understood, applied to all quantum states and it was only in this capacity that it became a fruitful part of quantum theoretical research. Its main importance was not to reestablish a connection between the old theories of mechanics and electrodynamics. Rather, it was a tool to solve specific problems, like the explanation of selection rules or the determination of transition probabilities.

This assessment provides a baseline for the discussion of the applications of the correspondence principle in the next chapters. As we have seen, physicists received the content of the correspondence principle in a homogeneous way. This provides an important piece for understanding the principle's *transformation through implementation*: Applying the principle to different problems, physicists used one tool, whose core idea was clearly understood throughout the quantum network. The emerging applications of the correspondence principle, which took widely different forms and led to very different conclusions, thus at least had a common point of departure. The diversification in the formulations of the principle can thus be thought of as a tree branching out in various directions.

The turn to transition probabilities and the institutional development of Copenhagen after World War I present two candidates for interpreting the overall pattern underlying the dissemination of the correspondence principle. The present discussion is not enough to decide whether one of these candidates or a combination of them is sufficient. In order to make sense of the turn from reception to adaptation, we need to look at case studies and analyze the motivations for making correspondence arguments and the role of interactions between the historical actors in detail. In other words, we need to probe the fine structure of the principle's pathways into the quantum network.

Chapter 4
Using the *Magic Wand*: Sommerfeld, Multiplet Intensities and the Correspondence Principle

This chapter discusses Arnold Sommerfeld's research on the intensity problem for multiplets and his application of the correspondence principle in the context of atomic spectroscopy, which he developed in collaboration with his students Werner Heisenberg and Helmut Hönl. The analysis is based on published research papers as well as on unpublished materials from the period 1921 to 1926. It discusses how Sommerfeld formulated empirical regularities governing the intensity of multiplets and at the same time sharpened the correspondence principle.[1]

This reconstruction can be read in two interconnected ways. It presents one of the case studies of the *transformation through implementation* of the correspondence principle. As such, it explicates what it meant for Sommerfeld to implement the principle in a new context, to recognize problems and to adapt the principle. At the same time, the chapter is a study of Sommerfeld's approach to quantum physics and aims to contribute to its recent reinterpretation by Suman Seth.[2] It aims in particular to challenge a prevalent characterization of Sommerfeld as a sharp critic of the correspondence principle, who rejected it as "a magic wand" and denied that it could play a foundational role in quantum theory.[3]

---

[1]In general, the development is reconstructed from the published papers, with a particular focus on Sommerfeld's papers on the correspondence principle, some of them written together with Heisenberg and Hönl: Sommerfeld and Heisenberg (1922a), Hönl (1924), Sommerfeld (1925), and Sommerfeld and Hönl (1925). The developments of Sommerfeld's first correspondence argument—discussed in Sect. 4.2 and 4.3—are also well documented in the correspondence of Werner Heisenberg and Alfred Landé (AHQP 6.2) as well as in a draft of Sommerfeld and Heisenberg's published paper in the Sommerfeld Papers (NL 89, 026). This rich material has no parallel in the later developments from 1924 to 1926, for which the available private correspondence provides limited insight into the development of the arguments.

[2]Seth (2010).

[3]See Darrigol (1992, 144), Tanona (2002, 73–78) and Kragh (2012, 210–211) for the standard presentation of Sommerfeld's position. For an alternative one, see Seth (2010, 232–241).

M. Jähnert, *Practicing the Correspondence Principle in the Old Quantum Theory*, Archimedes 56, https://doi.org/10.1007/978-3-030-13300-9_4

This interpretation has focused for the most part on Sommerfeld's programmatic statements in his book *Atombau und Spektrallinien* in its various editions, and his influential keynote address at the *Versammlung Deutscher Naturforscher und Ärzte* in Innsbruck in 1924.[4] The present chapter shows that Sommerfeld's position is not exhausted by his foundational critique in his programmatic remarks. Unimpaired by it, Sommerfeld frequently applied the correspondence principle in his research.[5] Taking these applications seriously, we can see that, as a user of the correspondence principle, Sommerfeld was no different from Franck and Hund or Reiche. He took up the correspondence principle as a tool and integrated it into his work. Thereby Sommerfeld recognized problems arising from this integration and eventually adapted it, thereby formulating his own version of the sharpening of the correspondence principle.[6]

My analysis of this *transformation through implementation* of the principle is divided into four sections. Sect. 4.1 introduces Sommerfeld's approach to atomic spectroscopy in the 1920s and its characteristic search for empirical regularities within spectral data. Following this approach in Sommerfeld's work on the intensity of multiplets, I reconstruct the problem which was addressed by Sommerfeld's first correspondence argument in 1922. The analysis of this argument in Sect. 4.2 then shows that Sommerfeld primarily used the principle to interpret the empirical regularities on multiplet intensities in terms of the state-transition model.

Section 4.3 analyzes how problems emerged within Sommerfeld's correspondence arguments. These were ultimately neglected at the time and played no

[4]See Darrigol (1992, 144) and Kragh (2012, 210–211) as well as Seth (2010, 226 and 241). Darrigol and Kragh have stipulated that Sommerfeld's rejection of the correspondence principle was rooted in a methodological demand for a mathematically well-defined framework and concrete physical models. Suman Seth has taken issue with this characterization of Sommerfeld as a "hyperrationalist." Instead, he has argued that Sommerfeld rejected the correspondence principle for conceptual rather than methodological reasons as it introduced classical electrodynamic concepts into quantum physics, leading to a "mixing of incommensurable worlds." Seth (2010, 241).

[5]These applications have not been subject to an extended analysis and were discarded by Olivier Darrigol on the assumption that Sommerfeld "distrusted the pervasive adaptability of the correspondence principle and [...] denied the 'sharpening of the correspondence principle' introduced by Bohr and Heisenberg in 1924–25." See Seth (2010, 232–241), Darrigol (1992, 144) and Kragh (2012, 210–211). See Darrigol (1992, 144) for the quotation. The present case study shows that this marginalization of Sommerfeld's applications is problematic. First, correspondence arguments played an important role in Sommerfeld's work on multiplet intensities from 1922 to 1925. Second and more importantly, he developed his correspondence arguments, discovered and solved problems within them, and last but not least, tried to obtain results significant for his work. The assertion that Sommerfeld did not lend much weight to his correspondence arguments does not help to understand why he engaged in these activities in the first place rather than discarding the principle.

[6]Darrigol claimed that Sommerfeld denied Heisenberg's and Bohr's sharpening of the principle. See Darrigol (1992, 236, 243 and 258). This may have been the case. As this chapter shows, he may have denied the idea of a sharpening of the correspondence principle in terms of the physical models like the *Zwischenbahn* or virtual oscillators, yet he used the term *Verschärfung* in an affirmative way in his work and developed his own approach to it.

substantial role in his subsequent work on the intensity problem from 1923 to 1926, which I discuss in the final Sect. 4.4. Following a radical change in Sommerfeld's work in this period, this section discusses the emergence of a new description of intensities and analyzes how Sommerfeld integrated and adapted his original correspondence argument to develop a new "theory of intensities."[7]

4.1 Formulating the Problem: The *Gesetzmäßigkeiten* of Multiplet Spectra and Their Model Interpretation

As Suman Seth argued in detail, Sommerfeld's work on multiplets was different from his previous work on the extension of the Bohr model for hydrogen and its mechanical framework. In the 1920s, Sommerfeld came to distrust mechanical models and eventually traded them for an approach in which "half-empirical *Gesetzmäßigkeiten*" governing the spectra played the central role.[8] The central idea of this new approach as Sommerfeld described it in the preface of the third edition of his book *Atombau und Spektrallinien* was that

> [t]he regularities [...] are primarily of an empirical nature, but their integral character from the outset demands a quantum theoretical clothing.[9]

This "clothing," Seth has stressed, did not have recourse to underlying dynamical models like the Bohr model. Instead it took "phenomenology-based equations as the starting point for an immediate discussion of the nature of atomic structure."[10]

Sommerfeld's discussion of the empirical regularities was indeed "immediate" insofar as it was not built upon dynamical models, as his previous work on the hydrogen spectrum had been. It would be a mistake, however, to consider this "immediate" discussion to be independent of the basic theoretical structure of the state-transition model. The "quantum theoretical clothing" Sommerfeld took for granted was based on the assumption of stationary states and the emission of spectral lines during transitions between these states. Only the implications of this approach, he thought, had to be made explicit in "a general remark, which does not say anything new to the spectroscopic expert":

---

[7]Sommerfeld (1925, 8).

[8]Seth (2010, 212). Seth's emphasis on empirical regularities follows Paul Forman's classic analysis of Alfred Landé's work on the anomalous Zeeman effect (Forman 1970), in which Forman argued that Landé followed an a posteriori approach that departed from the empirical line patterns and only afterwards couched the resulting regularities in terms of quantum theory.

[9]Sommerfeld (1922a, vi). "Die hier herrschenden Regelmäßigkeiten sind zunächst empirischer Natur; ihr ganzzahliger Charakter verlangt aber von Anfang an nach quantentheoretischer Einkleidung." Seth cites the translation by Henry L. Brose, which had introduced the metaphor of a "language of quanta" instead of Sommerfeld's use of the adjective "quantentheoretisch." See Sommerfeld (1923, v).

[10]Seth (2010, 213).

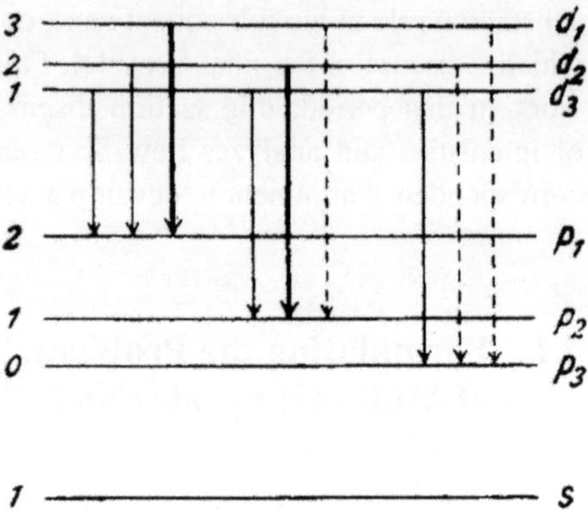

Fig. 4.1 Representation of *pd* transitions for alkaline earth metals in Sommerfeld (1920a, 231)

> The aim of spectroscopy is not so much the knowledge of the lines (energy differences), but the knowledge of the terms (the energy levels themselves), into which the frequencies of the lines can be resolved on the basis of the combination principle. These characterize the atomic states and are bearers of the *Gesetzmäßigkeiten* under consideration.[11]

Building on the state-transition model, Sommerfeld's "quantum theoretical clothing" of the empirical regularities identified spectroscopic terms with the stationary states and the combination principle with the frequency condition. In other words, it provided an interpretation of the term schemes used in spectroscopy in terms of quantum theory.

As Seth showed in detail, this identification led to the introduction of the inner quantum number in Sommerfeld's work on the spectra of alkaline metals, alkaline earth metals and other non-hydrogenlike elements. In contrast to hydrogen, whose spectral series consist of single lines, the spectra of these elements consist of a series of doublet, triplet, and generally multiplet lines. To describe them Sommerfeld used a spectroscopic term scheme, whose basic idea is shown in Fig. 4.1 for a composite triplet of the alkaline earth metals. As the term scheme shows, the composite triplet results from combining a threefold d-term with a threefold p-term. Of the nine possible combinations, only six correspond to lines that are actually observed, and are therefore represented by a full line. The three unobserved combinations, drawn with dashed arrows, must be excluded from the scheme. Interpreting this scheme within the state-transition model, Sommerfeld identified the s,p,d-terms with the azimuthal quantum number k ($s : k = 1$, $p : k = 2$, $d : k = 3$) in analogy to the theory of the hydrogen atom. He then introduced a new quantum number j to label

[11] Sommerfeld (1920a, 222). "Eine allgemeine Bemerkung, welche dem spektroskopischen Fachmanne nichts Neues sagt, ist für alles Folgende im Auge zu behalten: Das Ziel der Spektroskopie ist nicht so sehr die Kenntnis der Linien (Energiedifferenzen), sondern die Kenntnis der Terme (der Energiestufen selbst), in die sich die Schwingungszahlen der Linien nach dem Kombinationsprinzip zerlegen lassen. Diese charakterisieren die Atomzustände und sind die Träger der weiterhin zu betrachtenden Gesetzmäßigkeiten."

the different terms.[12] Finally, he observed that all that was necessary to reproduce the correct number of lines was to impose a selection rule on the new quantum number $\Delta j = \pm 1, 0$.[13]

After this brief discussion of Sommerfeld's general approach to multiplet spectra, we are in a position to discuss his approach to the intensity problem. While the interpretation of the term schemes within the state-transition model had to bridge a rather narrow gap for frequencies, the situation was quite different with respect to the intensities of spectral lines. As Sommerfeld pointed out as early as 1917, the intensity problem was one of the biggest lacunae in Bohr's quantum theory of the atom. It lacked the conceptual resources to describe the "form of oscillations" associated with the intensity or the polarization of light, and with the possibility of interference or radiation damping.[14] This deficiency was mirrored by spectroscopic observations: apart from singular cases, there was little more than "rough and qualitative estimates" on the intensity of spectral lines.[15]

While both the experimental knowledge on intensities and their theoretical description were rudimentary and deficient, Sommerfeld believed that the intensity of spectral lines was of vital importance for the description of atomic spectra. In any case, such a description would be necessary for a complete account of radiation. More importantly however, he argued, the intensity question was vital for obtaining the right number of spectral lines. Similar to Bohr, Sommerfeld was wrestling with this question from 1916 onwards and understood that it was necessary to restrict the number of possible transitions. To implement such a restriction he proposed a *Quantenungleichung*, according to which quantum numbers could only decrease or remain unchanged during the process of spontaneous emission. Such a restriction, he argued, implied that "certain lines have intensity zero" and thus that the *Quantenungleichung* was part of the "intensity question."[16]

[12]The letter assigned to the azimuthal and the inner quantum number differed considerably in 1922 and, to a lesser degree, also thereafter. Sommerfeld, Landé and Heisenberg used the letters j, k, n, n_j or J to denote the inner quantum number. The main text will use j except to avoid confusion in the discussion of quotes, which explicitly use a different nomenclature.

[13]Seth (2010, 204–210) and Sommerfeld (1920a, 231–232).

[14]Sommerfeld (1917, 83).

[15]Sommerfeld (1917, 84). This characterization of the state of affairs was rather moderate in comparison with other more critical statements like the ones made by Heinrich Kayser, who clarified in the introduction to his *Handbuch der Spectroscopie* that both the subjective judgements made by different observers and the lack of a standardized scale made it impossible to compare and discuss observations on spectral intensities. See Kayser (1900, Vol. 1, XXII).

[16]Sommerfeld (1916a, 26). Sommerfeld justified his *Quantenungleichung* on the assumption that energy did not increase during a transition. This assumption, however, remained disconnected from Sommerfeld's preliminary theoretical considerations on the intensity question. In them, Sommerfeld primarily discussed the number of systems in a particular state, and found that his *Quantenungleichung* could not be explained on the basis of statistical considerations. Rather, it required further "dynamical" assumptions about the transition process. Without making much headway in this direction, Sommerfeld realized in his treatment of the Zeeman effect that the *Quantenungleichung* did not suffice to restrict the possible transitions to a triplet (Sommerfeld

For Sommerfeld, the statement as to whether or not a certain line appeared in the spectrum provided merely a coarse-grained description for the intensity distribution. Even in his early work, Sommerfeld pondered the idea to go beyond a simple binary description. He distinguished between "unreal lines" with zero intensity, "very weak lines" with feeble intensity, and "real lines", implying that their intensity was correlated with his *Quantenungleichung*.[17] This gradual description of spectral intensities and the attempt to formulate an empirical regularity for them using the state-transition model became more important in Sommerfeld's work on the multiplet structure starting in 1920. In it, Sommerfeld worked with a distinction between strong *main lines* and weaker *satellites*. This distinction was commonly used in spectroscopy to characterize line structures and was well known to Sommerfeld as early as 1916. Back then, however, he had argued that it was something extrinsic and ill-defined. In 1920, he changed his position. In considering which transition frequencies were associated with main lines and satellites, he tacitly assumed that there was a typical order among them.[18] While these considerations entered into Sommerfeld's work in 1920 as an aside, he came to focus on the problem in the third edition of his book *Atombau und Spektrallinien* published in 1922. Considering the qualitative distinction between main lines and satellites as an indication for a *Gesetzmäßigkeit* governing multiplet intensities, he made an extended survey of the experimental material on different classes of multiplets. From it, he formulated an intensity rule that correlated the intensity with the "sense" in which the azimuthal and the inner quantum numbers changed during a transition:

> From the three transitions, that one that occurs in the same sense as the transition in the azimuthal quantum number n should appear with the maximum intensity; the more the nature of the transition in n_i [the inner quantum number, MJ] deviates from the one in n, the intensity should decrease. Hereafter, we speak of a "strong," a "slightly weak" and a "weak transition."[19]

1916b, 494). He found its "refined formulation" in Rubinowicz' explanation of selection rules based on energy and momentum conservation between the atom and the ether, which he initially defended as more suitable than Bohr's correspondence arguments, only admitting the superiority of Bohr's approach in the third edition of *Atombau und Spektrallinien* in 1922. For a more detailed discussion of Sommerfeld's position within the discussion of selection rules, see Borrelli (2009) as well as Tanona (2002, 73–90).

[17] Sommerfeld (1916a, 69).

[18] Sommerfeld (1916a, 68) and Sommerfeld (1920a, 226–228), in which Sommerfeld took the intensity relations found by Friedrich Paschen as typical and expected them to become apparent in different cases such as the lines analyzed by Meissner: "Wir erwarten nach Analogie mit Fig. 1 die kurzwelligste Linie als zu punktierende (Intensität Null), die langwelligste als Satelliten (Intensität klein). Nach Ausweis der Fig. 3 trifft beides nicht zu."

[19] Sommerfeld (1922a, 447–448). "Es soll von den drei Übergängen (4) derjenige mit der größten Intensität auftreten, der im gleichen Sinne geht wie der Übergang in der azimuthalen Quantenzahl n; und es soll die Intensität um so mehr abnehmen, je mehr die Art des Überganges in n_i von der in n abweicht. Wir sprechen hiernach von einem 'starken,' einem 'weniger starken' und einem 'schwachen Übergange'."

This qualitative rule, which emerged from the analysis of the empirical material in terms of the state-transition model, incorporated the distinction between main lines and satellites into Sommerfeld's interpretation of the spectroscopic term scheme based on the inner quantum number.

4.2 Implementing the Correspondence Principle: Heisenberg's Model Interpretation and Sommerfeld's Intensity Rule

After the previous introduction to Sommerfeld's thinking about the intensity problem, we are in a position to discuss his application of the correspondence principle in this context. As we have seen, the theoretical interpretation of the intensity rule remained entirely unclear in its initial formulation by Sommerfeld, and it was the search for such an interpretation that motivated Sommerfeld's turn to the correspondence principle.

In coming to use the principle for this purpose, Sommerfeld did not develop his correspondence argument from scratch. Rather, he adopted it from the work of Werner Heisenberg, who had already applied the correspondence principle to the motion of a multiplet atom. Heisenberg's application, as we will see, was intended initially as an argument in a debate with Landé on the physical interpretation of the inner quantum number. In it, Heisenberg considered the intensity distribution as a source of information for constructing a model of the multiplet atom. Sommerfeld abandoned this approach and adapted it for the interpretation of the intensity rule. To understand Sommerfeld's take on the correspondence argument, I analyze the transition from Heisenberg's to Sommerfeld's use of the correspondence principle, discussing Landé's and Heisenberg's initial correspondence arguments, and then turning to Sommerfeld's use of the principle in the interpretation of the intensity rule.

Both approaches to the correspondence principle, I show, were shaped by the debates on the model interpretation of multiplets and the search for empirical regularities within multiplet spectra. Landé, Heisenberg, and Sommerfeld followed the original correspondence arguments as they set up Fourier representations for the electronic motion and connected these representations with the respective transition processes. Due to the primacy of the empirical regularities and the uncertainty in the physical models, however, their correspondence arguments were interpretations of spectral data in terms of a physical model rather than deductive predictions of observable phenomena. This interpretative use of the correspondence principle, which was similar to Bohr's idea of obtaining *Fingerzeige*, relied on the general kinematic description of a precessional motion and the concept of space quantization. These elements formed the basis on which the regularities in multiplet spectra expressed within the state-transition model were interpreted in terms of an atomic model. Based on these central elements, Landé, Heisenberg, and Sommerfeld extended the original correspondence argument. Interpreting the empirical regularities as expressions of the spatial configuration of the atom in different stationary states, they advanced a new kind of correspondence argument, which resulted from the integration of the principle into their general approach.

Landé, Heisenberg and the Debates on the Model Interpretation of Multiplets

In the debate on the model interpretation of multiplets there were three different positions. First, there was Sommerfeld's initial interpretation of the inner quantum number from 1920, which associated it with a "hidden rotation" inside the atom. The axis of this rotation remained unspecified, as Sommerfeld continued to associate the azimuthal quantum number with the total angular momentum. The second position was formulated by Landé in 1921, who argued that the "hidden rotation" was nothing but the precession of the electronic orbit around the axis of the total angular momentum of the atom, which he now identified with the inner quantum number instead of the azimuthal quantum number. The third interpretation was proposed by Sommerfeld's young student Werner Heisenberg. He argued that the inner quantum number was not identical with the total angular momentum, but associated with some hitherto unspecified angular momentum that contributed to it vectorially.[20]

While Sommerfeld's initial interpretation dropped out of the picture, Landé and Heisenberg extensively debated the two latter possibilities in a series of letters in the fall of 1921.[21] While they disagreed on the details, Landé's and Heisenberg's interpretations were built on the same general model. Both considered an atom that precessed around the axis of the external magnetic field.[22]

As in Sommerfeld's and Debye's account of the normal Zeeman effect,[23] in which the description had first been proposed in 1916, the central quantity describing this precession was the angle Θ between the axis of the atom's total

[20]See Sommerfeld (1920a, 231) for Sommerfeld's hypothesis of a "hidden rotation." See also Landé (1921a, 234) for Landé's identification with the total angular momentum, and for Heisenberg's arguments against such an identification see for example Heisenberg to Landé, 11 October 1921 or Heisenberg to Landé, 16 October 1921 (AHQP 6.2).

[21]The Landé-Heisenberg correspondence was discussed by David Cassidy in his dissertation (Cassidy 1976) and a subsequent paper (Cassidy 1979). Cassidy focused mainly on the development of Heisenberg's rump model and the ensuing discussion between Landé and Heisenberg of the "deviations from established principles" inherent in the model. Landé's and Heisenberg's correspondence arguments, their relation to the model interpretation and importance for the genesis of the rump model did not play a central role in Cassidy's analysis.

[22]Both the basic model of the precessional motion of the atom's total angular momentum and the conception of space quantization were well established and taken for granted by Landé and Heisenberg.

[23]Sommerfeld (1916b) and Debye (1916a,b). For a discussion of Sommerfeld's earlier extensions of the Bohr model see Eckert (2013b, 2014) as well as Borrelli (2011).

angular momentum and the magnetic field. This angle determined the configuration of the atom in a stationary state and was space-quantized, as the projection of the total angular momentum on the axis of the magnetic field could only assume certain values. The issue at stake in the debate was whether the total angular momentum was to be identified directly with the inner quantum number, as Landé thought, or whether the total angular momentum was constructed by vectorial addition of the angular momentum of the valence electron and the angular momentum of the rump, as Heisenberg believed. In Landé's case, the different space-quantized configurations of the atom in a particular stationary state were specified by the relation:

$$\cos\Theta = \frac{m}{j},$$

where j is the inner quantum number and m is the magnetic quantum number, while Heisenberg did not propose a similar relation.

As they developed and argued for their interpretations, Landé and Heisenberg both turned to the correspondence principle and used it to determine the motions inside the atom from empirical regularities and the qualitative intensity distribution. This approach was made explicit in Heisenberg's first letter to Landé, in which he argued that the principle made it possible "to draw important conclusions on the oscillations [within the atom MJ]."[24] In another letter, he explained in an almost programmatic way that this approach was based on the inversion of Bohr's original correspondence argument:

> the correspondence principle initially states with Bohr: the initial and the final orbit are responsible for the intensity of a line. If we want to inversely infer from the intensity to the orbit, then it says: for the orbit, the intensities leading to and departing from such a state are responsible.[25]

Heisenberg's inference from the intensity distribution to the model interpretation was closely connected to the development of his rump model. It was also expressed in a first version of Sommerfeld and Heisenberg's paper to support Heisenberg's rump model.[26] However, they ceased to play a role in early 1922, as Sommerfeld and Heisenberg decided not to publish the first version of their intensity paper. Instead,

[24]Heisenberg to Landé, 11 October 1921 (AHQP 6.2).

[25]Heisenberg to Landé, 23 October 1921 (AHQP 6.2). "[D]as Korrespondenzprinzip sagt zunächst nach Bohr aus: Für die Intensität einer Linie ist Anfangs. u. Endbahn verantwortlich. Wollen wir von der Intensität umgekehrt auf die Bahn schließen, so [sagt] es: Für eine Bahn sind die zu diesem Zustand u. von diesem wegführende Intensitäten verantwortlich."

[26]The manuscript of the initial paper is not available; it is mentioned as the "unpublished part I of the intensity paper" in a letter from Heisenberg to Sommerfeld (Heisenberg to Sommerfeld, 17 October 1922 in Sommerfeld (2004, 126)) and connected to conclusions on the model interpretation. For the connection with Heisenberg's rump model see also Heisenberg (1922, 284), Heisenberg to Landé, 26 October 1921 (AHQP 6.2) and Heisenberg to Pauli, 17 December 1921 in Pauli (1979, 50–51). These sources indicate that Heisenberg's correspondence arguments were initially conceived in connection with the rump model and ready for publication in early 1922.

they submitted a second version, in which the intensity distribution was treated as a physical problem in its own right.

As David Cassidy has argued, the move from Heisenberg's initial argument to the Heisenberg-Sommerfeld paper was motivated by the growing opposition to the rump model from Bohr and Landé. Cassidy therefore interpreted the final paper as a "fallback" from Heisenberg's ambitious rump model to "the seemingly more 'qualitative' results obtainable from an application of the correspondence principle to the general 'kinematic description' of the orbits in the rump model."[27]

While Sommerfeld and Heisenberg certainly distanced themselves from the rump model in the published paper, the "qualitative results" mentioned by Cassidy were not a mere retreat marking the dead end of a young student's ambitious attempts. For Sommerfeld, as I will show, they were part of an attempt to provide a theoretical interpretation of his intensity rule. This discussion was built on Heisenberg's initial correspondence argument, but was essentially independent of its conclusion drawn with respect to the dynamics of the rump model.

Keeping the focus on the discussion of Sommerfeld's approach to the intensity problem, I will not discuss the correspondence arguments in relation to Heisenberg's work on the rump model. Rather, I will interpret the transition from Heisenberg's initial use of the principle to the Sommerfeld-Heisenberg paper as the separation of the dynamical model of the multiplet atom from the discussion of the intensity distribution. This separation remained in place until 1926 and is best illustrated by the fact that the discussions on the dynamics underlying the Zeeman effect had little impact on the description of the intensity distribution. Becoming a problem in its own right, as we will see, the latter was divorced from dynamic considerations and associated with the kinematic description of the multiplet atom.[28]

[27]Cassidy (1976, 142). In the same way, Mehra and Rechenberg stipulated that Heisenberg and Sommerfeld withheld their paper as they awaited Bohr's judgement on Heisenberg's rump model, which reached them at the Bohr *Festspiele* in Göttingen in June 1922. After Bohr's dismissal they stripped the paper of any controversial claims of Heisenberg's dynamical theory and ended up with their qualitative description. Mehra and Rechenberg (1982a, 44).

[28]In retrospect, the intensity problem was part of Heisenberg's initial attempt to formulate a dynamical model, but it ceased to function in this capacity in the continuing discussions. It played a role neither in the formulation of Goudsmit and Uhlenbeck's hypothesis of electron spin nor the problems that led to it. Conversely, spin turned out to be unimportant for the solution of the intensity problem in the new quantum mechanics. The intensity problem was first approached by Heisenberg in Heisenberg (1925a, 892) and then solved in general in Born et al. (1926, 603–605) within the discussion of the conservation of total angular momentum in matrix mechanics. As Born explicitly stated in his MIT lectures, one still needed to decide whether the quantum numbers were integers or half-integers. As such the intensity problem was independent of the introduction of a specific model, which included spin. See Born (1926a, 93–98) or the English version, Born (1926b, 106–112). The actual formulas found by Heisenberg and Jordan in their treatment of the Zeeman effect in matrix mechanics showed that the introduction of spin did not change the solution of the intensity problem. See Heisenberg and Jordan (1926, 270–272).

Landé's Correspondence Interpretation of the Intensity Distribution

The discussion of the intensity distribution departed from Landé's work on the anomalous Zeeman effect in 1921. He developed his first correspondence argument in a short paragraph at the very end of the paper.[29] In his short and tentative argument, he connected the qualitative aspects of the intensity distribution in the Zeeman effect with a model for the multiplet atom, which he used to justify an additional selection rule that he had found in his interpretation of the empirical data in terms of the state-transition model.

This short argument deviated considerably from Bohr's and Kramers' correspondence arguments. Whereas they had deduced consequences from a given physical model and compared them with available data, Landé tailored a model he thought would reproduce the empirical regularities through the correspondence approach.

In so doing, Landé departed from Sommerfeld's intensity rule. The correlation between the distinction between main lines and satellites, and the sense in which the azimuthal quantum number and the inner quantum number changed, Landé explained, would result "from Bohr's correspondence principle":

> if one represents the atom consisting of crossed orbits by one electron revolving in the invariable atomic plane with radius c, and a second electron oscillating axially with the small amplitude a ($a < c$).[30]

Proposing this first correspondence interpretation of Sommerfeld's intensity rule, Landé left open how the correlation between two quantum numbers emerged from his model and how he had come up with it. Instead he used the model to discuss the situation in the Zeeman effect. Considering the atom with two valence electrons in crossed orbits within an external magnetic field, he argued:

> The same substitute model then has x and z as mean values of the amplitudes within a magnetic field $H \parallel z$ with the equatorial component $m = k \cos \Theta$:
>
> $$\text{parallel to the field } \bar{z} = c \sin \Theta \quad + a \cos \Theta$$
>
> $$\text{perpendicular to the field } \bar{x} = c\sqrt{\frac{1+\cos^2 \Theta}{2}} + a \sin \Theta$$
>
> From this one can read from Bohr's correspondence principle: "the strength of the π-component is proportional to $\bar{z}(\Theta)$, the strength of a σ-component is proportional to $\bar{x}(\Theta)$, where Θ is an average value of the inclination Θ' and Θ'' of the atomic axis in the direction of the field in the initial and the final state."[31]

---

[29]Landé (1921a).

[30]Landé (1921a, 240). "Diesen Tatbestand würde man nach dem Bohrschen Korrespondenzprinzip erwarten, wenn man das aus gekreuzten Bahnen bestehende Atom repräsentiert durch ein Elektron, welches in der invariablen Atomebene mit dem Kreisbahnradius c zirkuliert, und ein zweites Elektron, welches mit geringer Amplitude a axial schwingt (a<c)."

[31]Ibid. "Dasselbe Ersatzmodell im Magnetfeld ($H \parallel z$) mit der äquatorialen Komponente $m = k \cos \Theta$ hat dann als Mittelwerte der Amplituden x und z

Again, the form in which Landé presented his correspondence interpretation makes it difficult to assess how he actually developed his considerations. From the short quote, however, we can see that Landé thought that the motions of the two electrons were projected onto the axes x, y, z of the coordinate system specified by the external magnetic field, and that one then summed up the different contributions to get the intensity of the perpendicular and parallel components. In his correspondence interpretation, Landé took the different coefficients in his Fourier series to be directly proportional to the "strengths," i.e. the intensities of the spectral lines. Taking Kramers' dissertation as the authoritative application of the principle, he assumed that these Fourier coefficients were mean values between the initial and the final state and naturally applied this idea to the angle Θ.[32]

Having implemented the correspondence principle, albeit in this tentative manner, Landé used it to draw conclusions about the intensity distribution. Following the initial assumption that the amplitude c was dominant with respect to a, he focused on the trigonometric terms, which he interpreted as space-quantized according to the equation $\cos\Theta = m/j$.[33] On the basis of this equation he concluded that the intensity of lines with circular polarization was highest for $\Theta = 0°$, whereas the intensity of lines with parallel polarization was highest for $\Theta = 90°$. This qualitative description of the intensity distribution of Zeeman components, he observed, was in agreement with the empirical intensity data, with one notable exception. It did not account for a new selection rule he had found in his interpretation of the empirical data within the state-transition model. This so-called *Zusatzregel* stated that the intensity vanished for transitions in which the magnetic quantum number was zero in the initial and the final state, and in which the inner quantum remained unchanged. Interpreting the *Zusatzregel* within his correspondence account, Landé argued:

> Only if $k - k' = 0$, the substitute radiator has to have the vanishing amplitude $c = 0$, so that the term a alone is responsible and an inversion of the dependence on Θ takes place. If it is also true that $m' = 0 = m''$, that is $\Theta = 90°$, $\overline{z} = 0$, through which the above mentioned *Zusatzregel* C' finds its justification.[34]

$$\text{parallel zum Feld } \overline{z} = c\sin\Theta + a\cos\Theta$$

$$\text{senkrecht zum Feld } \overline{x} = c\sqrt{\frac{1+\cos^2\Theta}{2}} + a\sin\Theta$$

Daraus lies man nach Bohrs Korrespondenzprinzip ab: 'Die Stärke einer π-Komponente ist proportional zu $\overline{z}(\Theta)$, die einer σ-Komponente zu $\overline{x}(\Theta)$, wobei Θ ein mittlerer Wert zwischen den Neigungen Θ' und Θ'' der Atomachse gegen die Feldrichtung im Anfangs- und Endzustand ist.' "

[32]See Landé and Back (1925, 48–53) for a more detailed exposition. Following the "method of the *Ersatzstrahler*," Landé discussed the intensity distribution in an "abridged argument" that did not take Sommerfeld and Heisenberg's extensive correspondence argument into account.

[33]Note that Landé and Heisenberg initially used the letter k to label the inner quantum number.

[34]Landé (1921a, 240). "Nur wenn $k'-k'' = 0$ ist, muss der Ersatzstrahler verschwindende zirkulare Amplitude $c = 0$ haben, so dass dann das a-Glied allein maßgeblich ist und eine Umkehrung der

At first glance this argument suggests that Landé was able to deduce the rule from his physical model, by arguing that the amplitude c vanished if the inner quantum number remained constant during a transition and then evaluating the angle Θ. The assumption that c vanishes, however, is rather problematic within Landé's model: it implies that the respective valence electron does not circulate the atom at all.

This inconsistency, which Landé did not discuss, again suggests that he had not come to his conclusion by considering the motion of the electron. It is much more plausible that he departed from the empirical regularity, which indicated that the total intensity of the $\mathfrak{z}$-component had to be zero for transitions $m = m' = 0$ if the inner quantum number did not change. For $m = 0$, the relation $m = k \cos \Theta$ then suggested that the component $a \cos \Theta$ would vanish, leading to the conclusion that the amplitude c is zero.

This reading is consistent with Paul Forman's characterization of Landé's a posteriori approach.[35] Moreover, it is supported by a comment Landé made three years later in his book *Zeemaneffekt und Multiplettstruktur der Spektrallinien*, coauthored with Ernst Back. In it, he remarked:

> It might also be mentioned that especially the intensity distribution of the Zeeman components gave the impetus, on the basis of considerations following the method of the substitute radiator, to interpret Sommerfeld's "inner" quantum number as the total angular momentum quantum J and thereby to prepare the understanding of the complex structure and its Zeeman components.[36]

Landé's assertion, according to which the interpretation of the inner quantum number as the total angular momentum emerged from considerations of the intensity distribution, indicates that the correspondence argument is best understood as an attempt to interpret empirical regularities in terms of a physical model.[37]

For the purposes of the present discussion, the question as to whether Landé's correspondence argument proceeded inductively or deductively is not essential. What is relevant here is that Landé took up Bohr's and Kramers' correspondence arguments and incorporated them into his work and the model he was working with. In this respect, it is telling that Landé made his correspondence argument in the

---

Abhängigkeit von Θ eintritt. Ist dazu noch $m' = 0 = m''$, d. h. $\Theta = 90°$, so wird $\mathfrak{z} = 0$, wodurch die obige Zusatzregel C' ihre Begründung findet."

[35] Forman (1970).

[36] Landé and Back (1925, 53). "Es mag noch erwähnt werden, dass gerade die Intensitätsverhältnisse der Zeemankomponenten der Anlass waren, auf Grund von Betrachtungen nach der Methode des Ersatzstrahlers die Sommerfeldschen 'inneren Quantenzahlen' als Gesamtimpulsquanten J zu deuten und dadurch das Verständnis der Komplexstruktur und ihrer Zeemaneffekte vorzubereiten."

[37] This was also the way in which Heisenberg described Landé's work in 1922. In an unpublished footnote to the Sommerfeld-Heisenberg paper, he clarified that "Landé has constructed an 'Ersatzmodell' to represent the intensities of the Zeeman components in accordance with experience." See Heisenberg's annotation in the draft of the Sommerfeld-Heisenberg paper: "*Landé* hat in seiner Arbeit [...] ein 'Ersatzmodell' konstruiert, um dadurch die Intensitäten der Zeemankompomenten der Erfahrung gemäß widerzugeben. Wie [gut?] die Analogie zwischen dem wirklichen Modell u. *Landé's* Ersatzmodell ist, ließe sich aus den resultierenden Intensitätsformeln ablesen." (Sommerfeld Papers (NL 89, 026, p. 25)).

form of a quote. This fact shows that Landé thought that his paper offered a mere recapitulation of an already established argument and that he did not perceive of what he was doing as anything new. Landé only realized that this was not the case when the young Werner Heisenberg pointed out to him that the quote in Landé's paper was nowhere to be found in the existing literature.[38]

Without knowing it, Landé had taken up the original correspondence argument and incorporated it into his work on the multiplet atom based on the notion of space quantization. This incorporation changed the original correspondence arguments considerably. The Fourier coefficients c and a describing the electronic trajectory, which were central for Bohr and Kramers, were constant and therefore irrelevant for understanding the varying intensities of the different Zeeman components. Instead, the intensity distribution was now governed by the angle Θ, which had formed the central notion within the attempts to understand the Zeeman effect since Sommerfeld's work in 1916. Thereby Landé had found that the intensity distribution first of all mirrored the spatial orientation of the atom with respect to the magnetic field, which was expressed by the equation $m = j \cos \Theta$.

Connecting the notion of space quantization within the correspondence approach with the empirical regularities in question, Landé also departed considerably from Bohr's original correspondence argument for the usual selection rules. As discussed in Chap. 2.3, the original correspondence argument for the selection rules depended on the absence of certain Fourier coefficients in all possible initial and final states. To restrict the change of a particular quantum number in a transition for all possible states, these arguments relied on considering the motions associated with the respective quantum numbers as independent of each other.

Yet Landé's *Zusatzregel* applies only to specific transitions with the initial and final states $m = m' = 0$. What is more, it is contingent on the transition of the inner quantum numbers j. These central differences in the empirical regularity alone made it impossible to interpret the *Zusatzregel* along the lines of the original correspondence argument. In his interpretation of the empirical regularity, Landé relied first of all on the notion of space quantization. Turning it into a central element with the correspondence approach allowed him to understand the new selection rule and the intensity of the Zeeman splittings as a result of the coupling between the inner quantum number and the magnetic field. While he was under the impression that he was recapitulating Bohr's and Kramers' arguments. Landé had thus integrated the correspondence principle into his work and developed Bohr's and

[38]See Heisenberg to Landé, 16 October 1921 as well as Landé's draft for a letter responding to Heisenberg in AHQP 6.2. Surprised that his correspondence argument had actually made an original contribution, Landé asked Heisenberg to acknowledge it. Heisenberg added a footnote to the Sommerfeld-Heisenberg paper, which did not make it into the final publication as the paper had already entered the second proofreading stage. See page 25 of the draft for the Sommerfeld-Heisenberg paper in the Sommerfeld Papers (NL 89, 026). For a discussion of this footnote see Heisenberg to Sommerfeld, 17 October 1922 in Sommerfeld (2004, 126) as well as a letter by Heisenberg to Landé, 13 November 1922 (AHQP 6.2).

Kramers' initial arguments into a new form, in which the idea of space quantization played the central role.[39]

For the subsequent discussion of the intensity problem in Sommerfeld's and Heisenberg's work, Landé's correspondence interpretation is important in two respects. First, it connected the intensity distribution with the spatial configuration of the atom and thereby opened up the possibility to draw conclusions in either direction. Second, as will be discussed below, Landé's tentative interpretation had its flaws when considered from the perspective of the original correspondence approach. Trying to "reproduce Landé's intensity distribution" in a way that fit more closely into Bohr's and Kramers' original arguments, Heisenberg and later Sommerfeld developed their implementation of the correspondence principle.

Heisenberg's Correspondence Argument for the Intensity Distribution

As already mentioned, Heisenberg developed his correspondence argument in a series of letters to Landé, in which he tried to argue for his own model of the multiplet atom. His initial correspondence argument, like his and Sommerfeld's later account of the intensity distribution, was based on the consideration of a precessional motion of an atom around a fixed axis, the axis of either the total angular momentum or an external magnetic field.

Developing this representation, Heisenberg proposed a model that was different from Landé's. He considered the description of Sommerfeld's theory of the fine structure of hydrogen. The atom had one (not two) valence electrons, which orbited the nucleus on an elliptic orbit with the fundamental frequency and its overtones ω_1.[40] In addition, the entire ellipse rotated in its plane with the frequency ω_2. On top of this motion, Heisenberg assumed, the entire system performed a precession with the frequency ω_3 around the z_0-axis, which was parallel to the axis of the

[39]In Bohr's original correspondence arguments, space quantization was absent from the discussion of spectral phenomena up to 1922. See Bohr (1918a,b). The idea began to play a role in Bohr's writings only in reaction to Landé's paper. In the 1922 Guthrie lecture, published as Bohr (1922), Bohr first considered the idea of the spatial orientation of the atom with respect to an external field. On this occasion, he neither discussed Landé's introduction of the angle Θ to the correspondence approach nor his interpretation of the intensity distribution and the *Zusatzregel* on its basis. See Bohr (1922, 296–302). In Kramers' dissertation the intensity distribution in the normal Zeeman effect had played a minor role. In his short discussion, Kramers had already considered the spatial orientation of the orbital plane with respect to the axis of the magnetic field. See Kramers (1919, 299–300). He had not used it, however, to account for features in the intensity distribution, nor had he anticipated the new kind of selection rules Landé was confronted with. In his estimates of the intensity of the Zeeman components, the spatial orientation of the atom was not a central physical property, rather it led to an additional numerical factor. See Kramers (1919, 378–384).

[40]Note that ω was consistently used to refer to frequencies and not angular frequencies, contrary to modern conventions.

angular momentum corresponding to the inner quantum number.[41] For this motion Heisenberg set up the Fourier representation:

$$z_0 = \sum_{\tau_1=-\infty}^{\infty} \sum_{\tau_2=\pm1} a_{\tau_1,\tau_2} \cos 2\pi((\omega_1\tau_1 + \omega_2\tau_2)t + \delta_{\tau_1,\tau_2})$$

$$x_0 \pm iy_0 = \sum_{\tau_1=-\infty}^{\infty} \sum_{\tau_2=\pm1} c_{\tau_1,\tau_2} e^{2\pi i((\tau_1\omega_1+\tau_2\omega_2+\omega_3)t+\gamma_{\tau_1,\tau_2})}.$$

This representation incorporated additional information on the electron's motion, which Heisenberg extracted from the established selection rules for the azimuthal quantum number and the selection rule for the inner quantum number. Like in Sommerfeld's theory of fine structure, he assumed that the azimuthal quantum number could only change by ±1, so that the corresponding rotation could only take place clockwise or counterclockwise with the fundamental ω_2; in other words, τ_2 could only be $+1$ or -1. Similarly, the selection rule for the inner quantum number $\Delta j = +1, 0, -1$ demands that the associated precession around the z_0-axis take place only clockwise or counterclockwise with the fundamental frequency ω_3.[42] As the motion along the z_0-axis is not affected by this precession, there is a non-zero Fourier coefficient corresponding to the overtone $0\ \omega_3$, and consequently, a transition $\Delta j = 0$.[43]

To describe the motion of this atom in a magnetic field, which he had depicted graphically in a previous letter (Fig. 4.2), Heisenberg introduced a precession of the entire system around the axis of the magnetic field. For a field that was weak enough not to alter the motion of an unperturbed system, this problem reduced to a standard transformation of the initial coordinate system into the coordinate system x, y, z, defined by the magnetic field based on the transformation equations:

$$x = x_0;\ y = y_0 \cos\Theta - z_0 \sin\Theta;$$
$$z = y_0 \sin\Theta + z_0 \cos\Theta,$$

[41] Heisenberg to Landé, 16 October 1921 (AHQP 6.2). Trying to undermine Landé's interpretation of the inner quantum number, Heisenberg's association of the inner quantum number with an angular momentum aimed to show that it was not necessary to identify it with the total angular momentum.

[42] It might be noted that the frequency $-\omega_3$ does not occur in Heisenberg's Fourier representation. We will see later how the corresponding transition was incorporated into Sommerfeld and Heisenberg's work.

[43] Heisenberg to Landé, 16 October 1921 (AHQP 6.2). Like Landé, Heisenberg used k instead of j for the inner quantum number.

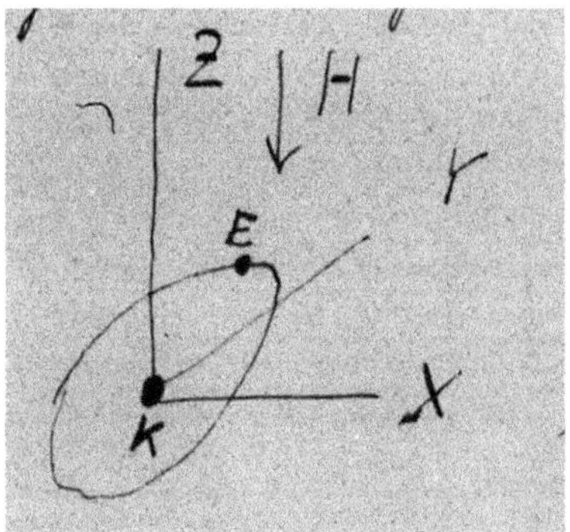

Fig. 4.2 Graphical representation of the precessional motion of an atom in a magnetic field in Heisenberg to Landé, 11 October 1921 (AHQP 6.2)

where Θ is the angle between the z_0-axis and the z-axis parallel to the magnetic field. Transforming the old coordinate system into the new one under these equations, Heisenberg obtained the new Fourier representation:

$$z = \sum_{\tau_1,\tau_2} c_{\tau_1,\tau_2} \sin\Theta \sin 2\pi (t(\omega_1\tau_1 + \omega_2\tau_2 + \omega_3) + \gamma_{\tau_1,\tau_2}) + \sum_{\tau_1,\tau_2} a_{\tau_1,\tau_2} \cos\Theta \cos 2\pi (t(\omega_1\tau_1 + \omega_2\tau_2) + \delta_{\tau_1,\tau_2})$$

$$x \pm iy = \sum_{\tau_1,\tau_2} c_{\tau_1,\tau_2} \cos^2\frac{\Theta}{2} e^{2\pi i (t(\omega_1\tau_1+\omega_2\tau_2+\omega_3)+\gamma_{\tau_1,\tau_2})} + \sum_{\tau_1,\tau_2} c_{\tau_1,\tau_2} \sin^2\frac{\Theta}{2} e^{-2\pi i (t(\omega_1\tau_1+\omega_2\tau_2+\omega_3)+\gamma_{\tau_1,\tau_2})} - \sum_{\tau_1,\tau_2} a_{\tau_1,\tau_2} \sin\Theta \cos 2\pi (t(\omega_1\tau_1 + \omega_2\tau_2) + \delta_{\tau_1,\tau_2}).$$

This representation, Heisenberg argued, had to be interpreted in a way that was quite different from Landé's. Whereas Landé had assumed that the c-term and the a-term both contributed to one and the same transition, Heisenberg clarified that the correspondence principle demanded a different reading:

> From this representation it follows that the coefficients *c alone* are responsible for the main lines $k - k' = \pm 1$, the coefficients *a alone* are responsible for the satellites $k - k' = 0$.[44]

[44]Heisenberg to Landé, 16 October 1921 (AHQP 6.2). "Aus dieser Darstellung folgt: Für die Hauptlinien $k - k' = \pm 1$ sind die Koefficienten *c allein*, für die Satelliten $k - k' = 0$ sind die Koefficienten *a allein* maßgebend."

The Fourier coefficients a and c, which he had introduced "to reveal the analogy to your *Ersatzstrahler*," belonged to the motion of one and the same electron. They had to be identified with different transitions as the two coefficients were associated with different frequencies (the c-coefficient with a frequency $\tau_1\omega_1 + \tau_2\omega_2 + \omega_3$, the a-coefficient with a frequency $\tau_1\omega_1 + \tau_2\omega_2$). Based on this new correspondence interpretation, there were two main lines, each associated with two satellites. Consequently, Landé's summation of intensities disappeared from the intensity distribution[45]:

$$\begin{array}{lll} & \text{main line} & \text{satellite} \\ I_{par} = \overline{z^2} = & c^2 \sin^2 \Theta & a^2 \cos^2 \Theta \\ I_{perp} = \overline{(x+iy)^2} = & c^2 \frac{1+\cos^2\Theta}{2} & a^2 \dfrac{\sin^2\Theta}{2}. \end{array}$$

In addition to the new distinction between main line and satellite, this distribution corrected two additional shortcomings Heisenberg had found in Landé's paper. First, he argued, the correspondence principle demanded the use of the mean squared amplitudes rather than the mean amplitudes adopted by Landé.

Moreover, he read from his Fourier representation that the intensity of the satellite I_{perp} was only half of the value Landé had given. This second deviation from Landé's description is of central importance for understanding Heisenberg's correspondence argument at this point and its relation to the intensity problem. As he argued, the factor $\frac{1}{2}$ necessarily followed "from the energetically deducible equation:"

$$I_{par} + 2I_{perp} = \text{const},$$

where "const" meant that the total intensity was "independent of Θ."[46] This conclusion followed directly from his intensity distribution, as the sum of the trigonometric terms was $\cos^2\Theta + \sin^2\Theta = 1$.

In October 1921, Heisenberg apparently had no doubt that the summation confirmed the empirical facts in a trivial way. This position is understandable if Heisenberg's derivation is taken into account. All that he had done up to this point was to perform a straightforward coordinate transformation and to set up a Fourier representation describing the motion of the electronic system in one particular state. Associating the Fourier coefficients of this state with the respective transitions, Heisenberg did not follow Landé in assuming that the angle in the intensity formulas

[45]Heisenberg to Landé, 16 October 1921 (AHQP 6.2). At this point, Heisenberg assumed that both satellites had the same intensity.

[46]Heisenberg to Landé, 16 October 1921 (AHQP 6.2). "Auch den Faktor $\frac{1}{2}$ im Satelliten bei I_{senkr} haben Sie wohl übersehen; daß er da sein muß, folgt schon aus der energetisch ableitbaren Gleichung $I_{parallel} + 2I_{senkr} = const$ (nicht von Θ abhängig)."

was associated with a *Zwischenbahn*. Rather, he assumed that the Fourier series in one particular state was associated with all possible transitions leading to and departing from this state, as he made explicit in a subsequent letter already quoted at the beginning of this section:

> to every oscillation in our "level" corresponds a quantum jump, which departs from this level (and also one that ends at this level).[47]

Heisenberg's interpretation of the correspondence principle as a relation between the Fourier coefficients of a particular state and the transitions leading to and departing from this state solved the initial-final-state problem in favor of the initial state. As a consequence, the summation argument was little more than a trivial observation.[48]

Considering the angle Θ as a parameter of a coordinate transformation, Heisenberg did not interpret it in terms of space quantization. He only began to consider the possibility and necessity of doing so when he realized that Landé's *Zusatzregel* did not follow from his Fourier representation "without the assumption that k is the total [angular] momentum."[49] Without this assumption, Heisenberg's initial correspondence argument involved only the correlation of the Fourier coefficients with the transition probabilities following Bohr's and Kramers' line of thought. Otherwise it remained a classical argument, which did not incorporate the notions that were central for Landé and Sommerfeld in their physical interpretation of the empirical regularities of multiplets within the state-transition model.

Sommerfeld's Adaptation of the Correspondence Argument

After this discussion of Heisenberg's initial correspondence argument, we are finally in a position to discuss Sommerfeld's approach to the principle. As has been mentioned, Sommerfeld and Heisenberg decided to publish a joint paper on the intensity problem based on Heisenberg's calculation. In it, they separated Heisenberg's considerations on the dynamical model of the multiplet atom from the discussion of the intensity distribution as a physical problem in its own right. The driving force behind this move was Sommerfeld, who wrote the paper on his own

---

[47] Heisenberg to Landé, 23 October 1921 (AHQP 6.2). "Aber wir können es auch so ausdrücken: Jeder Schwingung in unserem 'Niveau' [entspricht] ein Quantensprung, der von diesem Niveau ausgeht (ebenso noch einer, der in diesem Niveau endet)."

[48] As will be discussed in more detail in Sect. 4.3, Heisenberg and Sommerfeld recognized that the summation of the perpendicular and parallel components became problematic if one adopted Kramers' *Zwischenbahn* model and consequently interpreted the angle Θ in the intensity formulas as the average value involving the initial and the final state.

[49] Heisenberg to Landé, 16 October 1921 (AHQP 6.2).

and thereby developed Heisenberg's correspondence argument into a first theoretical interpretation of his intensity rule.[50]

Like Heisenberg's correspondence argument, the final argument in the Sommerfeld-Heisenberg paper was based on the Fourier representation of a precessional motion around a fixed axis. Building on this model, the motions of a multiplet atom appeared in a layered structure. First, there was the model of Sommerfeld's theory of the hydrogen atom, in which the electron moves around the nucleus on an elliptic orbit and rotates around the axis of the electron's angular momentum, as described by the quantum numbers n and k. Second, the plane of the electronic orbit itself rotates around the axis of the total angular momentum. This motion is associated with the doublets and triplets of the multipet spectrum. Thereby the inner quantum number j is identified with the total angular momentum. Finally, the entire atom precesses around the axis of the external magnetic field, resulting in the Zeeman effect described by the magnetic quantum number m. For Sommerfeld, this "general kinematic description" was independent of the underlying "*modellmäßigen* origin of the atomic trajectories," i.e., of the specific dynamics of a particular atom.[51]

Sommerfeld and Heisenberg extended Heisenberg's argument and built up their Fourier representation, mirroring the layered structure step by step. They started with the ellipse rotating in its orbital plane, describing it with a generic Fourier series in a coordinate system ξ, η:

$$(\xi + i\eta)e^{i\nu_k t} = \sum_{s=-\infty}^{s=\infty} a_s e^{i(s\nu_n + \nu_k)t},$$

where ν_n is the frequency associated with the principal quantum number n and ν_k is the frequency of the precession within the orbital plane associated with the azimuthal quantum number k. As in Heisenberg's initial argument, the precession of this system around the axis of the atom's total angular momentum is described by transforming the initial coordinate system into a new coordinate system x, y, z using the transformation equations:

$$x = \xi \cos \vartheta,\ y = \eta,\ z = \xi \sin \vartheta,$$

[50]The Sommerfeld Papers (NL 89, 026) contain the manuscript of the Sommerfeld-Heisenberg paper, which is written in Sommerfeld's hand. Sommerfeld, who anticipated that the publication process would not be finished before his departure to the U.S. in the fall of 1922, sent the manuscript to *Zeitschrift für Physik* with instructions that the proofs should first be sent to his assistant Gregor Wentzel, who should, if necessary, reformulate passages on his behalf, and then to Heisenberg for final proofreading. At this point, Heisenberg's only change had been to add a footnote on Landé's initial correspondence interpretation, which did not make it into the final publication as was discussed in Footnote 38.

[51]Sommerfeld and Heisenberg (1922a, 131). I will give quotes with reference to the published and hence more accessible version where it is identical to Sommerfeld's draft. In addition, I will use Sommerfeld's draft to trace some alterations made by Sommerfeld and Heisenberg during the reviewing process, which provide key insights into the development of the argument.

under the assumption that the z-axis is parallel to the j-axis, which forms the angle ϑ with the k-axis and thereby the orbital plane. Under the assumption that the precession does not alter the initial motion, this yields the following Fourier representation:

$$x + iy = \sum_{s=-\infty}^{\infty} \left(\frac{1+\cos\vartheta}{2} a_s e^{i(s\nu_n+\nu_k+\nu_j)t} - \frac{1-\cos\vartheta}{2} \overline{a}_{-s} e^{i(s\nu_n-\nu_k+\nu_j)t} \right)$$

$$z = \sum_{s=-\infty}^{\infty} \frac{\sin\vartheta}{2} \left(a_s e^{i(s\nu_n+\nu_k)t} + \overline{a}_{-s} e^{i(s\nu_n-\nu_k)t} \right),$$

or

$$x + iy = \sum_{s=-\infty}^{\infty} \sum_{r=\pm 1} \left(c_{s,1} e^{i(s\nu_n+r\nu_k+\nu_j)t} + c_{s,-1} e^{i(s\nu_n-r\nu_k+\nu_j)t} \right)$$

$$z = \sum_{s=-\infty}^{\infty} \sum_{r=\pm 1} \left(d_{s,1} e^{i(s\nu_n+r\nu_k)t} + d_{s,-1} e^{i(s\nu_n-r\nu_k)t} \right).$$

Applying the correspondence principle to this Fourier representation, Sommerfeld and Heisenberg associated the different Fourier coefficients with the different transitions and departed from Heisenberg's initial argument in an important way. Heisenberg had tacitly demanded that the Fourier coefficients c and a be real, finding identical values for the perpendicular components. Sommerfeld and Heisenberg operated with complex Fourier coefficients instead, obtaining a subdivision of the two intensities that became important in Sommerfeld's interpretation of his intensity rule.[52] Considering a spectral line with given Δn and Δk, they argued that the term $e^{i(s\nu_n+\nu_k+\nu_j)t}$ was associated with the component $\Delta j = 1$, the term $e^{i(s\nu_n+\nu_k)t}$ belonged to the transition $\Delta j = 0$, and the term $e^{i(s\nu_n+\nu_k-\nu_j)t}$ corresponded to $\Delta j = -1$. As the first two terms occur directly in Heisenberg and Sommerfeld's Fourier series, the Fourier coefficient can be read directly:

$$\Delta j = 1 : c_{s,1} = \frac{1+\cos\vartheta}{2} a_s$$

$$\Delta j = 0 : d_{s,1} = \frac{\sin\vartheta}{2} a_s.$$

[52]This result appears to have been developed by January 1922. See Heisenberg to Landé, 25 January 1922 (AHQP 6.2), in which Heisenberg gave the subdivision, indicating that his intensity formulas were "in need of improvement."

The term $e^{i(s\nu_n+r\nu_k-\nu_j)t}$, however, does not occur in the Fourier series. To get it, Sommerfeld and Heisenberg considered the complex conjugate of the Fourier series and replaced $+i$ with $-i$. They then considered the term $-s$ and $r=-1$ to obtain the Fourier coefficient associated with the corresponding term $e^{i(s\nu_n+\nu_k-\nu_j)t}$:

$$\Delta j=-1: c_{-s,-1}=-\frac{1-\cos\vartheta}{2}a_s.$$

Following this procedure Sommerfeld and Heisenberg obtained intensity ratios, which became central for the discussion of the intensity problem. I will therefore discuss them under the label Sommerfeld-Heisenberg formulas and give them here in full. For multiplets, Sommerfeld and Heisenberg obtained:

$$\begin{aligned} & J_{\Delta j=\pm1} : J_{\Delta j=0} : J_{\Delta j=\mp1} \\ k=\pm1: & \cos^4\tfrac{\vartheta}{2} : \cos^2\tfrac{\vartheta}{2}\sin^2\tfrac{\vartheta}{2} : \sin^4\tfrac{\vartheta}{2} \\ k=0: & \sin^2\vartheta : 2\cos^2\vartheta : \sin^2\vartheta. \end{aligned}$$

For the Zeeman effect, similar intensity ratios followed from the description of the motion in the external field based on the same ideas:

$$\begin{aligned} & J_{\Delta m=\pm1} : J_{\Delta m=0} : J_{\Delta m=\mp1} \\ j=\pm1: & \tfrac{1}{4}(1+\cos\Theta)^2 : \sin^2\Theta : \tfrac{1}{4}(1-\cos\Theta)^2 \\ j=0: & \tfrac{1}{4}\sin^2\Theta : \cos^2\Theta : \tfrac{1}{4}\sin^2\Theta. \end{aligned}$$

As in Landé's work, the variation of the intensities within a triplet is determined through the angle ϑ and Θ, i.e., the spatial configuration of the atom and its orientation with respect to the external field. The Fourier coefficients describing the undisturbed motion of the system, on the other hand, drop out of the respective ratios. Following Heisenberg's first correspondence argument, this result is obtained through a coordinate transformation, which introduces the angles and describes the precession of the undisturbed system around the fixed axis of the total angular momentum or the external magnetic field.

As has been shown, Heisenberg initially interpreted the angles as artifacts of the coordinate transformation and did not accept Landé's interpretation based on space quantization. This position was abandoned as Sommerfeld and Heisenberg embraced the fact that the *Zusatzregel* for the Zeeman effect was a direct result of space quantization and thereby converted to Landé's interpretation of the inner quantum number as the total angular momentum.[53]

[53] This position was in place as early as January 1922, as we can see from a letter by Heisenberg to Landé, in which Heisenberg expressed the angle as a fraction of the quantum numbers. Heisenberg to Landé, 25 January 1922 (AHQP 6.2).

This conversion dawned slowly on Heisenberg and Sommerfeld, however. Heisenberg continued to argue for his interpretation in his letters to Landé throughout the spring of 1922. Likewise, Sommerfeld remained undecided in his paper on Catalan's work in June. In it, he stated that the inner quantum number was associated with the total angular momentum or "was at least closely connected to it." Only in the final Sommerfeld-Heisenberg paper, did the two make explicit that the explanation of the *Zusatzregel* showed "unequivocally that the inner quantum number [...] has to be interpreted as the total mechanical moment of momentum."[54]

Sommerfeld and Heisenberg's conversion did not bring them closer to a numerical evaluation of the trigonometric terms. For the Zeeman effect, the angle Θ was determined from the equation $\cos\Theta = m/j$ for the simplest case of the hydrogen atom. However, this simple relation, Heisenberg and Sommerfeld argued, might not be applicable in all cases. For multiplets, the situation was even more problematic, since there was no equivalent equation for the angle ϑ. In general, Sommerfeld and Heisenberg assumed j and k to be related through the vectorial composition of the total angular momentum by the angular moment k and the angular momentum of the rump r, but they did not mathematize this idea and it is unclear whether they thought that the electronic orbit had a fixed orientation with respect to the total angular momentum.[55]

In this situation, and this is central for understanding Sommerfeld's use of the correspondence principle, Sommerfeld turned to the intensity rule and considered under which conditions it fit with the intensity ratios obtained from the correspondence principle. The basic idea of this argument was to order the components according to their empirical intensity rather than in the order suggested by the frequencies of the lines. This made little difference for the $\Delta k = \pm 1$ component, as the order remained the same. The strongest component in which j and k change in the same sense came first. It was followed by the transition in which j did not go in the same direction as k. Finally, there was the transition in which j and k changed in the opposite sense. For $\Delta k = \pm 1$, the Sommerfeld-Heisenberg formula thus became

$$J_{\Delta j=\pm 1} : J_{\Delta j=0} : J_{\Delta j=\mp 1} = \cos^4\frac{\vartheta}{2} : 2\cos^2\frac{\vartheta}{2}\sin^2\frac{\vartheta}{2} : \sin^4\frac{\vartheta}{2},$$

which preserved the order of the frequencies. For the $\Delta k = 0$ case,[56] however, Sommerfeld put the $J_{\Delta j=0}$ component first as the most intense, followed by two equally intense components $J_{\Delta j=+1}$, $J_{\Delta j=-1}$:

$$J_{\Delta j=0} : J_{\Delta j=+1} : J_{\Delta j=-1} = 2\cos^2\vartheta : \sin^2\vartheta : \sin^2\vartheta.$$

[54]See Sommerfeld (1922b, 35) and Sommerfeld and Heisenberg (1922a, 140) for the respective quotes.

[55]Within Landé's vector model, such a mathematization determined the angle as $\cos\vartheta = \frac{k^2+j^2-r^2}{2kj}$.

[56]Though forbidden according to the usual selection rule, these transitions had been observed under certain conditions by Götze and Franck.

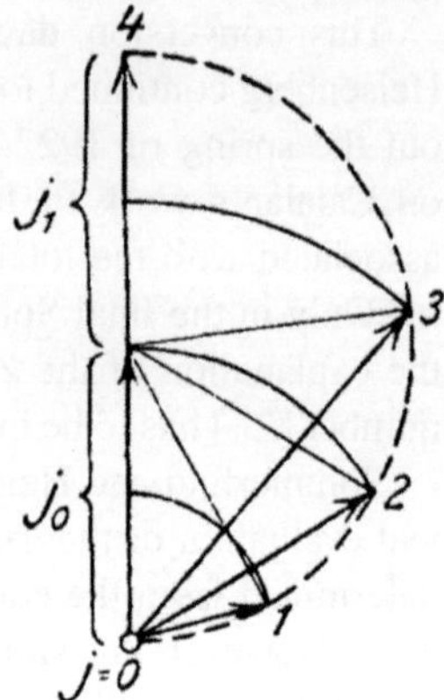

Fig. 4.3 Geometrical representation of the possible values for the inner quantum number in Sommerfeld (1922b, 54)

In this case, Sommerfeld clearly disregarded the order implied by the frequencies, in which the two weaker components were symmetric to the strong component.[57]

Setting up the representation with respect to the empirical intensities gave a clear indication for the ϑ-dependence. In all three cases ($\Delta k = +1, 0, -1$), the cosine term was associated with the strongest components, while the weaker satellites were connected with a sine term. Therefore ϑ had to be small if Sommerfeld's rule were to materialize. For small angles ϑ the intensity formula could be rewritten as

$$1 : \frac{1}{2}\vartheta^2 : \frac{1}{16}\vartheta^4 \text{ or } 1 : \frac{1}{2}\vartheta^2 : \frac{1}{2}\vartheta^2,$$

respectively, and showed the marked asymmetry between the strong main line and weaker satellites in the most pronounced way.[58] Focusing on this case, Sommerfeld analyzed the conditions under which the angle ϑ would be small according to the vectorial relation between the total angular momentum j and the angular momentum of the valence electron k.[59]

Sommerfeld connected this analysis with his work on the spectrum of manganese, which he had embarked upon following his reception of Miguel Catalan's work in March 1922. Catalan's extensive measurements and analysis of the spectrum of manganese had revealed that even the immensely more complicated spectra of manganese fit into the scheme of the inner quantum number and its selection rule, by simply assuming that the element had a d-term with five energy levels instead of a threefold d-term.[60] In this context, Sommerfeld developed a geometrical scheme (Fig. 4.3) to construct the possible values of the inner quantum numbers describing

[57] Sommerfeld and Heisenberg (1922a, 142–144).

[58] Sommerfeld and Heisenberg (1922a, 144).

[59] Sommerfeld and Heisenberg (1922a, 144–145).

[60] Sommerfeld (1922b). For the dating of Sommerfeld's reception see Eckert (2013a, 332–333).

the s,p,d...-levels of a given multiplet. This scheme was based on the assumption that:

> [the] total [angular] momentum [j] is determined vectorially from the angular momentum of the unexcited atom j_0 and the total angular momentum j_1 of the excitation [...] The largest value of j occurring in the respective line structure thus apparently corresponds to the concordant resolution [Auseinanderlegung] $j_0 + j_1$ of both momenta; the smallest value emerges when both partial moments superimpose in the opposite sense, and is equal to $|j_0 - j_1|$. Between these two extreme values only integer intermediate values are possible.[61]

When discussing the intensity distribution, Sommerfeld connected this geometrical scheme with the kinematic description underlying his and Heisenberg's correspondence argument. Following Heisenberg's rump model, he assumed that the angular momentum j_0 in the ground state should depend on the angular momenta of both the rump and the valence electron. In the excited state, the value for the rump remained the same, while the azimuthal quantum number k became larger. Consequently, the electron's angular momentum would dominate for larger values of k. This meant that the j-axis would have the same direction as the k-axis, or, in other words, that the angle between them became small. Therefore, Sommerfeld argued, the intensity rule should become manifest for transitions between terms with high k, like in transitions between d and b-terms, and appear less markedly for smaller ones like the sp-transition.[62]

Beyond the case of large k, Sommerfeld saw another possible prediction based on the vectorial decomposition of j. "The angle ϑ is also small," he observed, if the two components "add up algebraically and not vectorially."[63] Looking at the cases where algebraic addition did not hold, Sommerfeld formulated a different kind of regularity inherent in the correspondence formulas. The case of j_0 and j_1 adding up algebraically was realized, Sommerfeld argued, for the highest value of j within the multiplet. For triplets involving a term with smaller values of j, for which j_0 and j_1 added up vectorially, ϑ would not be larger and the intensity ratio in these triplets would not follow Sommerfeld's rule. Based on this argument, Sommerfeld

[61] Sommerfeld (1922b, 52–53). "Wir nehmen wie in §1 an, daß die innere Quantenzahl j das gesamte Impulsmoment des Atoms in dem betreffenden Anregungszustand bedeutet und daß dieses Gesamtmoment sich vektoriell zusammensetzt aus dem Impulsmoment j_0 des unangeregten Atoms und dem Impulsmoment j_1 der Anregung. Wir nehmen an, daß letztere in die Reihe der Terme S,P,D ... bzw. s,p,d ... =0,1,2 ... ist. Der größte in der betreffenden Linienstruktur vorkommende Wert von j entspricht dann offenbar der gleichsinnigen Auseinanderlegung $j_0 + j_1$ beider Momente; der kleinste Wert entsteht, wenn beide Teilmomente mit entgegengesetztem Sinn sich überlagern, und ist gleich $|j_0 - j_1|$. Zwischen diesen beiden Extremwerten sind alle ganzzahligen Zwischenwerte möglich."

[62] Sommerfeld (1922b, 145).

[63] Sommerfeld and Heisenberg (1922a, 145). "Der Winkel ϑ wird aber auch klein, wenn zwar nicht $k >> j_0$ ist, aber j_0 und j_1(bzw. k) sich algebraisch, nicht vektoriell addierend. Dieses findet statt im Hauptniveau jedes Terms, im Niveau der größten inneren Quantenzahl, welche zugleich im allgemeinen —wenigstens bei den gewöhnlichen Dublett- und Triplettsysstemen— das oberste Niveau, dasjenige des größten Termwertes darstellt."

formulated a second rule that described the relation of the intensity ratios of different triplets within a multiplet:

> We thus expect that our intensity rule will become manifest in different degrees for different energy levels; it will be most pronounced for the energy level of largest j, and is going to become blurred for smaller j.[64]

This second expectation was closely connected with Catalan's analysis of higher multiplets. Only the larger number of energy levels, and thereby the increasing number of triplets within these multiplets, made it possible to test and eventually confirm Sommerfeld's prediction about the intensity ratios of different j values.[65]

Most importantly for Sommerfeld, the application of the correspondence principle provided a theoretical interpretation for his intensity rule. The rule could now be understood as a direct consequence of the spatial configuration within the multiplet atom. This meant that "the opposition—satellite and main line—is formulated theoretically for the first time."[66]

Interim Conclusion

Before going on to the next section, I will briefly summarize the analysis of the implementation of the correspondence principle by Landé, Heisenberg, and Sommerfeld. As has been shown in this section, these implementations were closely connected to the search for the empirical regularities of multiplets and the Zeeman effect: Landé, Heisenberg and Sommerfeld used the correspondence principle to interpret the intensity distribution and its regularities in terms of a physical model and even tried to extract such a model from the empirical data.

With their correspondence arguments, Landé as well as Heisenberg and Sommerfeld thought that they were prolonging Bohr's original correspondence arguments in a straightforward manner. On the basis of this prolongation, they interpreted the qualitative features of the intensity distribution described by Sommerfeld's intensity rule and explained the additional selection rules. At the same time, their new applications of the correspondence principle differed remarkably from Bohr's

---

[64](Sommerfeld and Heisenberg 1922a, 145). "Wir erwarten also, dass unsere Intensitätsregel bei den verschiedenen Termniveaus verschieden stark zur Geltung kommen wird, am ausgesprochensten bei dem Niveau des größten j, und sich bei kleinerem j verwischen wird."

[65]Sommerfeld and Heisenberg (1922a, 148).

[66]Sommerfeld and Heisenberg (1922a, 146). "Durch unsere Intensitätsregel wird also erstmalig der Gegensatz—Satellit und Hauptlinie—theoretisch gefasst." As is clear from the unpublished draft, this result of the correspondence argument was so important that Sommerfeld initially thought about extending it to the Zeeman effect. He argued the Zeeman intensities could in principle be "formalized in the same clear manner" and only admitted that this extension was problematic in practice as the condition of a small angle was not realized. Without this precondition, the intensities had to be determined numerically using the equation $\cos\Theta = m/j$ or its equivalent in Heisenberg's rump model, so that Sommerfeld eventually decided to cut any reference to the extension.

original argument. Landé as well as Heisenberg and Sommerfeld brought the correspondence principle into contact with the attempts to interpret the structure of multiplets in terms of a physical model. These attempts, as has been shown, relied crucially on the idea that multiplet spectra were an expression of different space-quantized configurations of the atom. Integrating the correspondence principle into these attempts, the idea of spatial orientation became the central element in the correspondence approach within multiplet spectroscopy.

4.3 Recognizing Problems: Sommerfeld, Heisenberg and the Total Intensity of Split-up Lines

While Sommerfeld and Heisenberg had the overall impression that their application was a straightforward extension of the correspondence approach, they also encountered difficulties within their application. In his letter to Landé 25 January 1922, in which Heisenberg reported on the "improvement" of his intensity distribution due to the fully complex Fourier representation, Heisenberg communicated another revision of his initial considerations and the problem which resulted from it:

> It now turns out from the application of these formulas that for small quantum numbers (small n) the condition is no longer fulfilled that $I_\pi = I_\sigma$ holds throughout the separation image [Aufspaltungsbild, MJ] of a line. This condition has to be fulfilled, of course, because the original line is unpolarized. The reason for this error is located in the extrapolatory character of the correspondence principle."[67]

It is difficult to assess the extent to which Heisenberg had identified the origin of the discrepancy between the intensity distribution calculated from the correspondence principle and his physical expectation of an unpolarized parent line. His diagnosis that it resulted from the "extrapolatory character" of the correspondence principle did not clarify this point. A clue in this direction is given, however, by the manuscript of Sommerfeld and Heisenberg's final paper. In it, Sommerfeld added several comments on the implications of the initial-final-state problem for their discussion of the intensity problem:

> In general the intensities are determined by the correspondence principle as Fourier coefficients. This leads to the known difficulty that is characteristic for the correspondence principle: should one consider the Fourier series of the initial or of the final term, or of some appropriate average intermediate orbit?[68]

---

[67]Heisenberg to Landé, 25 January 1922 (AHQP 6.2). "Nun zeigt sich, bei Anwendung dieser Formeln, daß für kleine Quantenzahlen (kleine n), die Bedingung nicht erfüllt ist, daß für das Aufspaltungsbild einer Linie im ganzen $I_\pi = I_\sigma$ ist. Diese Bedingung muß natürlich erfüllt sein, da die ursprüngliche Linie unpolarisiert ist. Der Grund für diesen Fehler liegt im extrapolatorischen Charakter des Korrespondenzprinzips."

[68]Sommerfeld and Heisenberg (1922a, 140). "Dabei tritt die bekannte, für das Korrespondenzprinzip charakteristische Schwierigkeit auf: Soll man die Fourierreihe des Anfangs- oder Endtermes betrachten oder die einer geeignet gemittelten Zwischenbahn?"

In the case of multiplets, Sommerfeld explained, this ambiguity was not limited to the Fourier coefficients of the electron's ellipse, but extended to the space-quantized angle ϑ in the intensity ratios. Representing the configuration of the atom in a particular state, the angle ϑ was naturally different for the initial and final states of the atom. As a consequence, the determination of the intensity of an individual component in a multiplet or Zeeman triplet becomes ambiguous:

> Even if we are dealing with a transition, which departs from the same initial state, the final states to which they lead are different. Therefore the angles between the j and k-axis in three different intermediate states have to be considered as the angle ϑ.[69]

This assessment has major implications for the total intensity of a split-up line pattern. As each component involves the same initial but different final states, there are three different angles for the three *Zwischenbahnen*. Due to these different angles, the trigonometric terms no longer add up to one. Instead one arrives at an asymmetry between the different components and therefore at a non-vanishing polarization of the original line. In other words, the initial-final-state problem gave rise to predictions that were in conflict with hitherto unquestioned physical expectations.

It appears that Heisenberg alluded to this consideration in his letter to Landé when he diagnosed that the discrepancy between his physical expectation and the correspondence formulas was due to "the extrapolatory character of the correspondence principle." At the time, Heisenberg perceived this conflict as a challenge for the correspondence principle. As he argued:

> To remove the flaw one can go beyond the corresp[ondence] pr[inciple]. To begin with, the Voigtian Theory suggests itself as a point of reference, which says that I_π and $I_{\sigma_{+-}}$
>
> 1. are also quadratic forms of $\cos\theta$ resp. of m for $n = 2$
> 2. that in every case $2(I_{\sigma_+} + I_{\sigma_-}) + I_\pi$ is independent of m (this is immediately clear, if the levels are *equally* probable for all m)
>
> Moreover, necessary conditions for the formula are
>
> that I_{σ_+} and I_{σ_-} only differ in the sign of m,
> that for $m = n$, resp. $m^* = n^*$, $I_\pi = 0$ and for $m = n, n-1$ or $m^* = n^*, n^* - 1$ $I_{\sigma_-} = 0$,
> that finally for the total splitting $\sum I_\pi = \sum I_\sigma$.[70]

[69]Sommerfeld and Heisenberg (1922a, 142). "Auch wenn es sich um Übergänge handelt, die vom gleichen Anfangsniveau ausgehen, sind die Endniveaus, zu denen sie führen, verschieden. Als Winkel ϑ kommen dann die Winkel zwischen j und k-Achse in drei voneinander verschiedenen Zwischenzuständen in Frage."

[70]Heisenberg to Landé, 25 January 1922 (AHQP 6.2). "Der Grund für diesen Fehler liegt im extrapolatorischen Charakter des Korrespondenzprinzips. Um den Mangel zu beseitigen kann man über das Korresp[ondenz]pr[inzip] hinausgehen. Da bietet sich zunächst als Anhaltspunkt die Voigtsche Theorie diese sagt, daß I_π u. $I_{\sigma_{+-}}$ 1. auch für $n = 2$ quadratische Formen von $\cos\theta$ bzw. von m sind. 2. daß stets $2(I_{\sigma_+} + I_{\sigma_-}) + I_\pi$ von m unabhängig ist (das ist auch [...] klar, wenn die Niveaus für alle m *gleichwahrscheinlich* sind). Ferner sind notwendige Bedingungen für die [...] Formeln, daß I_{σ_+} und I_{σ_-} sich nur durch das Vorzeichen von m unterscheiden dürfen, daß für $m = n$ bzw. $m^* = n^*$ $I_\pi = 0$ ist und für $m = n, n-1$ oder $m^* = n^*, n^* - 1$ $I_{\sigma_-} = 0$ ist,

Trying to resolve the problem, Heisenberg turned to Voigt's theory of the anomalous Zeeman effect and formulated several formal constraints on the respective intensity formulas. Unfortunately, we know very little about how Heisenberg came to his conclusions, as his calculations were not published and neither Sommerfeld nor Heisenberg made further comments on the constraints. Nonetheless, some insights can be gleaned from the results, which Heisenberg gave in the letter to Landé:

> All these conditions can indeed be fulfilled by simple formulas. I write these formulas, i.e., I as a function of the quantum numbers of the initial orbit for e.g. the main line of the doublet and write the formulas of the correspondence principle on the side.
>
> | | extrapolated form | correspondence pr[inciple] |
> |---|---|---|
> | I_π | $1-(\frac{m*}{n_j*})^2$ | $1-(\frac{m*}{n*})^2$ |
> | $I_{\sigma+}$ | $\frac{1}{4}\left[(1+\frac{m*}{n_j*})(1+\frac{m*-1}{n_j*})\right]$ | $\frac{1}{4}(1+\frac{m*}{n*})^2$ |
> | $I_{\sigma-}$ | $\frac{1}{4}\left[(1-\frac{m*}{n_j*})(1-\frac{m*+1}{n_j*})\right]$ | $\frac{1}{4}(1-\frac{m*}{n*})^2$ |
>
> One sees at once that for large values of $n*$ both formulas coincide.[71]

Assuming that the intensities were described by quadratic functions, Heisenberg turned the quadratic expressions resulting from his correspondence argument into a product of two different factors. These two factors incorporated Heisenberg's requirement that the intensity vanishes for values of the magnetic quantum number ($m = n$ for I_π and $m = n, n-1$ for I_{σ_-}). As Heisenberg's intensity formulas were expressed in terms of the "quantum number of the initial orbit," this further ensured that the sum of the intensities of the different components was independent

daß endlich für die Gesamtaufspaltung $\sum I_\pi = \sum I_\sigma$ ist." The formatting is added to increase the readability of Heisenberg's text.

[71]Heisenberg to Landé, 25 January 1922 (AHQP 6.2). "All diese Bedingungen lassen sich nun tatsächlich durch einfache Formeln erfüllen. Ich schreibe diese Formeln, d. h. I als Funktion der Quantenzahlen der Anfangsbahn für z. B. für die Hauptlinien der Dubletts an und schreibe, zum Vergleich, die Formeln des Korrespondenzprinzips daneben.

	extrapolierte Form	Korrespondenzpr.
I_π	$1-(\frac{m*}{n_j*})^2$	$1-(\frac{m*}{n*})^2$
$I_{\sigma+}$	$\frac{1}{4}\left[(1+\frac{m*}{n_j*})(1+\frac{m*-1}{n_j*})\right]$	$\frac{1}{4}(1+\frac{m*}{n*})^2$
$I_{\sigma-}$	$\frac{1}{4}\left[(1-\frac{m*}{n_j*})(1-\frac{m*+1}{n_j*})\right]$	$\frac{1}{4}(1-\frac{m*}{n*})^2$

Man sieht auf den ersten Blick, daß für große Werte von $n*$ die beiden Formeln übereinstimmen."

of the magnetic quantum number, and that the total intensity of the components with different polarization was equal.[72]

While Heisenberg initially believed that the total intensity of a Zeeman pattern presented a challenge to the correspondence principle, which called for an adaptation of his intensity formulas, his considerations did not enter into the final Sommerfeld-Heisenberg paper. His intuitive assertion that the total polarization had to vanish became doubtful as the π-components indeed appeared to be more intense than the σ-components in Ernst Back's spectrographic photographs. In addition, Kramers had pointed to their asymmetry in his dissertation and Bohr had argued that this situation had to be expected on the basis of quantum theory, as a result of space quantization.[73] Accepting this situation as an empirical fact, Heisenberg and Sommerfeld dropped Heisenberg's diagnosis of a "flaw" in the correspondence principle and mentioned that Back's estimates and the theoretical predictions were quite "remarkable."[74]

Overall, Sommerfeld (and Heisenberg) had thus taken up the correspondence principle as a tool in 1922 and integrated it into his work on the *Gesetzmäßigkeiten* of multiplets. In this context, he used it to give a theoretical interpretation for his intensity rule and to formulate new regularities for the intensities of spectral lines. This implementation of the correspondence principle relied crucially on the notion of space quantization, which was central for the physical interpretation of multiplet atoms. As Sommerfeld and Heisenberg recognized, the application of the principle led to a new version of the initial-final-state problem, which manifested itself in a direct conflict between theoretical predictions and hitherto unquestioned physical intuitions. Due to the experimental evidence, however, this problem and its tentative solution were discarded. It appeared as a counterintuitive but empirically confirmed consequence of the correspondence principle.

4.4 Adaptive Reformulation: Sommerfeld, Hönl and the "Theory of Intensities"

Sommerfeld and Heisenberg completed their work on the intensity problem in the fall of 1922. Before leaving for Göttingen to work on his dissertation in hydrodynamics, Heisenberg sent the final version of the paper written by Sommerfeld

[72]Heisenberg to Landé, 25 January 1922 (AHQP 6.2). "All diese Bedingungen lassen sich nun tatsächlich durch einfache Formeln erfüllen. Ich schreibe diese Formeln, d. h. I als Funktion der Quantenzahlen der Anfangsbahn."

[73]See Sommerfeld and Heisenberg (1922a, 154) for the reference to Back's experiments as well as Bohr's position. Sommerfeld and Heisenberg might have discussed the issue privately with Bohr in Göttingen in June 1922. While the notes on Bohr's *Wolfskehl lectures* do not reflect any statements on the issue, it is clear that Bohr had formulated his position by March 1922. In his *Guthrie Lecture*, he had stated that "we should be prepared to find a resultant polarization of the total light of each triplet, even in weak magnetic fields." (Bohr 1922, 290).

[74]Sommerfeld and Heisenberg (1922a, 154).

to *Zeitschrift für Physik*. By that time, Sommerfeld was already aboard the S.S. George Washington headed to America, where he spent the academic year giving lectures on atomic theory.[75] Shortly after his return from the United States, the intensity problem for multiplets came back to Sommerfeld, albeit in an eminently different form. As we will see in this section, the qualitative description of intensities was replaced by quantitative regularities. Based on new experimental results, Sommerfeld and with him Ornstein, Burger and Dorgelo, who had conducted the experiments, formulated new empirical regularities and developed a description of the intensity of multiplets within the *Gesetzmäßigkeiten* approach. This approach culminated in the attempt to build intensity schemes, which completely determined the intensity of multiplets. While Sommerfeld initially thought that this new approach rendered the correspondence principle obsolete in its original realm of application, he soon returned to it. In so doing, he integrated and adapted the original Sommerfeld-Heisenberg formulas in order to resolve the problems with his "theory of intensities."

Within this process, the kinematic description and the problem within the correspondence approach discussed in the previous section resurfaced at the periphery, as the new experimental results revised the previous intensity estimates and reaffirmed Heisenberg's intuition. For Heisenberg and Pauli this turned the problem of split-up lines into a major challenge for the original correspondence approach. For Sommerfeld, however, the intensity problem had shifted away from the kinematics of a physical model, as it had become a prime example for his formal *Gesetzmäßigkeiten* approach and appeared in a different form.

The Intensity Problem in Sommerfeld's Gesetzmäßigkeiten *Approach: Intensity Schemes and the Utrecht Sum Rules*

In September 1923, Sommerfeld attended a lecture by Leonard Ornstein at the Deutsche Physikertag in Bonn entitled "Über photographische Photometrie." As Ornstein reported, the physicists at his institute in Utrecht had developed new experimental equipment and a new technique to measure the intensity of spectral lines.[76] This new technique, the Utrecht physicists would claim within the next

[75]See Heisenberg to Sommerfeld, 17 October 1922 in Sommerfeld (2004, 126) as well as Eckert (2013a, 343).

[76]The title of Ornstein's talk is given in the program of the Bonn conference. See "Zweiter Deutscher Physikertag in Bonn vom 16. bis 22. September 1923" in *Verhandlungen der Deutschen Physikalischen Gesellschaft im Jahre 1923* edited by Karl Scheel. For the contents of Ornstein's talk and Sommerfeld's reaction, see Sommerfeld (1925, 4) and Dorgelo (1924, 170). For a short discussion of the experimental setup and a contextualization of the Utrecht intensity measurements within the development of spectroscopic photometry, see Hentschel (2002, 276–283).

years, meant that "the measurement of the intensities of spectral lines has to be put on an equivalent footing with the measurement of wavelengths."[77]

One of Ornstein's Ph.D. students, Henk Dorgelo, had applied the new experimental procedure to the doublets of the alkaline metals and had found very promising initial results. The intensity ratio for the simple doublets was 2:1 for sodium, calcium and rubidium. Moreover, the triplets of the alkaline earth metals in the next subgroup of the periodic table all showed a ratio of 5:3:1. This occurrence of whole numbers caught Sommerfeld's attention. The number mysticist recognized a sequence within the intensity ratios of the different subgroups and understood that the intensity of the individual components was proportional to the inner quantum number of the final state involved in the respective transition. Sommerfeld's observation became the point of departure for an extended research cooperation between the Utrecht institute and the Munich school.[78]

Returning from Bonn, Sommerfeld wrote a letter to Dorgelo, explaining his interpretation and encouraging him to check the intensities for multiplets in which the inner quantum number assumed higher values, predicting ratios of 6:4:2 for a quartet system and 7:5:3 for a quintet system.[79] Confirming Sommerfeld's prediction, Dorgelo extended his measurements to composite doublets and triplets, i.e., to transitions between p- and d-levels in which the inner quantum number assumes different values for the initial and the final states.

Measuring the intensity when the composite doublets and triplets were not yet fully resolved, Dorgelo found intensity ratios according to Sommerfeld's prediction. As the non-separation of the different components was nothing but the result of low optical resolution, Sommerfeld and the Utrecht physicists concluded that the total intensity of lines was also proportional to the inner quantum number, and that the same had to be true if the multiplet was resolved into its components:[80]

---

[77]Burger and Dorgelo (1924, 258).

[78]The cooperation went beyond the exchange of letters and results. It led to a visit by Sommerfeld to Utrecht in April 1924 and to an extended research stay by Helmut Hönl at Ornstein's institute in 1926 and 1927 in connection with a Rockefeller fellowship. See Sommerfeld to Ornstein, 27 February 1926, Hönl to Sommerfeld, 12 June 1927, and Hönl to Sommerfeld, 13 August 1927 in the Sommerfeld Papers (NL 89, 009), indicating that the plan was for Hönl to go to Utrecht for one year starting on 1 October 1926. Although he did go to Utrecht, Hönl eventually broke off his stay due to nervous problems in August 1927. They affected him for some time. In November 1927 Sommerfeld tried to find a position for Hönl and wrote to Heisenberg that "a simple task like checking exercises or cooking tea would surely be good for him." Sommerfeld to Heisenberg, 15 November 1927 (Sommerfeld Papers (NL 89, 002)).

[79]Sommerfeld's letter to Dorgelo is not available. See (Dorgelo 1924, 174–177) and also Sommerfeld (1925, 4–5).

[80]This order of events is given explicitly in Sommerfeld (1925, 7) and is mirrored in the argument of Dorgelo and Burger's paper, see Burger and Dorgelo (1924, 259). The idea that the total intensity of split-up lines with different initial states is a significant property was absent in Dorgelo's work before he received Sommerfeld's rule, and came into view only when he observed that the unresolved complex doublets and triplets obeyed Sommerfeld's rule. See Dorgelo (1923, 209) and Dorgelo (1924, 177).

Tabelle 1.

	$1p_1 - 2d_3$	$1p_1 - 2d_2$	$1p_1 - 2d_1$	$1p_2 - 2d_3$	$1p_2 - 2d_2$	$1p_3 - 2d_3$
Wellenlänge	4456,61	4455,88	4454,77	4435,67	4434,95	4425,43
Intensität J .	1	18	100	19	54	25

Fig. 4.4 Table for the intensity of the *pd* transitions of calcium in Burger and Dorgelo (1924, 260)

Tabelle 2.

$1p_1\left(\frac{5}{2}\right)$	1	18	100
$1p_2\left(\frac{3}{2}\right)$	19	54	0
$1p_3\left(\frac{1}{2}\right)$	25	0	0
	$2d_3\left(\frac{3}{2}\right)$	$2d_2\left(\frac{5}{2}\right)$	$2d_1\left(\frac{7}{2}\right)$

Fig. 4.5 Intensity scheme for the *pd* transitions of calcium in Burger and Dorgelo (1924, 262)

> The sums of the intensities of the components of a multiplet, which correspond to jumps of the atom into the same final state, behaves like the inner quantum number J of the final states (Rule II).[81]

This proposition became known as the Utrecht sum rule. It was accompanied by a similar rule for the inverse process. Summing up all intensities with the same initial state yields ratios that are proportional to the ratios of the inner quantum numbers of the initial states.

To check these new relations simultaneously, Dorgelo and Burger began using a new representation. They transformed the tables, recording their experimental results (Fig. 4.4) into intensity schemes, which were analogous to term schemes for frequencies (Fig. 4.5). In this scheme the relevant intensity sums were given simultaneously. By looking at the rows or the columns one could check the consistency of the new rules immediately. For example, the sum in the first row is 119, in the second 73, in the third 25; their ratio is proportional to the ratio of the inner quantum numbers 5:3:1 labeling that row. The same holds for the columns.

[81] Burger and Dorgelo (1924, 259–260). "Die Summen der Intensitäten der Komponenten einer Mehrfachlinie, welche mit Sprüngen des Atoms in einem gleichen Endzustand korrespondieren, verhalten sich wie die inneren Quantenzahlen J dieser Endzustände (Regel II)."

	$k+1$	k	
k	$2k^2+k-1$	1	$k(2k+1)$
$k-1$		$2k^2-k-1$	$(k-1)(2k+1)$
	$(k+1)(2k-1)$	$k(2k-1)$	

Fig. 4.6 Intensity scheme for the *sp* transitions in Sommerfeld (1924a, 653)

Sommerfeld's "Theory of Intensities": The Ganzzahligkeitshypothese *and the Attack on the Correspondence Principle*

Independently, Dorgelo and Burger in Utrecht and Sommerfeld in Munich followed up on this new representation.[82] Going beyond the representation of empirical data, they developed intensity schemes like the one given in Fig. 4.6. These schemes no longer represented actual measurements but aimed to determine intensities theoretically using quantum numbers. They were constructed using the inner quantum number of the respective states as a label for the rows and columns, the selection rules and the sum rules. To determine the intensities of the individual components from such a scheme, Sommerfeld, Burger and Dorgelo used the sum rules to set up a system of linear equations. For simple doublets, this system could be solved, yielding intensities in agreement with experimental data.[83]

While successful for simple composite doublets, the new approach ran into a serious problem even in the case of simple composite triplets. The system of linear equations provided by the sum rules remained underdetermined, so Sommerfeld, Dorgelo and Burger needed to fix the intensity of one component by setting the smallest intensity equal to 1. Sommerfeld justified this move with the introduction of the so-called *Ganzzahligkeitshypothese*. According to it each line was assigned

[82] See Sommerfeld to Burger, 7 March 1924 (Sommerfeld Papers (NL 89, 001)) for the independence of the work. While he thought that Burger and Dorgelo deserved priority to publish the argument, Sommerfeld wrote to Burger that he had arrived at the same results. For the respective arguments see Burger and Dorgelo (1924, 262–265) and Sommerfeld (1924a, 651–653).

[83] See Burger and Dorgelo (1924, 262–264) and Sommerfeld (1924a, 652–653) for the essentially equivalent arguments and their slightly different schemes. Setting up the schemes in terms of quantum numbers consistently makes it necessary to introduce a normalization factor, as the sum rules otherwise lead to inconsistent equations. To resolve this issue, Burger and Dorgelo introduced this factor as a "relative quantum number," which is the inner quantum number of the respective state in a term divided by the sum of all values of the inner quantum numbers labeling the different states in this term. Sommerfeld introduced the normalization in a different way, multiplying each state by the total statistical weight of the states with which it combined. Both approaches are formally equivalent. At this point, neither Sommerfeld nor Burger and Dorgelo discussed that their arguments corresponded with the assumption that the total intensity of the unresolved lines is distributed among the resolved components according to their statistical weight.

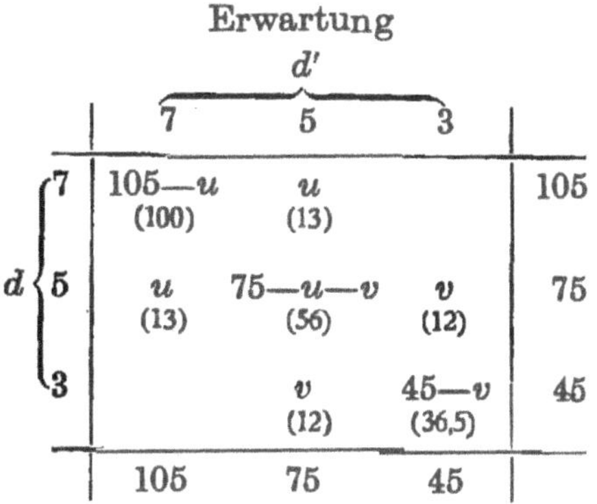

Fig. 4.7 Intensity scheme for *dd′* transitions of calcium in Sommerfeld (1924a, 655)

an integral number of what he called *Gewichtseinheiten*, with the weakest line receiving the smallest integral intensity possible.[84]

This hypothesis, which Sommerfeld admitted did not follow naturally from the intensity schemes, was an external assumption that could be justified empirically or on the basis of Sommerfeld's quantitative intensity rule. While it seemed to work for the most important cases, Sommerfeld did not regard it as a real solution to the problem. As he told Burger in a letter in April 1924, "whether one can indeed set the 'weak transitions' equal to one is in need of a more detailed investigation."[85] As he knew, for transitions between so-called primed and unprimed terms in which the azimuthal quantum number k remained the same, fixing the intensity on the basis of the *Ganzzahligkeitshypothese* would definitely fail to do the trick. The intensity schemes for these transitions (Fig. 4.7) have main lines on the diagonal, accompanied symmetrically by equally intense satellites. As a consequence one could not identify a faintest line to which an intensity of 1 could be assigned.[86]

By June 1924 at the latest, Sommerfeld was convinced that the *Ganzzahligkeitshypothese* was not a real solution. It merely determined the intensity from experience rather than reproducing it from a self-contained theoretical scheme. In a talk in Cologne in June 1924,[87] in which he described the development of the new approach, Sommerfeld made clear that this standard would have to be met by a future "theory of intensities":

[84]Burger and Dorgelo (1924, 264) and Sommerfeld (1924a, 654).

[85]Sommerfeld to Burger, 7 March 1924 (Sommerfeld Papers (NL 89, 001)).

[86]Sommerfeld to Burger, 7 March 1924 (Sommerfeld Papers (NL 89, 001)) and Sommerfeld (1924a, 655).

[87]Sommerfeld's hitherto little known talk, published as Sommerfeld (1925), was held on the occasion of the establishment of the *Ortsgruppe Rheinland der Deutschen Gesellschaft für technische Physik* and the *Gauverein Rheinland der Deutschen Physikalischen Gesellschaft*.

> Moreover, it cannot be denied that there is still a certain lacuna in our theoretical account. A complete [lückenlose] theory of intensities would have to determine all components unambiguously, without resorting to experience.[88]

Despite the problems the new approach was facing, however, Sommerfeld's overall perspective on the new research topic was quite optimistic, and justifiably so. Based on the new experimental possibilities and results, the discussion of the intensity problem had made immense progress within a single year. The qualitative intensity rule was replaced by intensity schemes in analogy with the term scheme for frequencies, which gave the prospect of developing a complete quantitative description of intensities. This goal, so Sommerfeld thought, could be achieved by analyzing new experimental results:

> There is no doubt that we will be able to amend our sum rule on the basis of greater material on intensity measurements, in particular for more composite multiplets [Fe, Cr, Mn], so that a complete determination of all components in any multiplet will emerge.[89]

Thus following the *Gesetzmäßigkeiten* approach, which operated with the inner quantum number as a label for the different states, its selection rules and the sum rules, atomic modeling did not play a constructive role in Sommerfeld's work, merely providing the basis for the interpretation for the new relations.

From this position, Sommerfeld reflected on his previous work on the intensity problem and his and Heisenberg's correspondence argument from 1922. In light of the new exact laws governing the intensity of multiplets, the previous considerations appeared to be clearly deficient: The vague formulation of the correspondence principle allowed only for qualitative conclusions and was therefore no longer suited to describe the new simple arithmetic relations. Believing that correspondence arguments would become obsolete in the future theory of intensities, Sommerfeld criticized the principle publicly in his address to the *Versammlung Deutscher Naturforscher und Ärzte* in September 1924 and in the fourth edition of his book *Atombau und Spektrallinien*.[90]

In his Innsbruck lecture, he acknowledged the major success of the principle in the explanation of the selection rules and its great importance "for every recent discovery by Bohr and his pupils." At the same time he confessed that:

> I cannot regard it as ultimately satisfying, because of its very mixture of quantum theoretical and classical points of view. I would like to regard the correspondence principle as an

[88] Sommerfeld (1925, 8). "Im übrigen [sic!] ist nicht zu verkennen, daß hier noch eine gewisse Lücke in der theoretischen Behandlung vorliegt. Eine lückenlose Theorie der Intensitäten müßte alle Komponenten eindeutig festlegen können, ohne auf die Erfahrung zurückzugreifen."

[89] Sommerfeld (1925, 8). "Es besteht kein Zweifel, daß wir auf Grund eines größeren Materials von Intensitätsmessungen, insbesondere an den mehr zusammengesetzten Multipletts [Fe, Cr, Mn], imstande sein werden, unsere Summenregel so zu ergänzen, daß daraus eine eindeutige Bestimmung aller Komponenten in beliebigen Multipletts hervorgehen wird."

[90] Sommerfeld had formulated his position by early July and communicated it to Kramers and Bohr in a letter to Kramers on 5 July 1924 (AHQP 8b. 11). Kramers' and Bohr's response did not reach Sommerfeld before the conference and will therefore be discussed in Chap. 7.

especially important consequence of a future, complete quantum theory, but not as its foundation.[91]

Sommerfeld's critique of the correspondence principle is well known within the historiography of quantum physics. As discussed in the introduction to this chapter, it has been generally interpreted as his final rejection of the principle as a "magic wand" and as an expression of his continued opposition against it as a foundation for quantum theory. As a consequence, this rejection appeared to be rooted solely in his general methodological convictions and his expectations for a future quantum theory. From this point of view neither Sommerfeld's work on the intensity problem nor his use of the correspondence principle received further attention.[92]

From Sommerfeld's Innsbruck lecture it is already clear, however, that his critique was closely connected to his work on the intensity problem and his attempt to establish arithmetic intensity schemes:

> What enforces my conviction are the Utrechtian intensity measurements of spectral lines. They show that the intensities of multiplets [...] are determined by the simplest arithmetic rules by means of certain integral quantum weights, which are derived from the possible states in phase space. From analogous rules, the intensities of the Zeeman splitting also occur as integers. A correspondence treatment of the intensity question gives only approximations, in a way that seems little suited to the arithmetical simplicity of the facts.[93]

In the fourth edition of *Atombau und Spektrallinien* Sommerfeld expanded this point. As he explained, the incompatibility of the correspondence principle and the arithmetic intensity schemes was rooted in the kinematic description of the atom underlying the correspondence approach:

> In the application of the correspondence principle we have to construct a general picture of the motion [...] and to infer the intensities from its Fourier representation. It is clear, however, that one introduces an element foreign to our problem. *The real arithmetic scheme is much simpler than such correspondence considerations. The problem is situated not in the*

[91] Sommerfeld (1924b, 1048). "Die Zauberkraft des Korrespondenzprinzips hat sich allgemein bewährt, bei den Auswahlregeln der Quantenzahlen, in den Serien- und Bandenspektren. Das Prinzip ist der Leitfaden geworden für alle neueren Entdeckungen BOHRs und seiner Schüler. Trotzdem kann ich es nicht als endgültig befriedigend ansehen, schon wegen seiner Mischung quantentheoretischer und klassischer Gesichtspunkte. Ich möchte das Korrespondenzprinzip als eine besonders wichtige Folge der zukünftigen, vervollständigten Quantentheorie, aber nicht als deren Grundlage ansehen."

[92] See Mehra and Rechenberg (1982a, 154–156), Darrigol (1992, 144), Kragh (2012, 210), and Seth (2010, 232–241) and the discussion in the introduction to this chapter.

[93] Sommerfeld (1924b, 1048). "Was mich in dieser Auffassung bestärkt, sind namentlich die Utrechter Intensitätsmessungen der Spektrallinien. Sie zeigen, daß die Intensitäten innerhalb eines Multipletts, also die Häufigkeiten der Übergänge aus einem Anfangs- in einen Endzustand, durch einfachste arithmetische Regeln bestimmt werden mittels gewisser ganzzahliger Quantengewichte, die sich aus den Zustandsmöglichkeiten im Phasenraum ableiten. Auch die Intensitäten in den Zeeman-Aufspaltungen ergeben sich nach analogen Regeln als ganzzahlig. Eine korrespondenzmäßige Behandlung der Intensitätsfragen liefert nur Näherungswerte, auf einem Wege, der der arithmetischen Einfachheit der Tatsachen wenig angemessen scheint."

realm of analytical mechanics and classical electrodynamics, but in the realm of arithmetic, discontinuous quantum theory.[94]

As Sommerfeld argued, the correspondence principle based on the kinematic description of the atomic motion introduced "foreign elements" to the description of atomic spectra. The inference from the "Fourier representation on the intensities" complicated things and prevented a clear understanding of the simple relations governing the empirical data. The success and superiority of the intensity schemes showed that this introduction was unnecessary.

In essence, Sommerfeld's argument thus constructed a dichotomy between the exactness and simplicity of the intensity schemes, on the one hand, and the conceptually flawed mixture of classical and quantum concepts and approximate character of the correspondence principle on the other. This argument consciously refrained from diagnosing a failure of the correspondence principle in a particular case. Instead it declared the issue a matter of methodology: either one worked with the exact *Gesetzmäßigkeiten* of the inner quantum number, or one adhered to the uncertainties involved in describing atoms on the basis of classical mechanics and electrodynamics.

The association of the correspondence principle with the ill-fated attempt to build atomic models was not set in stone, however. In parallel with his foundational critique, Sommerfeld introduced his students to the correspondence principle and its applications and pitfalls[95] and took up the principle again once he saw that it resolved the problems of his theory of intensities.

---

[94]Sommerfeld (1924a, 658, emphasis in the original). "Bei der Anwendung des Korrespondenzprinzip muß man sich ein allgemeines Bild von den Bewegungs-Vorgängen (Präzession der Elektronenbahnen um die Achse eines inneren Magnetfeldes) machen und muß aus der Fourier-Entwicklung dieser Vorgänge auf die Intensitäten schließen. Aber es ist klar, daß wir damit ein unserem Problem fremdes Element einführen. *Das wirkliche arithmetische Schema ist viel einfacher als solche Korrespondenz-Betrachtungen. Das Problem liegt nicht auf dem Gebiete der analytischen Mechanik und der klassischen Elektrodynamik, sondern auf dem der arithmetischen, diskontinuierlichen Quantentheorie. Gerade auf die Erfahrungen bei den Intensitätsfragen gründen wir die Überzeugung, daß das Korrespondenzprinzip nicht die endgültige Formulierung der Quantenprobleme sein kann.*"

[95]Sommerfeld dedicated his advanced course "Spektroskopische Probleme" to the discussion of the intensity problem in summer semester 1924 and winter semester 1925/26. The content of Sommerfeld's course is described by Helmut Hönl in "Memoirs of research on Zeeman effect in Munich in the early 1920's" written as a contribution to the AHQP (AHQP 66.10). As Hönl describes, Sommerfeld discussed Einstein's theory of emission and absorption, the correspondence principle and his and Heisenberg's account of multiplet intensities, along with his and Dorgelo and Burger's work on the intensity scheme. Helmut Hönl and Fritz London, who attended the lecture, subsequently worked on the sum rules in connection with the correspondence principle. Reporting to his advisor Edwin Kemble, Victor Guillemin gave a report on Sommerfeld's continued teaching activities on the intensity problem in 1925 and identified it as "the topic of chief interest at the institute for theoretical physics here at Munich." See Guillemin to Kemble, 3 August 1925 (AHQP 51.7).

Sommerfeld's Adaptive Reformulation: The Sharpening of the Correspondence Formulas

As will be discussed in the following, Sommerfeld's renewed use of the principle integrated the correspondence formulas Sommerfeld and Heisenberg had found in 1922 into Sommerfeld's intensity schemes and his arithmetic *Gesetzmäßigkeiten* approach. The resulting *sharpening* of the correspondence formulas resolved the problems of Sommerfeld's "theory of intensities." This marked a tremendous advance within Sommerfeld's formal approach to the intensity problem. At the same time it meant cutting the ties with the kinematic description of the atom that had underlain the original correspondence arguments.

Pauli, Sommerfeld and the "Requirement of the Correspondence Principle"

In Sommerfeld's next paper on the subject, Sommerfeld and his student Helmut Hönl summarized the development of the intensity problem in the fall of 1924. As they reported, the initial *Ganzzahligkeitshypothese* had been found to be unsatisfactory, and several parties had pointed to the correspondence principle as a possible alternative.[96] One of the first to do so was Wolfgang Pauli, whom Sommerfeld had asked for his opinion prior to the Innsbruck conference. In the answer to his former teacher, Pauli discussed the relation between the correspondence principle and the sum rules. Following the position of the Sommerfeld-Heisenberg paper, he assumed that the angle ϑ in the ratio

$$J_1 : J_0 : J_{-1} = \frac{(1+\cos\vartheta)^2}{2} : \frac{1}{2}\sin^2\vartheta : \frac{(1-\cos\vartheta)^2}{2}$$

was associated with different *Zwischenbahnen* for the transitions between one initial state and different final states. Consequently, the angle was different for all three components, and the total intensities would not be equal to that of the unresolved component. Without a physical link between the sum rules and the correspondence formulas, Pauli concluded, the sum rules would not be "deducible" from correspondence formulas.[97]

Without this physical link, Pauli compared the numerical values predicted by the sum rules with those resulting from the correspondence formulas in the limit of

---

[96] Sommerfeld and Hönl (1925, 141).

[97] Pauli to Sommerfeld, 29 September 1924 in Pauli (1979, 159). Pauli did not explicate this point in his letter to Sommerfeld; however, a letter by Heisenberg to Pauli, discussed in Chap. 7, makes it clear that Pauli thought along these lines. See Heisenberg to Pauli, 8 October 1924 in Pauli (1979, 167–168).

high quantum numbers, for which the initial-final-state problem did not arise. As the intensity ratio for the multiplet $\Delta k = 1$, $\Delta j = 1, 0, -1$,[98]

$$J_1 : J_0 : J_{-1} \sim 1 : \frac{1}{2}\frac{1}{k^2} : \frac{1}{16}\frac{1}{k^4}$$

showed, the intensity of the main lines always dominated the ratio, while the satellites vanished as $\frac{1}{k^2}$ and $\frac{1}{k^4}$. The main line, Pauli pointed out, also dominated the total intensity which thus became equal to 1. This meant, he concluded, "that the sum rules always satisfy the requirements of the correspondence principle (without being deducible from it)."[99]

As Pauli concluded, "it is thus *very little* which can be concluded about the intensity of the lines." At the same time, he did not accept that the correspondence approach failed entirely. Without a physical link between them, he argued, the two approaches still gave coinciding numerical predictions in the limit of large quantum numbers. This coincidence, Pauli stressed should be taken "for certain."[100]

This confidence in the principle was the starting point from which Pauli approached the relation between the intensity schemes and the correspondence principle in a constructive way. Taking the numerical coincidence seriously, he argued that one should amend the normalization factor of the intensity schemes, in which the weakest satellite was of the order $\frac{1}{k^2}$ whereas the correspondence ratios demanded a ratio of $\frac{1}{k^4}$. As he diagnosed, the reason for the discrepancy between the intensity schemes and the correspondence approach was Sommerfeld's *Ganzzahligkeitshypothese*. While the sum rules were empirically confirmed, this special hypothesis was not set in stone and the conflict with the correspondence ratios did not imply an inadequacy of the correspondence principle. On the contrary, Pauli argued:

[98] See Pauli to Sommerfeld, 29 September 1924 in Pauli (1979, 159). As in Sommerfeld's initial argument, Pauli treated the angle ϑ as small so that the Sommerfeld-Heisenberg ratios could be rewritten as $J_1 : J_0 : J_{-1} \sim 1 : \frac{1}{2}\vartheta^2 : \frac{1}{16}\vartheta^4$. For the interpretation of ϑ, Pauli went beyond Sommerfeld and Heisenberg's paper, which had not given an explicit quantization. He determined the angle ϑ from the by then standard vector model $\cos\vartheta = \frac{k^2+j^2-r^2}{2kj}$, so that the angle was set by the triangle between the momentum of the valence electron k, the total angular momentum j and the angular momentum of the rump r. In the limit of large quantum numbers k, Pauli used additional assumptions on the relation between the momenta to approximate ϑ^2 by $\frac{1}{k^2}$ and thus to arrive at the intensity ratio.

[99] Pauli to Sommerfeld, 29 September 1924 in Pauli (1979, 159). "Vergleicht man nun die Ergebnisse des Korrespondenzprinzips mit Ornsteins (natürlich viel weitergehenden) Regeln, so sieht man zunächst, daß die Summierungsregeln den Forderungen des Korrespondenzprinzips stets genügen (ohne etwa aus diesen ableitbar zu sein)."

[100] Pauli to Sommerfeld, 29 September 1924 in Pauli (1979, 159). "Es ist also *sehr wenig*, was man aus dem Korrespondenzprinzips über die Intensität der Linien schließen kann. Dieses wenige [sic!] möchte ich aber für sicher halten."

one can now, for triplets as a start, search for such expressions for the intensity ratios that fulfill the summation rules exactly and at the same time satisfy the requirements of the correspondence principle for large k.[101]

Amending the normalization factor of the intensity scheme in this way, Pauli still assumed that the intensity ratios were rational and set the weakest satellite equal to one.[102] While keeping the essential parts of the *Ganzzahligkeitshypothese*, Pauli departed from the "requirements of the correspondence principle" and thereby rendered the conflict between the correspondence formulas and the intensity schemes productive.

Sommerfeld received Pauli's evaluation quite positively and published it as an addendum to his Cologne talk.[103] As he put it, Pauli had shown that:

> The correspondence principle provides asymptotic values for the intensities for very high quantum numbers k [...] It prescribes, so to say, a *Größenordnungsrahmen* that is to be filled by the sum rules.[104]

While the correspondence principle remained unable to deduce the sum rules on the basis of its core idea of a connection between radiation and motion, Sommerfeld understood, it set the general *Größenordnungsrahmen*—the frame for the order of magnitude—which the sum rules had to fill out and, as such, provided a numerical constraint on the sum rule schemes.

Sommerfeld, Hönl and the Sharpening of the Correspondence Formulas

Following this positive evaluation in October 1924, Sommerfeld took a next step towards integrating the principle into his approach when he received the work of Helmut Hönl. His student used and adapted the correspondence principle to establish the intensity schemes of the Zeeman effect and had thereby convinced

[101]Pauli to Sommerfeld, 29 September 1924 in Pauli (1979, 159–160). "Man kann nun zunächst bei Tripletts solche Ausdrücke für die Intensitätsverhältnisse suchen, die sowohl die Summierungsregeln exakt erfüllen als auch für große k den Forderungen des Korrespondenzprinzips genügen."

[102]Pauli knew that this adaptation could be achieved consistently in many different ways, as the requirement of the correspondence principle was rather weak. The simplest choice was to introduce a factor of $\frac{1}{k^2}$ into the scheme so that the weakest satellite became equal to $\frac{1}{k^2(4k^2-1)}$. However, setting the weakest satellite equal to 1 was now consistent with the requirements of the correspondence principle. See Pauli to Sommerfeld, 29 September 1924 in Pauli (1979, 159–160).

[103]Sommerfeld (1925, 8). The addendum is dated October 1924.

[104]Sommerfeld (1925, 9). "Das Korrespondenzprinzip liefert asymptotische Werte der Intensitäten bei sehr hoher Quantenzahl k, weil nur in diesem Falle die Quantengesetze in die Gesetze der klassischen Optik übergehen. Es schreibt sozusagen einen Größenordnungsrahmen vor, der durch die Summenregeln auszufüllen ist."

Sommerfeld that the principle could play a constructive role within the theory of intensities.

Working on the intensity schemes for the Zeeman effect in his hometown of Offenburg during the semester break, Hönl had found that the situation was analogous to the one for multiplets.[105] Whereas the simplest cases of the simple doublets and triplets could be treated using the empirically confirmed sum and polarization rules, it was necessary to introduce further assumptions to establish intensity schemes for arbitrary Zeeman splittings. Approaching this task, Hönl turned away from the *Ganzzahligkeitshypothese* and instead applied the correspondence principle.

Hönl's approach to the problem was a formal one. He departed from the Sommerfeld-Heisenberg formulas and expressed the angles Θ in terms of the inner quantum number j and the magnetic quantum number m. Fitting them into the structure of the intensity schemes, he argued that the main obstacle blocking the use of the original correspondence formulas was the initial-final-state problem:

> The application of the correspondence principle presents the known difficulty that it leaves undetermined whether its formulas are to be applied to the initial, final or any intermediate state.[106]

Hönl resolved this ambiguity of the original formulation in a two-step process.[107] First, he considered the original correspondence formulas for the π-components, for which the initial and the final state of m are the same so that "one might expect definite statements" from the original formulas.[108] He then compared the intensities of the correspondence formulas with those of the alkaline doublet established from the sum and polarization rules, changing the correspondence formulas to yield the same result, where necessary. Without giving further indications about the adopted procedure, he then extrapolated these formulas for all Zeeman intensities and took them as the sharpened formula for the π-components.[109]

---

[105] Hönl's absence from Munich is clear from a letter from Sommerfeld on 20 October 1924. In it, Sommerfeld informed Hönl of parallel developments in Munich and Utrecht. See Sommerfeld to Hönl, 20 October 1924 (Hönl Papers E14/23).

[106] Hönl (1924, 343). "Die Anwendung des Korrespondenzprinzips bietet die bekannte Schwierigkeit, daß es unbestimmt läßt, ob seine Formeln auf den Anfangs-, End- oder irgend einen Zwischenzustand des Atoms anzuwenden sind."

[107] The fact that Hönl's paper presented the result in this two-step argument, which was based on a rather bold extrapolation, makes it highly plausible that Hönl had also approached the problem in this way in his original manuscript. See also Hönl's "Memoirs of research on Zeeman effect in Munich in the early 1920's" (AHQP 66.10).

[108] Hönl (1924, 343).

[109] Ibid. For the Zeeman splitting of the component $\Delta j = 0$, Hönl kept the Sommerfeld-Heisenberg formula $J_{\Delta m=0} = m^2$, which predicts an intensity of 0 for the transition $m = m' = 0$ in accordance with the intensity schemes and Landé's *Zusatzregel*. For the Zeeman splitting of the component $\Delta j = \pm 1$, on the other hand, he found that the original formula $J_{\Delta m=0} = j^2 - m^2$ gave the intensity ratio predicted by the sum rules if he introduced the higher of the two values of the quantum number j involved in the transition. For $\Delta j = 1$, this was the quantum number of the initial state j_a; for $\Delta j = -1$ it was the quantum number j_e of the final state.

Schema I.

m	j	$j-1$	$j-2$	$j-3$	$j=4$ ...
j	j^2	j			
$j-1$	j	$(j-1)^2$	$2j-1$		
$j-2$		$2j-1$	$(j-2)^2$	$3j-3$	
$j-3$			$3j-3$	$(j-3)^2$	$4j-6$
$j-4$				$4j-6$	$(j-4)^2$ ...
...					...

Fig. 4.8 Table for the intensity of the transitions with $\Delta j = 0$ in Hönl (1924, 345)

After tentatively adapting, if not guessing, these intensity formulas, Hönl's adaptation followed a much clearer prescription in the second part of the argument. Turning to the σ-components with $\Delta m = \pm 1$, for which the initial-final-state problem could not be neglected, he introduced the already sharpened formula for the π-components into the respective intensity scheme for arbitrary Zeeman splittings. Filling the main diagonal of the intensity scheme with these components, he determined the intensity of each row and column from the sum and polarization rule and then determined the off-diagonal σ-components by iteratively applying the sum rules.[110] Arriving at the intensity scheme for the case $\Delta j = 0, \Delta m = \pm 1, 0$ shown in Fig. 4.8, he found the sharpened correspondence formulas

$$J_{\Delta m=-1} : J_{\Delta m=0} : J_{\Delta m=+1}$$
$$\frac{1}{2}(j+m)(j-m+1) : \quad m^2 : \quad \frac{1}{2}(j-m)(j+m+1),$$

in which the quantum numbers j, m refer to the initial state and which return the original Sommerfeld-Heisenberg formulas for large quantum numbers.

While Hönl's procedure yielded this promising result, it had a serious drawback. The stepwise determination, which established the intensity formula for one class of transitions ($\Delta m = 0$) and then determined the remaining transitions ($\Delta m = \pm 1$), relied crucially on the polarization rule. This limitation meant that Hönl's procedure could be applied only to the Zeeman effect and was consequently replaced by a different approach in Sommerfeld and Hönl's ensuing work on multiplets.

While the specific method of obtaining the sharpened correspondence formulas fell into oblivion due to its limited applicability, Hönl's idea of fitting the correspondence formulas to the intensity schemes led to immense progress within Sommerfeld's approach to the theory of intensities. When he received Hönl's work in October 1924, Sommerfeld agreed with the new "application of the correspondence principle," informing his student that he had "considered the entire problem [of multiplet intensities] pretty extensively in the last days" in the process

[110] Hönl (1924, 345–350).

of writing the aforementioned addendum to his Cologne talk.[111] Compared to what Sommerfeld could do on the basis of the *Ganzzahligkeitshypothese* and Pauli's correspondence argument on the normalization factor of the scheme, Hönl had taken a big step forward, as he could determine the intensity schemes for all relevant cases, including the ones in which the intensity scheme is symmetric with respect to the main diagonal. Whereas the *Ganzzahligkeitshypothese* had failed in this case, the sharpened correspondence formulas solved it without additional difficulties and reproduced the results that had been obtained successfully from the *Ganzzahligkeitshypothese*.

Welcoming Hönl's approach, Sommerfeld nevertheless requested that Hönl's initial manuscript be revised considerably "in tone and composition" before publishing it as a "*short* note."[112] In this revised and shortened version, Sommerfeld and Hönl commented on Hönl's approach, summarizing it as the attempt:

> [...] to sharpen the conclusions of the Bohrian correspondence principle for the present special case of the Zeeman effects in such a way that they fit into the present rules on the one hand, and that on the other hand allow a general scheme extending beyond these rules to be set up for the number of jumps in arbitrary Zeeman splittings.[113]

In the secondary literature, the central term *Verschärfung* is generally associated with Heisenberg's paper on the polarization of fluorescence radiation written in Copenhagen in 1924. The fact that the term was absent from private discussions between them, however, strongly indicates that Heisenberg and Sommerfeld and Hönl were unaware of each other's work and came to use it independently.[114] Both

---

[111] Sommerfeld to Hönl, 20 October 1924 (Hönl Papers E14/23).

[112] Sommerfeld to Hönl, 20 October 1924 (Hönl Papers E14/23). Most importantly, Sommerfeld argued, Hönl had to give priority to Ornstein and Burger, who had independently found the extension of sum rules to the Zeeman intensities in simple cases. See Ornstein and Burger (1924). The scope and character of these changes cannot be reconstructed, as Hönl's original manuscript is not part of Sommerfeld's or Hönl's papers.

[113] Hönl (1924, 342). "Es soll nun hier ein Vorschlag gemacht werden, die Folgerungen aus dem Bohrschen Korrespondenzprinzip für den hier vorliegenden Sonderfall der Zeemaneffekte derart zu verschärfen, daß sie sich einerseits dem Rahmen der bisherigen Regeln einfügen. andererseits [sic!] aber darüber hinausgehend zusammen mit den vorstehenden Regeln zugleich ein allgemeines Schema für die Sprungzahlen bei beliebigen Zeemanaufspaltungen aufzustellen gestatten."

[114] There is only indirect evidence indicating their independence. The term was not mentioned in the dense private correspondence on the subject among Heisenberg, Pauli, Sommerfeld and Hönl from September until the publication of the two papers at the end of November 1924. See Heisenberg to Pauli, 30 September 1924 and 8 October 1924, Pauli to Sommerfeld, 29 September 1924, Sommerfeld to Hönl, 20 October 1924 and Heisenberg to Sommerfeld, 18 November 1924. Sommerfeld's letter to Hönl does not mention Heisenberg's paper, while pointing out that Heisenberg had already found Hönl's sharpened correspondence formulas in 1922 to some extent. Heisenberg's letter to Sommerfeld announces Heisenberg's paper as a work on the polarization of fluorescence radiation, mentioning neither the sum rules, nor the conflict with Sommerfeld's position, nor a relation to Hönl's work. This shows that the two sides did not discuss their respective approaches and apparently did not know of each other's work at the time. This reading is supported by the fact that Heisenberg's paper and Hönl's paper, both of which were published in the same issue of *Zeitschrift für Physik*, were received on 30 November (Heisenberg) and 26 November (Hönl) 1924 without referencing each other explicitly.

used the term *Verschärfung* to describe the attempt to draw exact conclusions from the correspondence principle. However, they approached this general idea quite differently. On the one hand, Heisenberg, whose arguments will be discussed in Chap. 7, kept the connection to the kinematics of a physical model with quantum transitions as the core physical idea of the principle and argued for its consistency with the sum rules. Hönl and Sommerfeld's sharpening, on the other hand, left such physical assumptions aside and cut the ties with the original correspondence principle in a radical way. At the same time, they bracketed Sommerfeld's foundational criticism of the correspondence principle. Rather than diagnosing the principle's conceptual flaws, they accepted that the correspondence *formulas* could be integrated and adapted in such a way that they fit into the intensity schemes and at the same time completed them.

This idea of *sharpening* was in keeping with Sommerfeld's previous idea that the principle prescribed the realm of the order of magnitude for the intensities in the high quantum number limit. It meant extending this prescription from the limiting case to all intensities:

> The correspondence principle appears in essence to be in need of a sharpening by additional rules. By itself it can only prescribe the general, individually not sharply defined distribution of intensities. Within the leeway [Spielraum], which it gives in its application, the above defined [sum and polarization, MJ] rules specify such values for intensity and/or number of jumps that are distinguished by especially simple arithmetic relations.[115]

The correspondence formulas now set up the *Spielraum* for the intensity schemes in general, so that the numerical coincidence in the high quantum number limit had become a formal constraint for intensity schemes in general.

In a joint paper, Sommerfeld and Hönl discussed the intensity problem for multiplets from this new perspective of the sharpening of the correspondence formulas. Extending their reinterpretation of the correspondence principle as a formal constraint, they now assumed that the principle prescribed the "analytic form of the expressions, which represent the intensities for small quantum numbers."[116] In the concrete case of multiplet intensities, this meant that the intensities would be determined by quadratic functions $P(j)$ and $Q(j)$, just like the original Sommerfeld-Heisenberg formulas, which read:

$$\begin{aligned} J_{-1} : J_0 : J_{+1} &= \cos^4 \tfrac{\vartheta}{2} : \cos^2 \tfrac{\vartheta}{2} \sin^2 \tfrac{\vartheta}{2} : \sin^4 \tfrac{\vartheta}{2} \\ &= \frac{P(j)^2}{(4jj_a)^2} : \frac{2P(j)Q(j)}{(4jj_a)^2} : \frac{Q(j)^2}{(4jj_a)^2}. \end{aligned}$$

[115]Hönl (1924, 345). "Das Korrespondenzprinzip scheint seinem Wesen nach einer Verschärfung durch zusätzliche Regeln zu bedürfen. Es selbst kann nur den allgemeinen, im einzelnen nicht scharf definierten Verlauf der Intensitäten vorzeichnen. In dem Spielraum, welchen es bei seiner Anwendung noch gewährt, werden nun durch die aufgestellten Regeln solche Intensitäts- bzw. Sprungzahlwerte hervorgehoben, welche durch besonders einfache arithmetische Zusammenhänge ausgezeichnet sind."

[116]Sommerfeld and Hönl (1925, 141).

Schema H.

		d					
		9	7	5	3	1	
p	7	135	$40-u$	u			175
	5		$65+u$	$60-u-v$	v		125
	3			$15-v$	$60-v-w$	w	75
	1				$-15+w$	$40-w$	
	−1					-25	
		135	105	75	45	15	

Fig. 4.9 Intensity scheme with negative intensity in Sommerfeld (1925, 11)

To determine the sharpened correspondence formulas explicitly, Sommerfeld and Hönl fit the sharpened form of Q and P into the intensity scheme.

Sommerfeld and Hönl's paper, which provides the only source of information, is difficult to analyze. The logic of the paper alternates between a demonstration that the already sharpened formulas $P(j)$ and $Q(j)$ are consistent with the sum rules, and a presentation of the construction of these formulas that had predated this consistency check. The reconstruction of this latter process is necessarily problematic and difficult due to the pivotal character of this paper. Following the traces which were not erased in the process of writing the paper, I will nonetheless attempt to give a temporal order to the various arguments and to show how Sommerfeld and Hönl integrated the correspondence formulas into the intensity schemes.

This attempt takes off from Sommerfeld and Hönl's description of the "first ansatz of our intensity formulas,"[117] which is given in Paragraph 5 of their paper. In it, Sommerfeld and Hönl state that their considerations centered on the intensities at "the edges" of the intensity schemes. Such considerations had already played a role in Sommerfeld's work prior to Hönl's sharpening. In his addendum to the Cologne talk in October 1924, he had considered an intensity scheme with rows and columns that did not correspond to existing states, and therefore led to transitions that were not observed (Fig. 4.9). Considering this situation in a concrete example, he had applied the sum rules to both the actually observed intensities and the fictitious intensities, finding that transitions between the j-values that corresponded to existing states gave the correct intensity ratios, "although our scheme was not designed to give them." At the same time, transitions involving non-existing states had negative "unreal" intensities and were therefore physically meaningless.[118]

[117] Sommerfeld and Hönl (1925, 155).

[118] Setting the highest azimuthal quantum number to $k = 2$ in the 5x5 scheme, he found that the intensity associated with the non-existent terms in the threefold p-level $k = 1, k = -1$ gave negative and therefore "nonsensical" values for the intensity of the non-existent lines.

Schema 1. $k \to k-1$.

	0								
	0	0							
j_a+j_s-1	J_{-1}	J_0	J_{+1}						
j_a+j_s-2		J_{-1}	J_0						
j_a+j_s-3			J_{-1}						
j_a-j_s+1					J_{-1}	J_0	J_{+1}		
j_a-j_s						J_{-1}	J_0	0	
j_a-j_s-1							J_{-1}	0	0
j	j_a+j_s	j_a+j_s-1	j_a+j_s-2		j_a-j_s+2	j_a-j_s+1	j_a-j_s		

Fig. 4.10 Intensity scheme with vanishing intensity at the edges in Sommerfeld and Hönl (1925, 143)

While Sommerfeld did not see a specific theoretical significance for this kind of consideration in October 1924, similar arguments were central for Sommerfeld and Hönl's determination of the sharpened correspondence formulas in early 1925. Fitting the correspondence formulas into their intensity schemes, they also considered the total intensity of the lines actually observed and the unreal lines beyond the edges of their scheme (e.g. Fig. 4.10). Whereas these intensity sums had been identical with the sum rules in Sommerfeld's earlier argument, this was no longer the case in Sommerfeld and Hönl's work:

> In the proof of the sum rules we have up to now always taken our intensity sums over all three components; in reality, however, the opinion of the sum rules is of course that we have to sum up only the actually existing lines. In the upper and lower corner of our schemes 1 and 2 there are not three but only two or one line. The contradiction is resolved only if our formulas automatically give the value 0 for the components that are illegally included in the summation.[119]

Sommerfeld and Hönl thus separated the sum rules from the intensity sums of a given triplet. The former referred to the intensity of the actual existing lines and was empirically testable. The latter referred to the intensity formulas to be constructed

[119]Sommerfeld and Hönl (1925, 154). "Wir haben bisher beim Nachweis der Summenregeln unsere Intensitätssummen stets über alle drei Komponenten J_{-1},J_0 und J_{+1} erstreckt, während natürlich in Wirklichkeit die Meinung der Summenregeln die ist, daß nur über die tatsächlich vorhandenen Linien zu summieren ist. An der oberen und unteren Ecke unseres Schemas 1 und 2 sind dies nicht drei, sondern nur zwei oder eine Linie. Der Widerspruch löst sich nur dann, wenn unsre Formeln für die bei der Summation mit Unrecht mitgezählten Komponenten automatisch den Wert 0 ergeben."

by sharpening the correspondence formulas. These sums involved both the existing and the unreal lines and thus presented purely theoretical expressions.

The task, Sommerfeld and Hönl argued, was to resolve the contradiction between these two statements by demanding that the intensity of the unreal lines at the edges of the schemes vanish. This demand was the key to adjust the correspondence formulas to the structure of the intensity schemes:

> We can, however conclude inversely: because we want to (in the interest of the generality of our sum calculations) let the components that are labeled 0 vanish at the edges of our schemes, we have to construct the factors in J_0 and J_+ in such a way that they have the zeros (24) and (25). Indeed, we proceeded in this way in our very first ansatz for our intensity formulas.[120]

As Sommerfeld and Hönl explicitly stated, this approach had been their point of departure. The conflict between the sum rules as a statement derived from experience and the total intensity of the sharpened correspondence formulas at the edges of the scheme had provided them with a condition for sharpening the correspondence formulas: the assumption of the vanishing intensity "at the edges" provided the zeros of the quadratic intensity formulas and thereby made it possible to construct the sharpened correspondence formulas from the product of these zeros.

This prescription, which Sommerfeld and Hönl described as the *Methode der Nullstellen*, was built on a general characteristic of the intensity schemes, which was independent of the sum rule argument that had led to the consideration of the edges of the scheme in the first place. The edge of the scheme is defined by the fact that the inner quantum number j always ranges from a maximum value j_{max} to a minimum value j_{min}. These extreme values are determined by the sum and difference of the azimuthal quantum number, labeled j_a, of the respective term and the quantum number of the rump, labeled j_s, according to the vector model. This definition of the edge is independent of Sommerfeld and Hönl's sum rule argument.

It appears that Sommerfeld and Hönl realized the importance of this fact when they first applied their *Methode der Nullstellen* in the scheme (Fig. 4.10) for $\Delta k = -1$. This scheme, which they discussed in their paper as the first example for their initial ansatz, fits the approach based on the sum rule argument. Considering the edges of the schemes in the top left and lower right corner, Sommerfeld and Hönl constructed their intensity formulas from the zeros provided by the final state. For the rows in which the component J_{+1} vanishes independently of the component J_0, this leads to the sharpened intensity formula:

$$\begin{aligned} Q(j-1) = &\; (j_a - j_s - j)(j_a + j_s - j + 1) \\ &= j_s(j_s+1) - (j - j_a)(j - j_a - 1). \end{aligned}$$

[120] Sommerfeld and Hönl (1925, 155). "Wir können nunmehr umgekehrt schließen: Weil wir (im Interesse der Allgemeingültigkeit unserer Summenberechnung) die mit 0 bezeichneten Komponenten an den Ecken unseres Schemas zum Verschwinden bringen wollen, haben wir die Faktoren in J_0 und J_+ so einzurichten, daß sie die Nullstellen (24) und (25) haben. Dadurch ist der Zähler von J_+ vollständig, diejenigen von J_0 zur Hälfte bestimmt. Tatsächlich sind wir bei dem erstmaligen Ansatz unserer Intensitätsformeln in dieser Weise vorgegangen."

Schema 3. (*pd*) im Septettsystem.

	j	d: 5	4	3	2	1	0	−1
p	4	J_{-1}	J_0	J_{+1}				
	3		J_{-1}	J_0	J_{+1}			
	2			J_{-1}	J_0	J_{+1}		
	1				o	o	×	
	0					o	×	×
	−1						×	×
	−2							×

Fig. 4.11 Intensity scheme with vanishing intensity at the edges in Sommerfeld and Hönl (1925, 155)

For the rows in which J_0 and J_{+1} are both zero, the same approach gives:

$$\begin{aligned} Q(j) = &\quad (j_a + j_s - j)(j_a - j_s - j - 1) \\ &= j_s(j_s + 1) - (j - j_a + 1)(j - j_a). \end{aligned}$$

The product of these two functions $Q(j)Q(j-1)$ replaces the term Q^2 from the original correspondence formula and thereby fixes the J_+-component (as well as half of the intensity J_0). At the same time, Hönl and Sommerfeld realized, their approach did not allow them to determine all sharpened correspondence formulas on the basis of their sum rule argument. The intensity J_- was left completely undetermined, as there is no intensity sum at those edges which involves real intensities and unreal intensities for J_-.

To obtain the missing intensity formula, they extended their initial ansatz to a different type of scheme (Fig. 4.11).[121] In this scheme, the vanishing intensities at the bottom of the scheme provide zeros for the components J_- and J_0 on the basis of the sum rule argument. At the same time, the vanishing intensities in the top left corner are associated with the components J_+ and J_0, just as in the previous case. This new situation had important implications for the sum rule argument. As Hönl and Sommerfeld realized, only half of the zeros necessary to construct the intensity

[121]This scheme corresponds to the one considered by Sommerfeld in his addendum in October 1924. Whereas the scheme of the former type represents cases in which j_a is always larger than j_s, the inverse is true in the latter type.

Schema 4. *pd* im Triplettsystem

j	3	2	1	0	−1	−2	−3	−4
4	o							
3	o	o						
2	J_{-1}	J_0	J_{+1}					
1		J_{-1}	J_0	o				
0			J_{-1}	o	o			
−1				o	o	·		
−2					o	·	·	
−3						·	·	·
−4							o	o
−5								o

Fig. 4.12 Intensity scheme with vanishing intensity at the edges in Sommerfeld and Hönl (1925, 158)

formulas were actually subject to the sum rule argument. The zeros of the already sharpened correspondence formulas Q marked with "x," on the other hand, could no longer be "controlled" by the sum rules. Including them thus implied dropping the initial sum rule argument.

In addition, they found that it was necessary to construct the other half of the zeros for the function P in a different way. As the intensity of J_- decreases steadily from top to bottom, they thought that the second pair of zeros could not be located at the top of the scheme, but rather even further within the unreal part. To construct them, they employed a mathematical trick. They added $+1$ on each side of the zero $j + j_a = j_s$ and then multiplied the two equations to get a formal expression that closely resembled the already sharpened function Q:

$$P(j) = (j + j_a)(j + j_a + 1) - j_s(j_s + 1)$$
$$P(j - 1) = (j + j_a)(j + j_a - 1) - j_s(j_s + 1).$$

These equations, which followed from the qualitative intensity distribution within the scheme and would have led to different results if 2, 3, 4 etc. had been added, could be rewritten as a linear combination and showed that the missing zeros ($j = -j_s - j_a - 1$) and ($j = -j_s - j_a$) were not situated at the edges of the scheme in any case. This is shown in Fig. 4.12, which combines all necessary zeros in one scheme. At the same time, some of the zeros clearly cannot be justified from the original sum rule argument, thus undermining the initial justification for the *vanishing at the edges* argument. Nonetheless, the *Methode der Nullstellen* remained the main heuristic technique Sommerfeld and Hönl used to construct their intensity schemes.

In short, this reconstruction suggests that Sommerfeld and Hönl found a condition for sharpening the correspondence formulas through their sum rule argument. They realized, however, that their initial approach had its limitations and was insufficient to obtain all relevant intensity formulas. They completed them initially by sticking to the argument and working with intensity schemes representing different types of transitions. Establishing the sharpened correspondence formulas in this way, they understood that the assumption that the intensity vanishes at the edges was independent of the sum rule argument. Dropping the initial justification, they adopted it as the *Methode der Nullstellen*.

Having found the sharpened correspondence formulas through their argument of vanishing intensities, Sommerfeld and Hönl considered whether these formulas fulfilled the sum rules in general. As they found that this was not the case, they manipulated the way in which the sum was taken, and modified the denominator of the intensity ratios.[122] In contrast to the construction of the sharpened formulas for Q and P from the assumption of the vanishing intensities, these changes did not follow from considerations on the structure of the intensity scheme, but were rather tailored to meet the requirements of the sum rules.

For the time being, however, Sommerfeld and Hönl accepted this ad-hoc approach, concluding that "[t]he communicated formulas should solve the intensity problem for multiplets for all practical purposes."[123] Sharpening the correspondence formulas, Sommerfeld and Hönl had integrated the principle into the *Gesetzmäßigkeiten* approach and had found a way to set up all kinds of intensity schemes for multiplets without resorting to the *Ganzzahligkeitshypothese*. With this

[122]Without further changes the sum of the correspondence formulas is:

$$J_{-1} + J_0 + J_{+1} = 2j(P(j)(2j_a - 1) + Q(j)(2j_a + 1)).$$

While the sum rules demand that the total intensity of the three components be proportional to $2j + 1$, the above sum cannot be factorized in this way. In order to obtain the factor, Sommerfeld and Hönl realized, the argument of the intensity always has to be the larger value of the quantum number j, so that

$$\begin{aligned} J_{-1}(j+1) + J_0(j) + J_{+1}(j) &= P(j)[P(j+1) + Q(j)] + Q(j)[P(j) + Q(j-1)] \\ &= P(j)(2j_a + 1)2(j+1) + Q(j)(2j_a + 1)2j. \end{aligned}$$

They then changed the denominators of the correspondence formulas by replacing the $(4jj_a)^2$ of the Sommerfeld-Heisenberg formulas with $4jj_a$ and exchanging the factor $\frac{2}{j}$ by $\frac{1}{j} + \frac{1}{j+1}$ for the intensity J_0. In this manner, the factors $2j + 1$ and $2j$ canceled out and it was possible to factorize the sum as

$$\begin{aligned} J_{-1}(j+1) + J_0(j) + J_{+1}(j) &= P(j)\frac{[P(j+1)+Q(j)]}{2j_a(j+1)} + Q(j)\frac{[P(j)+Q(j-1)]}{2j_a j} \\ &= (2j_a + 1)(2j + 1). \end{aligned}$$

[123]Sommerfeld and Hönl (1925, 161). "Durch die hier mitgeteilten Formeln dürfte das Intensitätsproblem der Multiplets nach der praktischen Seite erledigt sein."

integration, they dropped the kinematic underpinning of the original correspondence formulas and thereby the physical core of the principle. Departing from the original correspondence approach in this radical way, Sommerfeld and Hönl consciously stayed away from a discussion of their adaptation and its consequences for the correspondence approach:

> On the other hand, we have not touched upon the question of the actual theoretical significance of our formulas. This, as for other applications of quantum theory, remains dark for the moment.[124]

Conceptual clarification was admittedly beyond the scope of the *Gesetzmäßigkeiten* approach, whose prime objective was to establish a complete and consistent description of empirical regularities.

At this stage Sommerfeld concluded the study of multiplet intensities, which had been a focus of his work from 1922 until 1925. He left it to Hönl to earn his doctorate by expanding the new method using techniques from function theory. This new approach did not change the situation significantly with respect to the theoretical understanding of the formulas. In the published version of Hönl's dissertation, he admitted in passing that "we are still a long way from a real theoretical understanding of the intensity of spectral lines."[125] While Sommerfeld and Hönl had left it at this characterization, Hönl tried to go further in his dissertation, addressing the conceptual puzzles and insights he saw arising from the new intensity scheme.

As Hönl saw it, the intensity scheme based on the sharpened correspondence formulas gave insight into the underlying physical structure of the intensity problem. The sharpened correspondence formulas, Hönl found, could be written as the "*product of two factors*, the first of which is associated with the initial and the second with the final state of the atom."[126] This meant, he argued, that the intensity resulted from the combination of two "multiplicative terms" of the atom, which were analogous to the terms involved in the description of spectral frequencies:

> In this way the representation of intensities comes to parallel the term representation for frequencies. As frequencies result from two "terms," one of which is associated with one stationary atomic state while the other is associated with another, the same now also holds for intensities.[127]

---

[124] Sommerfeld and Hönl (1925, 161). "Andererseits haben wir die Frage nach der eigentlichen theoretischen Bedeutung unserer Formeln nicht berührt. Diese bleibt, wie in anderen Anwendungen der Quantentheorie, bis auf weiteres dunkel."

[125] Hönl (1926, 273–274). "Abgesehen davon, daß wir von einem eigentlichen, theoretischen Verständnis der Intensitäten der Spektrallinien noch weit entfernt sind, fordern doch auch die gewonnenen Intensitätsgesetze dazu auf, nach allgemeineren Zusammenhängen formaler Art zu suchen."

[126] Hönl (1926, 289). "Wir können demnach die Gln. (6′) und (16′) so deuten, daß sich [...] die Intensitäten der Zeemankomponenten als Produkt zweier Faktoren darstellen lassen, von denen der eine dem Anfangs-, der andere dem Endzustande des Atoms zugeordnet ist."

[127] Hönl (1926, 289). "Damit tritt die Darstellung der Intensitäten in eine gewisse Parallele zur Termdarstellung der Frequenzen. Ebenso wie sich die Frequenzen aus zwei "Termen""

Though similar in some sense, the differences between the terms of spectral frequencies and the terms of intensities were considerable. While the combination relation for frequencies was a difference equation, which made it possible "to distinguish between an 'initial' and 'final state'," Hönl argued, this possibility did not exist for his "multiplicative terms," as these played entirely symmetrical roles. Therefore, "the representation of the intensities does not allow for the same conception" of "discrete atomic states, between which transitions take place." Therefore it was also impossible to use the correspondence principle to associate the intensity with the "initial, final, or some intermediate state of the atom."[128]

Hönl's notion of a "multiplicative term," which was the conceptual result of Sommerfeld and Hönl's work on the intensity scheme, remained an illustration and did not allow him to develop a deeper understanding of the puzzles that had arisen from the work on the intensity problem. For example, Hönl had realized that the factor $2j + 1$, which had been interpreted as the statistical weight of a particular state, resulted from his approach without introducing statistical assumptions. This, so Hönl, was "a particularly characteristic trait of the intensity problem" and pointed towards an "intimate fusion, which the constituents of the correspondence principle (trigonometric intensity expressions) and of quantum statistics (quantum weights) formed through the sharpening." However, this fusion remained enigmatic. The two had "entirely different origins" and it was "apparently only the constraint of mathematical form that unifies the heterogenous elements."[129]

Hönl had come to these insights into the intensity problem and to the formulation of its puzzles without using the new quantum mechanics of Heisenberg, which he knew was being developed in parallel to his work. As he thought, the new theory might provide the "deeper understanding" of the intensity problem which he and Sommerfeld had left untouched. As will be shown in Chap. 7, the intensity problem not only received some elucidation from the newly developed theory of

zusammensetzen, von denen der eine zum einen, der andere zum anderen stationären Atomzustand gehört, gilt nunmehr auch für die Intensitäten."

[128]Hönl (1926, 289–290). "Dabei besteht jedoch ein charakteristischer Unterschied. Während nämlich die Terme, welche die Frequenzen bestimmen, nicht miteinander vertauscht werden können, da sie mit entgegengesetzten Vorzeichen in die Darstellung der Schwingungszahlen eingehen—hierauf beruht die Möglichkeit, zwischen 'Anfangs-' und 'Endzustand' zu unterscheiden—, besteht in Bezug auf die multiplikativen 'Terme' der Intensitäten eine vollkommene Symmetrie, da bei der Produktbildung diese miteinander vertauscht werden können. [...] Wie wohl nun die Termdarstellung der Frequenzen die Vorstellung diskreter Atomzustände, zwischen denen Übergänge stattfinden, nahe legt, so ist doch a priori die Möglichkeit nicht von der Hand zu weisen, daß die Darstellung der Intensitäten dieselbe Vorstellung (Versinnlichung) nicht mehr zuläßt."

[129]Hönl (1926, 311). "Als ein besonders charakteristischer Zug des Intensitätsproblems der Multiplets und der Banden sei die innige Verschmelzung hervorgehoben, welche bei der quantentheoretischen Verschärfung die vom Korrespondenzprinzip (trigonometrische Intensitätsausdrücke) und von der Quantenstatistik (Quantengewichte) herrührende Bestandteile eingehen. Dieser Zug ist um so merkwürdiger, als der Ursprung dieser beiden Bestandteile ein ganz verschiedener ist und es offenbar nur der Zwang der mathematischen Form ist, welcher die so verschiedenartigen Elemente vereinigt."

quantum mechanics, it also played a crucial role in its formulation. In Copenhagen, Heisenberg, Pauli, and Kronig understood the sharpened correspondence formulas—as Pauli summarized it—as providing insight into the formal structure of the "yet unknown, general quantum kinematics, which will take the place of the classical kinematics operating with definite electron orbits."[130] Trying to find this new kinematics, they searched for the connection between the physical core of the correspondence principle and the formal structure of the intensity problem, which Sommerfeld and Hönl (and in parallel Kronig and Goudsmit) established through their sharpening.

4.5 Conclusion

Within the historiography on quantum physics, Arnold Sommerfeld's work on the intensity of multiplets does not play a role and even Suman Seth and Sommerfeld's biographer Michael Eckart skipped this part of his work in their book-length studies. So why is his work on multiplet intensities important? A simple answer would be that it was important from Sommerfeld's own perspective, not just during the 1920s but also in retrospect: In an autobiographical sketch, he wrote that the intensity problem was his central research problem from 1922 to 1926.[131] This chapter has presented the first detailed study of this work. Filling this temporal or thematic gap in Sommerfeld scholarship, I have challenged the topos that Sommerfeld was first and foremost as a critic of the correspondence principle and showed that his foundational critique was paralleled by his use of the principle as a research tool.

Sommerfeld's use of the correspondence principle, however, is significant beyond his personal trajectory in the 1920s. As an application of the correspondence principle outside of Copenhagen, it is central for the main argument of this book. As we have seen in this chapter, Sommerfeld's work followed the pattern of *transformation through implementation.* Taking it up in its original formulation in 1921, Sommerfeld integrated the correspondence principle into his work on the empirical regularities governing multiplet intensities. This integration led to a combination of Bohr's and Kramers' correspondence arguments and Landé's, Heisenberg and Sommerfeld's models for multiplet atoms based on space quantization. The resulting implementation of the correspondence principle allowed him to give a first theoretical interpretation for his qualitative intensity rules.

Initially, Sommerfeld (and with him Heisenberg and Landé) perceived this new correspondence argument as a straightforward extension of Bohr's work. As he realized only later, this was not the case: the implementation of the principle to

[130] Pauli (1925, 68). "Man hat es hier mit einem Spezialfall der uns noch unbekannten allgemeinen Quantenkinematik zu tun, welche an die Stelle der mit eindeutig definierten Elektronenbahnen operierenden klassischen Kinematik treten wird."

[131] Sommerfeld (1968b, 678).

multiplets led to a new version of the initial-final-state problem, in which it extended to the total intensity of each triplet. In this form, the correspondence argument became a challenge for the *Zwischenbahn* model. Recognizing this challenge, Heisenberg initially tried to resolve it by adapting the correspondence argument. As we have seen, however, this approach was abandoned as Back's experimental estimates on the intensity of split-up lines appeared to confirm the predictions obtained from the correspondence principle. Instead of calling for an adaptation of the correspondence principle, the challenge turned into a confirmation of the correspondence approach.

Sommerfeld's work on the intensity problem changed drastically from 1923 onwards as he developed intensity schemes to represent and predict the intensity of multiplets within his *Gesetzmäßigkeiten* approach. These intensity schemes, Sommerfeld initially thought, rendered the correspondence principle obsolete. In a sense, they did: Sommerfeld's intensity schemes discarded the kinematic description of atoms, the geometric interpretation of space quantization, and the initial application of the principle once and for all. Yet, Sommerfeld returned to the correspondence principle: together with Pauli and Hönl, he integrated the original correspondence formulas into his work by adapting them to the formal structures of the intensity schemes. Thereby he arrived at a complete description of multiplet intensities from a "practical point of view." This solution to the intensity problem, Sommerfeld knew full well, left the theoretical significance of his intensity schemes untouched. Cutting the ties to the former correspondence arguments based on the kinematic description of the atom, the implications of Sommerfeld's adaptive reformulation for both the study of atomic structure and the correspondence principle remained unclear.

Sommerfeld's work on multiplet intensities thus bears the central characteristics of *transformation through implementation*: the integration of the correspondence principle into an existing framework, the implementation of it in its original formulation, the emergence and recognition of challenges arising from this implementation, and the adaptive reformulation of the correspondence principle. At the same time, this reconstruction is more than an illustration or a proof of principle. It allows us to analyze the principle's *transformation through implementation* by exploring each step in more detail and highlighting aspects that stand out in the individual case.

Adoption: Motivations for Taking up the Correspondence Principle

With respect to the motivation for taking up the correspondence principle, Sommerfeld's case is remarkable. As we have seen, his applications were driven by the research problem: the intensity of multiplets. To tackle this problem, Sommerfeld initially adopted the correspondence principle as the only available tool for the task. Even more importantly, he returned to it within his work on the "theory of intensities" which sought to render the principle obsolete.

This points to the remarkable feature of Sommerfeld's case: the persistence, if not irresistibility, of the correspondence principle. Turning on its head the old topos of Sommerfeld as a critic, we can see that he used the correspondence principle despite rejecting it as a foundational idea of quantum theory. Moreover, as he focused on his formal *Gesetzmäßigkeiten* approach and abandoned atomic models, his research moved away from the core ideas of Bohr's original correspondence arguments. In other words, without close ties to the Copenhagen community, Sommerfeld had no clear personal motivation for applying the correspondence principle; at the same time there was no strong conceptual affinity between the correspondence principle and the framework of Sommerfeld's research, yet he integrated the principle into his work and adapted it.

Implementation: Preconditions for Making Correspondence Arguments

Second, Sommerfeld's case tells us something about the preconditions for making a correspondence argument. As we have seen, Sommerfeld and Heisenberg's initial correspondence argument did not touch on the dynamical problems encountered in the anomalous Zeeman effect, and did not employ the elaborate techniques of the quantum theory of multiply periodic systems. Instead, their argument relied on the state-transition model, the notion of space quantization and a general kinematic description of atomic motion.

At first sight, this kinematic description of the atom appears to vanish in Sommerfeld and Hönl's "sharpening" of the correspondence formulas. As the state-transition model found a new explication in the form of intensity schemes, this kinematic description no longer served as a representation of the stationary states. At the same time, Sommerfeld and Hönl still relied on the kinematic description and the original correspondence argument, as these provided the formulas that were to be adapted to the structure of the intensity schemes. In other words, the core idea of the correspondence principle—the relation between radiation and motion—still formed an essential part of the argument, even though Sommerfeld and Hönl rejected a spatiotemporal description of the atom and interpreted the correspondence principle as providing a formal constraint for a conceptually independent quantum theory.

Recognizing Problems: How Challenges Arise

Third, Sommerfeld's case is illuminating with respect to how problems arose within the correspondence approach. As is evident from Sommerfeld's transition from atomic models to formal *Gesetzmäßigkeiten*, this depended decisively on the existing framework. Having integrated the correspondence principle into a description of the atom based on space quantization, Sommerfeld and Heisenberg initially recognized that the initial-final-state problem for multiplets presented a challenge for the *Zwischenbahn model*. This challenge arose from the disagreement between the theoretical predictions of the *Zwischenbahn* model and the physical expectation on the intensity of a multiplet.

This challenge pertained to the model representation of the correspondence principle and thus had an impact on the conceptualization of the principle's physical core. Tellingly, the problem only resurfaced in 1923 in a short, heated debate between Pauli and Heisenberg. Meanwhile, such conceptual issues did not play a significant role in Sommerfeld's work, as he developed his intensity schemes within his formal *Gesetzmäßigkeiten* approach. In this context, the problem was not whether or not the correspondence principle was conceptually compatible with the relation between resolved and unresolved lines. Rather, Sommerfeld aimed to establish the empirical regularities of multiplet intensities in a quantitative way. This meant adjusting the correspondence formulas to the structure of the intensity schemes. In this attempt the former conceptual challenge presented a key constraint for resolving the ambiguity inherent in the original correspondence formulas and obtaining a quantitative description of the intensities.

Adaptive Reformulation: Implications for the Formulation of the Correspondence Principle

Finally, Sommerfeld's case provides an example of what *transformation through implementation* meant for the formulation of the correspondence principle. As we have seen, Sommerfeld and Hönl's solution was a formal one. It was on this formal level that Sommerfeld's work came to a close, for he believed that solving the intensity problem meant establishing intensity schemes—if possible, in a unique way.

Due to this focus on the formal character, the quantitative description of intensities was central for Sommerfeld, while the physical core of the correspondence principle—the relation between radiation and motion—ceased to play a role. In Copenhagen, as we will see in Chap. 7, Sommerfeld and Hönl's successful description amplified the challenge to the original correspondence principle brought about by Sommerfeld's *Gesetzmäßigkeiten* approach and called for a new kinematics embedding the new intensity formulas. As we have seen, such questions about the theoretical significance of the "sharpening" of the correspondence formulas remained almost untouched by Sommerfeld and Hönl. This should not be considered a step which Sommerfeld and Hönl refused to take. Rather, it shows very clearly that research problems and the approaches taken to them shaped the application of the correspondence principle decisively. This imposed limitations on the development of the tool itself and thereby also shows how far an approach searching for *Gesetzmäßigkeiten* and solutions "from a practical point of view" was able to carry the conceptual development of quantum physics in the 1920s.

Chapter 5
Fertilizer on a Sandy Acreage: Franck, Hund and the Ramsauer Effect

The applications of the correspondence principle to multiplet spectroscopy, discussed in the previous chapter, presented an extension and rather radical adaptation of the correspondence principle in its original domain of atomic spectroscopy. The present chapter focuses on a research field that was well beyond the principle's initial scope and studies an application of the correspondence principle to electronic collisions, which was developed in Göttingen between 1922 and 1923 by James Franck and Friedrich Hund to account for the newly discovered Ramsauer effect. The chapter reconstructs how the two physicists came to apply the principle to a phenomenon, which initially, as we will see, had no connection to the principle, be it on a phenomenological or on a conceptual level. As such, it shows how the application of the correspondence principle turned the Ramsauer effect into a quantum phenomenon and eventually led to a first conceptualization of collision processes in terms of the state-transition model.[1]

This reconstruction is based on hitherto unused archival material and published papers and it presents the first detailed reconstruction of this episode in the history of quantum physics, which has received little attention within the historiography of quantum physics of the 1920s.[2] Indeed, the long-term history of the Ramsauer effect and the quantum theory of collisions could be written largely without taking recourse to the developments discussed in this chapter. Franck and Hund's work on the Ramsauer effect did not lead to insights that were directly relevant for later

[1]This chapter was published in a much abridged form as Jähnert (2015).

[2]One reason for this might be that collision processes did not play an extensive role in the development of quantum physics in the 1920s. As such, Roger Stuewer's work on the Compton effect and Bruce Wheaton's work on the empirical roots of wave-particle duality remain the only studies to deal extensively with collision processes connected with X-ray spectroscopy. Comparable studies are lacking for the history of collision processes in gases, which have been studied only exemplarily by Clayton Gearhart in the case of the Franck-Herz experiment, and in the work of Geyong Soon Im on the experimental formation of the Ramsauer effect. See Stuewer (1975) and Wheaton (1983) as well as Gearhart (2014) and Im (1995, 1996).

M. Jähnert, *Practicing the Correspondence Principle in the Old Quantum Theory*, Archimedes 56, https://doi.org/10.1007/978-3-030-13300-9_5

quantum theories of collision processes. Yet, even as a marginal episode in the history of quantum physics, Franck and Hund's case is important. It shows how the Göttingen community came to perceive the Ramsauer effect as a quantum phenomenon in the first place and developed a first qualitative conception of scattering in terms of the state-transition model.

The work of Franck and Hund is even more important as an application of the correspondence principle. As this chapter shows, Franck and Hund initially aimed to account for the Ramsauer effect on the basis of classical electron theory. They only came to perceive it as a quantum phenomenon in the fall of 1922 after Franck had spent time with Bohr and Kramers in Copenhagen and returned with a new, qualitative explanation of the effect based on the correspondence principle. Franck and Hund then went on to expand this explanation into a full correspondence argument and encountered several challenges arising from the attempt to implement the correspondence principle in the case of aperiodic motions. These challenges led them to adaptively reformulate the correspondence principle and eventually to developed a new conception of scattering.

Franck and Hund's case points towards several key aspects of the transfer and adaptation of the correspondence principle. First, their application brought the correspondence principle in an entirely different research field and did not make use of the larger technical apparatus of Bohr's quantum theory of multiply periodic systems. It thus shows how much work had to be done to push the principle's conceptual boundaries in order to transfer and extend the principle beyond its initial domain. At the same time, it highlights, more than any other case, that making correspondence arguments relied on conceptual preconditions, which were largely independent from the framework of Bohr's quantum theory.

Second, Franck and Hund's application of the correspondence principle is illuminating with respect to the importance of personal communication for the transfer and adaptation of the principle as a research tool. The fact that personal discussions with Bohr and Kramers in Copenhagen played an important role makes it possible to probe "what exactly was being transferred"—to quote David Kaiser—when physicists took up the correspondence principle through personal communication and to which extent this predicated the outcome of the resulting correspondence arguments.

Last but not least, Franck and Hund's work is documented exceptionally well in published and unpublished material: Franck and Hund's correspondence arguments are presented in their most detailed form in Hund's dissertation and a subsequent paper published in *Zeitschrift für Physik*. The central source for the reconstruction of the development of this argument, however, is Hund's scientific diary.[3] In it, Hund kept track of the development of his work almost daily, at times even to the exact hour. The diary's analysis shows how conceptual, technical, and social problems arose, and how they were solved in order to put the principle to work.

[3]Hund's diary is part of the Hund Papers held at Handschriften und Nachlässe, Niedersächsische Staats- und Universitätsbibliothek Göttingen.

Section 5.1 reconstructs how the problem of the Ramsauer effect was first received by Franck and Hund in Göttingen and how they tried to account for it on the basis of classical electron theory. Section 5.2 analyzes how Franck and Hund came to develop a new explanation based on the correspondence principle. In it, I will discuss the role of the aforementioned transfer of the correspondence principle from research on the continuous X-ray spectrum in Copenhagen to the study of electronic collisions in gases in Göttingen. Section 5.3 then discusses the problems that arose from the implementation of the correspondence principle in Hund's work, leading to the temporary conclusion that the correspondence principle was inapplicable in the case of scattering. Finally, Sect. 5.4 analyzes Franck's adaptive reformulation of the correspondence argument based on a new conception of scattering in terms of the state-transition model, and the way in which Franck and Hund's work came to an end.

5.1 Formulating the Problem: Franck, Hund and the "Argon Effect"

The Ramsauer effect became a research topic for physicists in Göttingen following the first *Deutsche Physikertag* held in Jena in September 1921. There Franck attended a talk by Carl Ramsauer about his experiments on the passage of very slow electrons through noble gases.[4] Working in Lenard's *Radiologisches Institut* in Heidelberg at the time, Ramsauer had developed an experimental setup shown in Fig. 5.1. In it, electrons were emitted from a metal plate Z and sent through a system of metal diaphragms B_1, $B_2 \ldots B_6$ by means of a magnetic field, which forced them to move on circular orbits with a certain diameter depending on their velocity. Selecting the circular orbits in this way, Ramsauer controlled the velocity and direction of the electrons, which passed through the gas under study before they were detected in the metal cages A, a, or A_2.

This setup was a manifestation of the rationale behind Ramsauer's consideration: Electrons that collided with gas atoms effectively would change their velocity, either in magnitude or in direction. They would thus be forced onto different circular trajectories by the magnetic field and would not reach the detector. This allowed Ramsauer to discriminate between the electrons that had been affected by elastic or inelastic collisions with a gas atom and those that had not.[5]

Studying the number of electrons passing through the gas at different velocities, Ramsauer—and in parallel with him Townsend and Bailey, Mayer and later Hertz, Minkowski and Sponer—reported an unexpected relation between the velocity of

---

[4]Ramsauer's talk was published as Ramsauer (1921a) and was followed by a series of papers; Ramsauer (1921b,c, 1923).

[5]See Ramsauer (1921a,b,c, 1923) for descriptions of the apparatus and the argument that was built upon it.

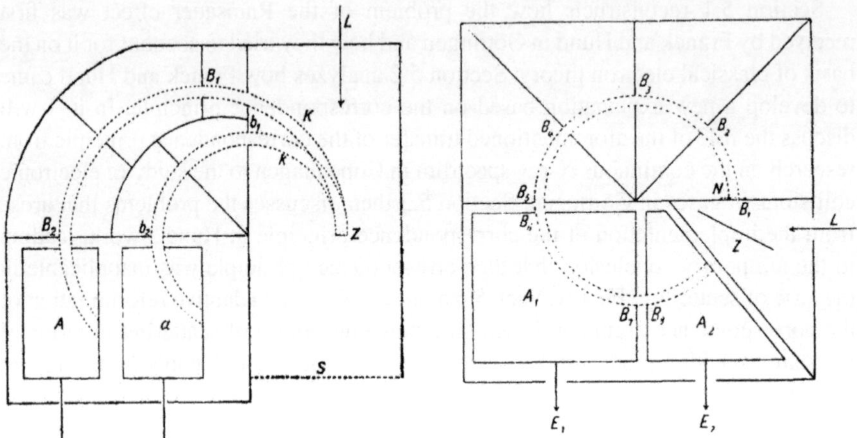

Fig. 5.1 Schematic of Ramsauer's apparatus in Ramsauer (1921a, 517) (left) and in Ramsauer (1921c, 547) (right)

slow electrons and the permeability of different noble gases, which is expressed through the cross section of the atom or the mean free path of the electron.[6] Ramsauer's results—summarized in Fig. 5.2—showed that the atomic cross section was more or less constant for electrons of different velocities passing through neon and helium. Electrons passing through argon, however, behaved quite differently. For slow electrons, the atomic cross section nearly vanished. As indicated by the horizontal dash G, it amounted to only about one-tenth of the value expected from kinetic gas theory. The cross section then increased to a maximum at 12 eV and decreased again for higher voltages. For Ramsauer, this behavior implied that very slow electrons could pass through argon without being disturbed by the gas atoms.[7]

It was this interpretation which struck Franck as both interesting and provocative. If anything, he thought, very slow electrons should be strongly affected by atomic force fields. The atomic cross section should therefore increase rather than decrease for lower electron velocities. For the time being, he therefore assumed that an experimental error on Ramsauer's part was the most likely explanation. Bringing the problem to the classroom of his *Proseminar* in Göttingen, Franck and Max Born set students to work on the experimental and theoretical refutation of "Ramsauer's crazy assertion."[8]

---

[6] See Im (1995) for a discussion of these different approaches and their role in the formation of the Ramsauer effect.

[7] Ramsauer (1921a, 614). Ramsauer further discussed this interpretation in his following papers and discarded alternative explanations. See Ramsauer (1921b,c, 1923).

[8] For the reception in Göttingen, see Franck to Bohr, 25 September 1921 in Bohr (1987, 689) and Born to Einstein, 29 November 1921 in Born et al. (1969, 91–93). As Born stressed in his letter to Einstein, Ramsauer's results implied that "the atoms are passed through freely by slow electrons!"

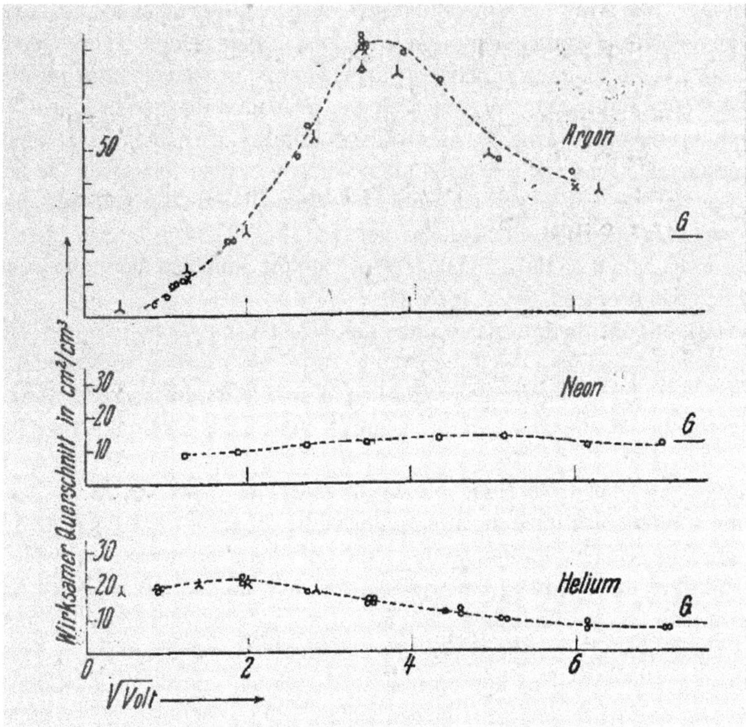

Fig. 5.2 Ramsauer's graphs for the atomic cross sections of helium, neon and argon in Ramsauer (1921a, 614)

The Göttingen community passed the stage of denial in March 1922, after Franck's old friend and colleague Gustav Hertz confirmed the strange behavior of electrons in argon.[9] Nonetheless, the idea that electrons actually passed through argon without any disturbance, i.e., somehow ceased to interact with atoms, remained unacceptable to Franck.

Considering an explanation for the "argon effect," as he called it, Franck did not identify Ramsauer's newly discovered effect with quantum theory let alone seek an explanation in terms of the correspondence principle. This was entirely consistent with the prevalent approach to scattering at the time. While the Franck-Hertz effect had shown that the transfer of energy from a scattering electron to the atom was quantized, this quantization relied solely on the fact that the atom could absorb energy only when making a transition to a higher state. In general, however, there was no general conviction that the interaction between free electrons and atoms had

[9]For the first indications of this confirmation, see Hertz to Franck, 15 December 1921 (Franck Papers [Box 3, Folder 13]). See also Hertz (1922a,b) and Franck to Bohr, 21 February 1922 in Bohr (1987, 693).

to be described in terms of quantum theory. Consequently, Franck did not consider a quantum theoretical explanation of the collisions in Ramsauer's experiments which did not involve such a transfer of energy to the atom.[10] Instead, he turned to classical electron theory and tried to come up with an explanation for it on this basis.

From this perspective, he considered the possibility of an anomalous interaction between the electrons and the argon atoms. This was plausible given the fact that argon was the only gas known to exhibit the effect and Franck could thus speculate that argon had a special force field that deflected slow electrons into an angle of 2π. The electrons would go through one loop around the atom and then leave it as if no interaction had occurred.[11]

The task of constructing such a force field was taken over by Friedrich Hund in his doctoral dissertation.[12] Well prepared by his *Examensarbeit* on potential theory written with Richard Courant, he produced a force field for the desired 2π deflection and thereby turned Franck's hypothesis into a classical explanation of the Ramsauer effect.

As we can gather from Hund's scientific diary, this work started in May 1922 with an exploration of the deflection produced by different force fields and atomic configurations. This exploration aimed to identify the conditions under which slow electrons were deflected into an angle of 2π and left the atomic force field on the same path along which they had entered it. With his background in potential theory, Hund treated the deflection of the electron into an angle of 2π as a problem of "conic sections."[13] This approach to scattering as a problem of mathematical physics was quite different from the mathematically less sophisticated one that had been entertained in the discussion of the Franck-Hertz effect. While the latter was based on the kinematics of billiard balls and the application of energy conservation, Hund conceptualized the scattering process as a dynamical problem governed by classical mechanics and electrostatics. Like in Rutherford's scattering experiments with α particles, the electron approached the atom at the distance ϱ—today's impact parameter—and a certain velocity v. It was then forced onto a hyperbolic trajectory as it interacted with the atomic force field. As it left the field, the electron approached an asymptotic straight line again.

[10]For an overview of scattering theory, see Franck (1923) and Landé (1926, 21–23). For an account of the Franck-Hertz experiment and its context, see Gearhart (2014).

[11]Franck to Bohr, 21 February 1922 in Bohr (1987, 693). Note that Franck initially proposed a deflection of 180°. The electron would make a U-turn and hence would require two such deflections to stay on its path. It is highly doubtful whether Franck seriously proposed this explanation; a careless mistake is much more likely. In any case, Franck and Hund soon adopted the 2π deflection hypothesis.

[12]Friedrich Hund had come from Marburg to Göttingen with the aim of finishing his studies and becoming a teacher of physics, mathematics and geology. He already worked at the local *Gymnasium* and switched over to research only when Born offered him a position as his assistant in the course of his dissertation.

[13]Hund (1922, 3). "In einem Kraftfeld dessen Potential durch $\frac{c}{r}$ wiedergeben wird, laufen die Elektronen auf Kegelschnitten um den Mittelpunkt (den Atomkern), in besonderen [sic!] beschreiben die von aussen kommenden Elektronen Hyperbeln."

With this general setup, Hund developed an expression for the deflection in a potential of the form $\frac{c}{r^n}$. From energy and angular momentum conservation, he first deduced the differential equation governing the scattering angle φ of the trajectory of the electron as:

$$d\varphi = \frac{v\varrho\, dr}{r\sqrt{v^2r^2 + U(r)r^2 - v^2\varrho^2}},$$

where the velocity of the electrons v and ϱ are the key parameters determining the scattering process and $U(r)$ is the potential of the atomic force field. Integrating this equation, Hund determined the deflection function $\chi(v, \varrho)$:

$$\chi = -\pi + 2\arcsin\varrho + 2v\varrho \int_{r_{min}}^{\infty} \frac{dr}{r\sqrt{v^2r^2 + U(r)r^2 - v^2\varrho^2}}.$$

According to this formula electrons are deflected at different angles, depending on their initial velocity and the perihelion. These angles are generally different from 2π and Hund spent May and June experimenting with atoms consisting of positively charged nuclei and one or two negative shells to get the desired deflection. The most promising results could be obtained for a positive nucleus and a spherically symmetric negative charge distribution, as Hund noted in his diary:

> If the atom consists of a +-core and a spherical symmetrically distributed −-charge, there exists a deflection function $\chi_v(d)$ for every initial velocity v. The ~~curves~~ functions for smaller v in general have a maximum. For the v, where the maximum of χ is equal to 2π, the cross section has a minimum. In general, the less deflection is still allowed, the weaker the minimum will be.[14]

Sticking to this configuration, Hund soon found a deflection function χ for a potential $\frac{c}{r^n}$ depending on v^2 and ϱ^n. For electrons near the axis, the dependence on the distance and the velocity of the electrons was negligible and the angle of deflection depended only on the exponent n:

$$\chi = \frac{n}{2-n}\pi.$$

The desired 2π deflection would thus be possible only for $n = \frac{4}{3}$. Establishing this result, Hund noted the research question for his subsequent investigation in his diary: "Is there a force field where $\chi_v(d) = \text{const}$ holds?" In July, he was able to answer this question affirmatively for a potential, which generalized his initial solution. Shortly before and after his summer vacation, Hund investigated

[14]Hund's scientific diary, 18 June 1922 (Hund Papers). "Besteht das Atom aus +-Kern und kugelsymmetrisch verteilter −=Ladung, so gibt es für jede Anfangsgeschw. v eine Ablenkungsfunktion $\chi_v(d)$. Die ~~Kurven~~ Funktionen für kleineres v haben i.a. ein Maximum. Für das v, wo das Max. von χ gleich 2π ist, hat der Wirkungs=querschnitt ein Minimum. Das Minimum wird i.a. um so schwächer, je geringer Ablenkung noch zugelassen wird."

whether his results were consistent with known experimental facts. He found that his result was not in conflict with the results on the atomic cross section of Lennard, Mayer and Ramsauer. Moreover, he investigated whether the exponents needed for the 2π deflection hypothesis matched the value for the exponents extracted from spectroscopic data. While he found sufficient agreement between these two values, another parameter, the effective charge of the argon atom, did not match.[15] Summarizing this situation in his dissertation, Hund concluded that "our knowledge of the argon atom thus gives no confirmation but also no definite refutation of our conception."[16]

With this somewhat inconclusive result, Hund's work on the "argon effect" came to a close in the fall of 1922. Hund reported on his progress repeatedly in the proseminar and gave a synopsis to Franck, hoping for a final discussion with his advisor before handing in his dissertation at the university.[17] Upon Franck's return to Göttingen, however, the 2π deflection hypothesis became increasingly problematic for various reasons and was finally abandoned by Franck and Hund.

The 2π deflection hypothesis was confronted by new empirical results obtained for krypton by Rudolf Minkowski and Herta Sponer. Their measurements indicated that the element "also showed the large permeability for low velocities" exhibited by argon.[18] This result implied that the "argon effect" could not be explained as a peculiarity of a single element. Thereby, the assumption that the effect was due to a specific atomic force field became increasingly improbable. Indeed, Hund found that krypton could not possibly have a force field allowing for a 2π deflection as the exponents in the potential implied by spectroscopic data on krypton were taken into consideration.[19]

Taken individually, Sponer and Minkowski's results for krypton might have been considered a problem for the 2π deflection hypothesis. The value Hund calculated for krypton was too high for a 2π deflection. It squared reasonably well, however, with values for a 6π or 8π deflection. Along with the new results for krypton, however, Hund and the classical explanation of the Ramsauer effect were confronted with a second challenge. Returning from Copenhagen, Franck proposed a new explanation of the effect that was no longer based on the 2π deflection hypothesis. Instead, he argued, Ramsauer's results should be explained on the basis of the

[15]See Hund's scientific diary, 26 June to 9 September 1922 (Hund Papers), and also Hund (1922, 29–34) for his discussion of experimental data.

[16]Hund (1922, 34). "Unsere Kenntnisse über das Argon-Atom geben also keine Bestätigung, aber auch keine sichere Widerlegung unserer Auffassung [. . .]."

[17]Hund's scientific diary, 9 September 1922 (Hund Papers).

[18]Hund's scientific diary, 20 October 1922 (Hund Papers). "Krypton zeigt für geringe Geschw[indigkeiten] ebenfalls große Durchlässigkeit."

[19]Hund (1922, 34–37). In his dismissal of the 2π deflection hypothesis in his dissertation, Hund mentioned additional problems. The classical approach, he argued, could account for neither the low value of the minimum of the cross section nor for the fact that the cross section did not depend on the experimental setup. These difficulties also applied to Hund's original consideration on argon and had been tolerated before. See Hund (1922, 27–29).

correspondence principle. Together, Minkowski and Sponer's results for krypton and Franck's new explanation led to the abandonment of the classical approach.

Flirting with Quantum Theory: Hund and the Bohr Festspiele

As we have seen in this section, Franck and Hund's attempt to explain the "argon effect" was based on a classical description of scattering in terms of potential theory. Quantum theory did not play a decisive role in this approach before Franck returned from Copenhagen in October 1922. Yet, Hund had considered a connection with quantum theory before. Inspired by Bohr's Wolfskehl lectures given in Göttingen from 12 June to 24 June 1922, he tried to impose quantum conditions on the trajectories of the scattering electron and formulated a quantized version of the 2π deflection hypothesis:

> Assumption: The radiation damping on the moving electron is so high that it very rapidly enters a quantum orbit. The quantum orbits in the inner atom are the same as those that correspond to the spectral terms. The quantum conditions [...] allow only for few discrete orbits. Especially for slow v the selection is small. The deflections for different d thus more or less coincide there. The possibility for sharp minima is given.[20]

According to Hund's tentative conceptualization of scattering in quantum theory, the scattering electrons moved around the atom on the same trajectories as bound electrons. Entering the atomic force field, however, they were not captured by the atom and instead left the atom again at angles of 2π or its multiples. Just as in Bohr's quantum theory of the atom, the quantum conditions selected a particular set of classical orbits and thereby secured the straight path of the electrons.

This qualitative description, which Hund expressed in terms of the initial Bohr model from 1913, did not play a decisive role in Hund's work. It was obviously ad-hoc, conceptually problematic from the perspective of the state-transition model, and not up to the standard of the quantum theory of multiply periodic systems and its sophisticated techniques of Hamilton-Jacobi theory and perturbation theory.[21] As

---

[20]Hund's scientific diary, 21 June 1922 (Hund Papers). "Annahme: Die Strahlungsdämpfung auf das bewegte Elektron ist so stark, dass es sehr rasch auf eine Quanten=bahn kommt. Die Quantenbahnen sind im inneren Atom die gleichen, die den Spektraltermen entsprechen. Die Quantenbedingungen

$$vd = kc$$

$$v^2 = f(n, k)$$

lassen nur wenige diskrete Bahnen zu. Besonders für kleine v ist die Auswahl gering. Die Ablenkungen für verschiedene d stimmen da also ziemlich überein. Die Möglichkeit für scharfe Minima besteht.

[21]As Hund's diary indicates, Hund had not mastered these techniques at this point. He only learned them after completing his dissertation. See Hund's scientific diary (Hund Papers). "9.1.–4.3. [1923 MJ] Hamilton-Jacobische Theorie, Störungs-Rechnung." Even if Hund had been an expert on these

such, it is not surprising that Hund's idea seemed to be little more than toying around with quantum formulas for the visitors from Copenhagen and Hund's two advisors. After a "meeting with Bohr's assistants and the *Bonzen*," Hund noted in his diary, "the quantum theoret[ical] treatment [was] rejected, because the assumptions are too artificial."[22] After another meeting with Bohr's assistants on the next day, the quantum theoretical version of the 2π deflection hypothesis ceased to play a role in Hund's work on the "argon effect" and was mentioned only briefly as an ill-fated approach in his unpublished dissertation.[23]

Although it presents a small, ultimately peripheral episode within Hund's dissertation, Hund's attempt and its rejection have interesting implications for the discussions in this chapter. It shows very clearly that both the Copenhagen and the Göttingen communities were undecided as to whether the Ramsauer effect was a quantum effect, and did not buy into the speculative explanation of a young student. In addition, one can see from Hund's quantized version of the 2π deflection hypothesis that the correspondence principle was not the only place to start in the search for a quantum theoretical explanation of the Ramsauer effect. Even more so, it was not an obvious one: While they rejected Hund's attempt, neither the Göttingen nor the Copenhagen physicists proposed an alternative based on the correspondence principle.

5.2 Implementing the Correspondence Principle: Franck, Hund and the Non-deflection Hypothesis

While quantum theory had not played a substantial role throughout Hund's dissertation, it took center stage in October 1922. At this point Franck integrated the correspondence principle into his and Hund's approach to electron scattering and gave a new interpretation of Ramsauer's results. As we will see in this section, this new interpretation departed radically from his previous work: Franck abandoned the 2π deflection hypothesis and moved on to the idea that "electrons [...] really

techniques and their application to atomic systems, he would have had to recognize that these tools were virtually useless for treating aperiodic motions. Action angle variables were not useful in this case, as the energy could not be written as a function of the angle variables alone. In addition, the quantum condition could not be applied to aperiodic motions because their system trajectories do not enclose a finite area in phase space, and thus do not allow for quantization in the sense of the old quantum theory. Other, more experienced historical actors like Kramers or Wentzel did not apply or even discuss them in their work.

[22]Hund's scientific diary, 23 June 1922 (Hund Papers). "[11 Uhr Bespr[echung] mit Bohrs Assistenten und den Bonzen; die quantentheoret[ische] Behandlung wird abgelenht, da die Annahmen dabei zu künstlich sind]."

[23]Hund (1922, 40–41).

pass the atoms without deflection."[24] Franck thus dropped the idea of manipulating atomic force fields. Instead he introduced a new hypothesis to explain the Ramsauer effect. As summarized by Hund in his diary, Franck now argued that:

> [...] an electron [would pass] straight through an atom if the radiative loss corresponding to the frequency were greater than the kinetic energy of the electron.[25]

Electrons stopped interacting with the atom, so Franck argued, if they would have to radiate with a frequency corresponding to an energy higher than its kinetic energy during the scattering process. From this position, as we will see, Franck's and Hund's attempts to explain the Ramsauer effect came to focus on the occurrence or absence of radiation during scattering and relied on an application of the correspondence principle.

As mentioned in the introduction to this chapter, this major shift in Franck's thinking emerged after his visit to Copenhagen in the fall of 1922. In his discussions with Bohr and Kramers at the *Universitetes Institut for Teoretisk Fysik*, he received the central input for his hypothesis. To understand how Franck arrived at this point, we thus need to consider his discussions with Bohr and Kramers. As Franck learned in Copenhagen, Kramers had also studied collision processes albeit in a context which was very different and hitherto unrelated to his and Hund's work on the Ramsauer effect. Rather than considering the passage of electrons through gases, Kramers worked on radiation phenomena of X-ray spectroscopy which resulted from the collision of electrons with metal plates. In this context, as we will see, he developed an application of the correspondence principle for scattering processes and arrived at a description of *bremsstahlung*, e.g., the continuous X-ray spectrum as well as the emission of X-ray lines due to the capture of electrons.

Adapting Kramers' correspondence argument, Franck developed his new interpretation of the Ramsauer effect. As he put it in December 1922, when he presented Hund's final paper to Bohr:

> In my opinion it is essential that this process [Ramsauer effect] occurs as soon as an electron would have to radiate more energy at its entrance into the atom than it possess. The problem is surely totally akin to that which Kramers discussed in the continuous X-ray spectrum.[26]

At this point, Franck was already certain that the continuous X-ray spectrum provided the clue for understanding the Ramsauer effect. In order to arrive at his new explanation of the Ramsauer effect and thus to reconceptualize it as a

---

[24]Franck to Bohr, 23 December 1922 (BSC 2.4). "Wir sind nämlich jetzt sicher geworden, dass die Elektronen im Argon aber auch im Krypton wirklich die Atome ohne Ablenkung durchqueren können."

[25]Hund's scientific diary, 18 October 1922 (Hund Papers). "Franck meint ein Elektron geht glatt durch ein Atom durch, wenn der der Frequenz entsprechende Strahlungs-Verlust größer wäre als die kinet[ische] Energie des Elektrons."

[26]Franck to Bohr, 23 December 1922 (BSC 2.4). "Wesentlich ist nach meiner Meinung, dass dieser Prozess stattfindet, sobald ein Elektron bei seinem Eintritt ins Atom quantenmässig mehr Energie ausstrahlen müsste, als es besitzt. Das Problem ist sich ganz verwandt mit dem was [sic!] Kramers beim kontinuierlichen Röntgenspektrum behandelt hat."

quantum phenomenon, however, Franck needed to take several steps. Connecting Kramers' correspondence argument with his own work on the Ramsauer effect, Franck combined two physical phenomena that were observed under vastly different circumstances. The electrons producing the continuous X-ray spectrum had energies a thousand times higher than Ramsauer's slow electrons. The *bremsstrahlung* of these slow electrons had not been observed even in a qualitative way. Rather than building on similar observations, Franck thus presupposed the existence of *bremsstrahlung* for slow electrons on the basis of the general analogy to the case of high-energy electrons emitting the continuous X-ray spectrum.

To connect these two phenomena, he moreover relied on a shared conception of the scattering process: in both cases, he assumed that electrons approached the nucleus on a straight line. They were then deflected by the atom's force field and emitted radiation during this process and finally left the atom approaching a straight line again. This shared description allowed Franck to bridge the gap between two research areas focusing on different phenomena and to integrate Kramers' correspondence argument into his work on the Ramsauer effect.

In December 1922, Franck believed that this integration led to the new interpretation of Ramsauer's results in a straightforward way. He even suggested that his new interpretation was implicit in Kramers' work or at least "pretty much suggested" by it. As such, Franck and Hund's work had not introduced any new conceptual ideas and the work in Göttingen had more or less redone Kramers' calculations, which Franck had not seen during his stay in Copenhagen.

Given this characterization, it is remarkable that Bohr and Kramers reacted quite differently to Franck's announcement. Rather than being "pretty much suggested" or even anticipated in Kramers' work, Franck's interpretation came as a surprise to both. It led Bohr to question the applicability of spatiotemporal descriptions of collisions and was welcomed by Kramers as a new idea, which he mentioned as an addendum to his work.[27]

Given this contrast between Franck's self-evaluation and the reaction in Copenhagen we should be skeptical about the idea that Franck had come up with his new interpretation of the Ramsauer effect by simply taking up Kramers' correspondence argument or that the argument developed in Göttingen was nothing more than a recapitulation of Kramers' work; indeed the following analysis will show that this was not the case. On the basis of their shared conception of the scattering process, as we will see, Franck took up the most tentative speculation of Kramers' correspondence argument and developed it further into his new explanation of the Ramsauer effect.

[27]For Bohr's reaction see Bohr to Franck, 21 April 1925 (BSC 10.4) as well as Hendry (1984, 57) and Darrigol (1992, 251) for its role in the demise of the BKS theory. For Kramers' reaction, see Kramers to Franck, 8 January 1923 (Franck Papers [Box 4, Folder 9]) and Kramers (1923, footnote on p. 856).

Input from Copenhagen: Kramers' Correspondence Approach to Scattering

In order to see where Franck departed from Kramers' work and how he eventually arrived at his new interpretation, we need to discuss Kramers' work in more detail and consider it in relation to Franck's perception of it. In this respect, it is essential that Kramers' main research goal was quite different from what Franck believed it to be. Whereas Franck saw him developing a correspondence argument for electron scattering leading to an account of the continuous X-ray spectrum, Kramers' main goal was to account for a phenomenon that was related to the scattering problem in an indirect way.[28]

This phenomenon was the absorption of X-rays and the associated emission of an electron from the atom.[29] In terms of the state-transition model, this process presented a prime example of an electron making a transition from a stationary state in the atom to a free state.

These free-bound transitions constituted one of the conceptually most challenging cases for developing a correspondence argument: Since there was no clear spatiotemporal description of the motion of the electron ejected from the atom, the possibility to associate the transition frequency with a certain harmonic component in the electron's motion did not exist; nor could one construct a state lying in between the bound and the free state.[30]

---

[28]Analyzing Kramers' approach is difficult because sources on it are limited. Apart from Franck's brief description in his letter to Bohr, the content and scope of his work in the summer and fall of 1922 is available only in the form of a short abstract for a talk titled "Om Absorption af Røntgenstraaler" [On the absorption of X-rays] which Kramers held at the *2nd Nordic Physicist Meeting* in Uppsala on 24–26 August 1922, shortly before Franck's visit to Copenhagen. The only substantial source for reconstructing Kramers' correspondence argument is his final paper "On the Theory of X-Ray Absorption and of the Continuous X-Ray Spectrum." See Kramers (1922) for Kramers' abstract as well as Kramers (1923) for the published paper. These admittedly limited sources hardly allow a historical analysis of Kramers' work in its own right. They do provide the central clue for understanding Kramers' work in its relation to Franck's new interpretation and therefore for understanding Franck and Hund's integration and implementation of the principle.

[29]The only source on the content of Kramers' argument is his final paper. As Kramers' paper was published only in 1923, it is doubtful whether Kramers had already established the full-fledged argument in October 1922. However, Franck's letter as well as the abstract of Kramers' paper suggest that the central elements of the argument were already in place. The former suggests that Kramers' consideration was based on his correspondence argument on the continuous X-ray spectrum. The latter specifies that Kramers' argument was based on Bohr's correspondence principle and produced a theoretical value for the absorption coefficient α of an atom for X-rays. Kramers (1922, 131). For understanding the transfer of the correspondence argument from Copenhagen to Göttingen, it is not essential to which extent Kramers had already developed his argument. Since Franck had not seen actual calculations, all we need to assume is that Kramers had already developed the argument in a qualitative form.

[30]As we will see below, even if Kramers had been able to describe the necessarily aperiodic motion of the free state, it would not have been clear at all how the continuous range of frequencies in this motion would correspond to a set of discrete transition frequencies.

Trying to circumvent this problem, Kramers considered the inverse process, in which a free electron is bound by the atom under the emission of an X-ray line. In this case, there was again no description for the electron's motion during the binding process and thus a correspondence argument faces similar problems. In this version, however, Kramers could at least identify the motion of an electron in the free state. It would be nothing but the hyperbolic trajectory resulting from a treatment of collisions in classical electron theory. Interested in bound-free transitions, Kramers thus came to consider collisions in terms of the state-transition model and applied the correspondence principle in this case.

The capture of an electron by the atom is, however, only one of two possible outcomes of a collision process. In addition, the electron could also remain in a free state after the collision, being scattered in a particular direction through the interaction with the atom. In this case, corresponding to transitions between two free states, the electron emits the continuous X-ray spectrum ranging from frequency 0 up to a maximum frequency ν_0 depending on the kinetic energy of the scattering electron as $\nu_0 = E_{kin}/h$. This maximum frequency and its dependence on the kinetic energy was well known in X-ray spectroscopy as the Duane-Hunt limit. First established by William Duane and Franklin Hunt in 1915, it was almost trivially understood as a consequence of energy conservation and the Planck relation.[31]

For Kramers, the Duane-Hunt limit played an important role in delineating the two possible outcomes of the collision process. As he interpreted it, the continuous X-ray spectrum up to the Duane-Hunt limit was associated with electrons making transitions between free states. Beyond the limit, they could only make transitions from a free state to a bound state in the atom under the emission of a spectral line:[32]

> The statistical result of a large number of collisions with free electrons of the same velocity will therefore be the emission of a continuous spectrum, extending from very small frequencies to a limiting frequency ν_0 determined by $h\nu_0 = mv^2$, and of a spectrum of discrete lines, extending from this limit towards larger frequencies. The line of highest frequency will correspond to a transition by which the electron is bound in a one-quantum orbit.[33]

With this delineation, Kramers effectively split his correspondence argument into two parts. On the one hand, he was concerned with the continuous classical spectrum

[31] Duane and Hunt (1915).

[32] Kramers (1923, 852). It should be noted that this delineation did not resemble the conditions in X-ray scattering experiments, in which the binding of the electron did not occur. Rather than addressing actual experiments, Kramers considered a highly idealized situation designed to make the argument on the binding process. In it the atom was thought of as completely ionized, so that the scattering electron could be bound in any possible state. This setup of the scattering / binding process only related to observation in an indirect way: it allowed Kramers to determine the transition probabilities for the inverse process of the emission of electrons in the photo effect. As such, Kramers' description, and even more so his quantum spectrum for binding, present a manifestation of his expectations based on the state-transition model and the known experimental facts on the scattering of electrons.

[33] Kramers (1923, 852).

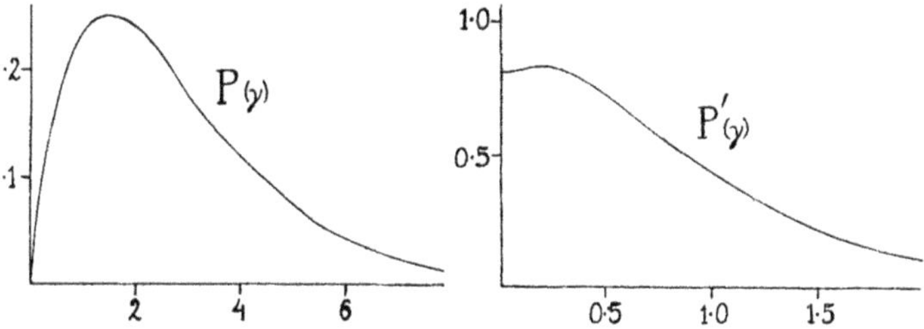

Fig. 5.3 Kramers' classical spectrum for hyperbolic orbits in Kramers (1923, 855), $P(\gamma)$ describes parabolic orbits, $P'(\gamma)$ describes nearly rectilinear orbits

up to the Duane-Hunt limit, corresponding to free-free transitions. On the other hand, he discussed the discrete set of spectral lines beyond the Duane-Hunt limit. To account for both of these processes, Kramers considered the description of scattering in classical electron theory and developed his correspondence argument to which we now turn.

Kramers' Correspondence Argument for Scattering

Implementing the correspondence principle in these two cases, Kramers followed the general structure of a correspondence argument: First he Fourier-analyzed the classical motion and established the classical spectrum of the radiating system. Then he mapped this spectrum on the quantum theoretical radiation spectrum. For the present discussion of Kramers' correspondence argument, the first part of this argument is less problematic. Establishing this classical account, Kramers followed a mathematically complicated, but clearly defined argument: He developed a Fourier representation for the hyperbolic trajectory of a scattering electron, arriving at the classical energy spectrum of *bremsstrahlung*, which, as shown in Fig. 5.3, ranges from frequency 0 to ∞ with a maximum intensity at frequency ν_{max}.[34] As we have seen, Kramers' extensive calculations remained unknown to Franck and we can assume that Kramers, Bohr, and Franck discussed only the qualitative aspects of the resulting spectrum, and the way in which Kramers mapped it onto the quantum theoretical spectrum.

As such, the analysis of Kramers' mapping procedure and his underlying conceptualization of scattering in the state-transition model in relation to the correspondence principle is essential for understanding the transfer and adaptation of the correspondence argument by Franck and Hund. From it we can see what Franck picked up in the discussions in Copenhagen, which steps he took to arrive at

[34] Kramers (1923, 845–847). The calculation leads to a spectral distribution in terms of Hankel functions $H(\gamma)$, where the argument γ depends on the angular momentum of the scattering particle.

his new interpretation, and how Hund implemented the correspondence principle in Göttingen.

Kramers justified his mapping by taking recourse to the original formulation of the correspondence principle: He argued that, just as for atomic systems, "the intensity distribution in this spectrum may be estimated from the intensity distribution in the radiation which the colliding electron would emit on the classical theory."[35] At the same time, he knew very well that an extension of the correspondence principle for atomic systems to scattering was not unproblematic:

> Thus with reference to the application of the correspondence principle to series spectra, we should expect that every possible quantum transition corresponds to a certain frequency present in the motion of the electron [. . .] Here we meet immediately the question, Which [sic!] frequency in the motion corresponds to a given transition? In the present case the answer cannot be given in the same unambiguous way as was possible in the quantum theory of simple and multiple periodic systems, where the motion can be analysed in discrete harmonic components.[36]

As Kramers pointed out, the application of the correspondence principle in the case of atomic systems meant that the classical spectrum was determined from the frequency and Fourier coefficients of the motion in the stationary state or the *Zwischenbahn*. One could thus associate a particular harmonic component in the motion with a particular quantum transition. Mapping the classical spectrum on the quantum theoretical counterpart in the case of the scattering or the binding process thus led to the question: "[w]hich frequency in the motion corresponds to a given transition" in these cases?

Considering the centrality of this question in the case of atomic systems, Kramers' take on it in the cases of scattering and binding is surprising. In the case of scattering, he assumed that the correspondence between the classical spectrum and the quantum theoretical counterpart was a one-to-one correspondence:

> the very simple assumption offers itself that the corresponding frequency is just equal to the frequency of the emitted radiation. As well known, such is not in general true for simple or multiple periodic systems where we have to do with discrete stationary states, but here there is at first sight nothing which prevents the introduction of such an assumption.[37]

This solution of a one-to-one correspondence, Kramers knew very well, was at odds with the correspondence approach for atomic systems, in which the frequency in the motion of the initial or final state was not identical to the radiation frequency. Yet he adopted it on the grounds of feasibility. In analyzing Kramers' considerations, we need to keep in mind that his consideration of the scattering process was an auxiliary consideration. Discussing the intricacies of identifying and establishing the corresponding motion, the conceptualization of the scattering process within the state-transition model or its relation to the correspondence principle did not play a major role for Kramers. Rather, he adopted an intuitive conception of scattering

[35] Kramers (1923, 852).

[36] Kramers (1923, 852–853).

[37] Kramers (1923, 853).

and adopted rather pragmatic solutions when making his correspondence argument without insisting on its consistency with the original correspondence principle.

For the binding process, Kramers realized, a similar one-to-one correspondence was not possible. The simple connection between the quantum theoretical radiation spectrum and the corresponding motion could not be given simply because there was no explicit description of this motion to begin with. Moreover, Kramers knew quite well how problematic it would be to associate any kind of aperiodic motion, hyperbolic or spiraling, with a discrete transition:

> As mentioned above, the emission spectrum resulting from such collisions will consist of separate lines, and there can be no question of a simple correspondence with the frequencies occurring in the original motion of the electron.[38]

From the perspective of the correspondence principle for atomic systems, the discrete set of transitions associated with the binding of the electron corresponded to a single frequency in the motion of the electron. As Kramers saw it, this meant that the usual correspondence relation would not be able to associate the discrete set of transitions with the continuous range of frequencies in the motion of the electron.

Engaging in tentative speculation, he pondered the possibility of resolving this issue. He considered the motion of an electron that was scattered rather than bound as the initial state. Next, he abandoned the correspondence between the frequency in the motion and a given transition. Instead, he stipulated that one could map the classical radiation spectrum onto the discrete set of transitions by assuming the correspondence of a certain range of frequencies in the continuous classical spectrum and the discrete set of transition frequencies:

> It is tempting nevertheless [...] to assume that *a certain frequency interval in the radiation emitted on the classical theory corresponds with a process by which the electron is bound in a certain stationary state.*[39]

Without addressing how such a relation could be understood on the basis of the core idea of the correspondence principle, Kramers stipulated, the continuous classical spectrum could be connected with the discrete set of spectral lines. This association allowed him to determine the intensity of the X-ray line from the integral intensity of a range of frequencies $\nu + d\nu$ in the cut-off part of the spectrum. In the graphical representation shown in Fig. 5.4, the area corresponding to the cut-off part of the classical spectrum was transformed into the area corresponding to the spectral line. Somehow, it seems, the classical intensity should be concentrated around the frequency of the spectral line. Building on this speculation, Kramers used his description of the classical spectrum to determine the transition probability for bound-free transitions and finally arrived at a value for the absorption cross section.

---

[38] Kramers (1923, 854).

[39] Kramers (1923, 854, emphasis in the original).

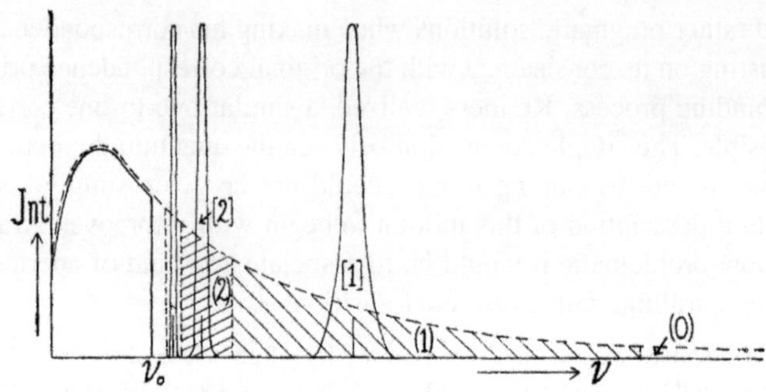

Fig. 5.4 Kramers' qualitative representation of the continuous spectrum in Kramers (1923, 855)

Franck and the Correspondence Interpretation of the Ramsauer Effect

The previous discussion puts us in a position to analyze what Franck took up during his discussion with Bohr and Kramers and how he developed his new interpretation of Ramsauer's results. As we have seen, the main idea of Kramers' correspondence argument for scattering was the mapping of the continuous radiation spectrum in classical electron theory onto the spectrum expected for transitions of a free electron to another free state or to a bound state. Through this mapping, the classical spectrum became relevant within quantum theory as a description of both the continuous X-ray spectrum up to the Duane-Hunt limit and the cut-off part beyond it. As such, the cut-off part was clearly not an appendix that ought to vanish in a correct treatment based on a future quantum theory of radiation.[40] Instead it played a constructive role in his approach.

It was this main idea of mapping the classical onto the quantum theoretical spectrum and Kramers' tentative speculation associated with it which Franck took up during the discussions in Copenhagen. Integrating Kramers' argument into his own approach to scattering, Franck focused on the radiative loss of energy during scattering and Kramers' speculation regarding the cut-off part beyond the Duane-Hunt limit and considered its implications for the explanation of the Ramsauer effect. The Ramsauer effect, Franck and with him Hund realized, was different from the scattering process discussed by Kramers. Since noble gases had no electron affinity, the electron's collision would not lead to its binding to the atom under the emission of a spectral line. Without the possibility of being captured by the atom,

[40]Such a position had already been formulated by Einstein at the 1911 Solvay Conference. For the discussions at the conference, see Eucken (1914, 308–310). In light of the limited amount of energy that could be radiated by an electron with a definite energy, Einstein had argued against the tenability of a classical understanding of the continuous x-ray spectrum based on a Fourier analysis of an aperiodic process. Bergen Davis also recognized this discrepancy independently of Einstein (Davis 1915).

electrons that would have to radiate with frequencies corresponding to the cut-off part had to remain free after the collision. As Hund put it in his doctoral dissertation, this argument implied that:

> the incoming electron can thus not describe a classical trajectory for these low velocities; for atoms without electron affinity it can also not reach one of the quantum orbits; it has to pass the atom on a trajectory without radiation.[41]

Taking a huge step, Franck connected this assessment with Ramsauer's experiments and conjectured that the absence of radiation meant nothing but the absence of interaction in general: electrons associated with the cut-off part do not radiate *and* pass through the atom without being deflected. In accordance with Hund's diary entry of October 1922, he described this argument in his letter to Bohr:

> We are namely now certain that electrons in argon but also in krypton really pass through the atoms without deflection and we believe that this is even a requirement of quantum theory as soon as electrons have low velocities [...] In my opinion it is essential that the process occurs as soon as an electron would have to radiate more energy, quantum-wise, upon its entrance into the atom, than it possesses.[42]

Based on this assumption, the velocity dependence in Ramsauer's experiments received a new and strikingly simple explanation: For slow electrons, the cut-off part of the spectrum becomes larger and larger so that the atom becomes transparent for more and more electrons.

While this new interpretation was "pretty much suggested" to Franck at a later point, it was by no means inherent in Kramers' original argument. It required several non-trivial assumptions on Franck's part. To begin with, as we have seen, he assumed that the description of electrons impinging on metal plates at about $1keV$ could be transferred to the description of electrons colliding with gas atoms at about $1eV$. Taking this step, Franck took off, as far as we can construct it, from Kramers' central idea of mapping the classical onto the quantum theoretical spectrum and the speculation about the cut-off part. Combining this speculation with the experimental facts on the electron affinity of noble gases and the free passage of slow electrons through gases, he arrived at his new interpretation and thereby gave new meaning to the cut-off part.

Making this argument, it appears, Franck did not consider how one could conceptualize scattering as a spatiotemporal process in the state-transition model. This is quite understandable given Kramers' discussion of a corresponding motion.

[41]Hund (1922, 42). "Das von aussen kommende Elektron kann bei diesen geringen Geschwindigkeiten also nicht die klassische Bahn beschreiben; es kann auch bei Atomen ohne Elektronen-Affinität nicht auf eine Quantenbahn kommen; es muss das Atom auf einer Bahn ohne Strahlung durchsetzen."

[42]Franck to Bohr, 23 December 1922 (BSC 2.4). "Wir sind nämlich jetzt sicher geworden, dass die Elektronen im Argon aber auch im Krypton wirklich die Atome ohne Ablenkung durchqueren können und meinen, dass das sogar eine Forderung der Quantentheorie ist, sobald die Elektronen kleine Geschwindigkeiten haben [...]. Wesentlich ist nach meiner Meinung, dass dieser Prozess stattfindet, sobald ein Elektron bei seinem Eintritt ins Atom quantenmässig mehr Energie ausstrahlen müsste, als es besitzt."

In it, as we have seen, he intuitively identified the hyperbolic trajectory of a scattering electron as the initial state and identified the frequency of this hyperbola with the radiation frequency of a transition to another such hyperbola. In the case of binding, he associated a multitude of frequencies in the motion with a single transition without considering any kind of corresponding motion. As such, the conception of scattering remained either intuitive or completely unclear. For Kramers, however, this was no reason for concern. He explicitly acknowledged that his corresponding motions were at odds with the correspondence principle for atomic systems and left the question of the corresponding motion an open one while using his intuitive, classical conception of scattering to establish a classical spectrum.

Following Kramers' approach, Franck also left aside how scattering could be conceptualized in terms of the state-transition model. As we will see in the following, this problem was only discussed by Hund in his attempt to substantiate Franck's new interpretation. Tellingly, the conception developed by Hund and the intricacies of making a correspondence argument for scattering on its basis were strikingly different from Kramers' argument. These differences and the implicit character of Kramers' description suggests that the question of the corresponding motion was not discussed by Franck, Kramers, and Bohr in Copenhagen. In other words, Franck took up the general idea of Kramers' correspondence argument for scattering. However, by integrating this argument into his work on the Ramsauer effect, and even more so, by implementing the correspondence principle in this context, Franck and Hund developed their own correspondence argument.

Hund's Implementation: Adaptive Reformulation in a Nutshell

As is clear from his letter to Bohr from December 1922, Franck thought that Hund's work on the new interpretation of the Ramsauer effect was more or less identical to Kramers'.[43] In any case, he thought Hund's work was little more than a technical elaboration of his hypothesis. Receiving the new interpretation of the Ramsauer effect after Franck's return from Copenhagen, it appears, Hund had more or less redone Kramers' Fourier analysis, mapped the Fourier representation of the scattering electron onto the radiation spectrum, and introduced the cut-off at the maximum frequency to substantiate Franck's new interpretation.

As will be shown in the following, the situation had been quite different. From Hund's diary and his unpublished dissertation we can see that the implementation of the correspondence principle in the case of scattering was anything but unproblematic for Franck and Hund. Trying to make the correspondence argument, Hund realized that he first needed to describe scattering in terms of the state-transition model in order to identify the initial and final states in scattering. The description

[43] Franck to Bohr, 23 December 1922 (BSC 2.4).

adopted by Hund was not only quite different from Kramers', it also led to a first adaptation of the correspondence principle.

These considerations developed within a short period of time. Two days after receiving Franck's new interpretation, Hund made the following entry in his diary, which summarizes his reaction to Franck's hypothesis:

The radiation loss for small v in a finite Coloumb field is

$$W = c\frac{z^4}{\varrho^5 v^5}$$

and the cross section, in which class[ical] deflection is impossible, is

$$\pi \varrho^2 = c\, Z^{8/5}\, v^{-7/5}$$

for $z = 8$ and 0.7 volt ca. atom=cross section. Here W was calculated classically. If one calculates ν classically (Fourier terms), $h\nu$ becomes considerably larger. *The modified Franckian conjecture (calculate W classically) appears to be able to explain the permeability of argon* [krypton also shows high permeability for small velocities].[44]

This entry summarizes and mixes several steps in Hund's work on Franck's new interpretation. On the one hand, it encapsulates the result of his implementation of the correspondence principle in scattering and his recognition of problems in the approach by noting that "[i]f one calculates ν classically (Fourier terms), $h\nu$ becomes considerably larger." On the other hand, it sketches a "modified Franckian conjecture" as an alternative to the correspondence approach. Finally, it indicates Hund's awareness of Minkowski and Sponer's experiments on krypton, which—as already discussed—Franck and Hund took as indications for the untenability of the 2π deflection hypothesis. In the following, I will discuss the first two steps in logical order and assume that Hund implemented the correspondence approach, discovered its inadequacy, and then developed the "modified Franckian conjecture."

This reading is based on Hund's unpublished dissertation. The dissertation ultimately presents a counterargument to the correspondence approach, however, Hund's line of argument is a constructive one. I will therefore interpret it as the final result of Hund's implementation in light of Franck's hypothesis, which still

[44] Hund's scientific diary, 20 October 1922 (Hund Papers). "In einem begrenzten Coulomb-Feld ist der Strahlungsverlust für kleine v

$$W = c\frac{z^4}{\varrho^5 v^5}$$

und der Querschnitt, in dem klass[ische] Ablenkung unmöglich ist,

$$\pi \varrho^2 = c\, Z^{8/5}\, v^{-7/5}$$

bei z=8 und 0,7 Volt etwa Atom=Querschnitt. Dabei war W klassisch gerechnet. Rechnet man ν klassisch (Fourierglieder), so wird $h\nu$ wesentlich größer. *Die modifizierte Francksche Vermutung (W klassisch rechnen) scheint Durchlässigkeit des Argon erklären zu können.* [Krypton zeigt für geringe Geschw[indigkeit] ebenfalls große Durchlässigkeit]".

reflects the stages of its development. Interpreted in this way, Hund's quote shows how he implemented the correspondence principle, developed his first *adaptive reformulation*, and arrived at the conclusion that it was inapplicable.

In full, Hund's argument on the implementation of the correspondence principle in scattering reads:

> In the application of this principle in atomic physics the frequencies which are radiated in a quantum jump relate to the frequencies contained in the motion on the initial and on the final orbit; the transition probabilities are determined by the coefficients associated with the respective frequencies in the Fourier series of the motions in the initial and final orbit. In our case the initial and final orbits are straight [...]; they have no frequency; we could thus only let the frequency of the classically continued initial and final trajectory relate to the emitted frequencies. For the case that $h\nu$ is larger than the energy of the colliding electrons, we have to assume that the corresponding transition does not occur, but that the electron instead flies straight on.[45]

On the assumption that this quote mirrors the development of Hund's correspondence argument, we can see that he started from the formulation of the correspondence principle for atomic systems, which he paraphrased in accordance with Bohr along the same lines discussed in Chap. 3.1. While physicists working on periodic systems could all implement the correspondence principle in its original form, Hund was pressed to extract a prescription for applying the principle to aperiodic motion in the case of scattering. In this situation, he focused on the fact that the principle associated the frequencies and Fourier coefficients of the stationary states with the frequency and the probability of a quantum transition.[46] To extend the principle to scattering, he concluded, it was necessary to identify the initial and final orbits of the scattering electron. Only this identification made it possible to think about a correspondence relation between the Fourier representation and the radiation frequency and transition probability. In other words, the application of the correspondence principle required him to conceptualize the physical system in terms of the state-transition model.

Even this first step was quite problematic, as aperiodic motions could not be described within the framework of Bohr's quantum theory of multiply periodic

[45]Hund (1922, 43). "Bei der Anwendung dieses Prinzips in der Atom-Physik entsprechen die Frequenzen, die bei einem Quantensprung gestrahlt werden, den Frequenzen, die die Bewegungen auf der Anfangs-und [sic!] Endbahn enthalten; die Uebergangswahrscheinlichkeit wird durch die Koeffizienten gegeben, die die betreffende Frequenz in den Fourierreihen der Bewegungen auf der Anfangs-und[sic!] Endbahn hat. In unserem Fall sind die Anfangs-und [sic!] Endbahnen gerade (wenigstens wenn das Atom aussen neutral ist); sie haben keine Frequenzen; Wir könnten also nur die Frequenzen der klassisch fortgesetzten Anfangs-und[sic!] Endbahnen den ausgestrahlten Frequenzen entsprechen lassen. Für den Fall, dass $h\nu$ grösser als die Energie des auftreffenden Elektrons ist, müssen wir annehmen, dass der betrffende [sic!] Uebergang nicht vorkommt, sondern stattdessen das Elektron gerade ausfliegt."

[46]Hund (1922, 43). Hund's understanding did not change significantly in the subsequent paper (Hund 1923, 250–251). In it, he specified that the Fourier series of the initial state had to be taken into account instead of the initial and final states. This difference did not affect the argument or the calculations.

systems. The lack of a mechanical framework, however, did not put an end to Hund's attempt to use the correspondence principle. He intuitively identified the electron approaching the atom on a straight line with the initial state, and—vice versa—the electron moving on a straight line after the collision with the final state. Hund left open what he thought about the transition. By definition, however, it would have to take place somewhere in between the two states and would thus be connected with the deflection of the electron.

This conceptualization was not only considerably different from Kramers' identification of the initial and final states with two different hyperbolae characterized by different energies, it also led to the peculiar situation that the correspondence principle was not applicable at all. As Hund realized, the newly defined states had no frequencies and could therefore not possibly be connected with a continuous radiation spectrum. Finding the new description of scattering to be incompatible with the correspondence principle, Hund adapted it for the first time, adopting a novel solution to the initial-final-state problem: He concluded that the only way to make the correspondence argument in the case of scattering was to "relate the frequency of the classically continued initial and final trajectory to the emitted frequencies."[47] Explicitly, these "classically continued trajectories" did not represent the initial or the final state. Rather, they described the electron during its deflection and were therefore implicitly associated with the transition process. Effectively but unconsciously adapting the reference system of the correspondence principle, Hund recognized that his subtle change made it possible to obtain the desired continuous spectrum from the Fourier representation and to follow Franck's hypothesis.

Interim Conclusion

Before discussing the problem Hund subsequently encountered, a short review of the analysis of the implementation is in order. As we have seen in this section, Franck and Hund abandoned their initial approach to the Ramsauer effect in the fall of 1922 and developed a new interpretation on the basis of the correspondence principle. Instead of explaining the effect by manipulating the classical force field of the atom, this new interpretation assumed that slow electrons actually ceased to interact with the atom.

This interpretation was based on considerations about the spectrum radiated during scattering, which had emerged from a transfer of Kramers' correspondence argument for the continuous X-ray spectrum to Franck's and Hund's work on the passage of electrons through gases. While Franck later portrayed this transfer as a mere takeover and explication of ideas inherent in Kramers' work, we have seen that this was not the case. Taking up Kramers' correspondence argument,

[47] Hund (1922, 43).

Franck integrated it into his work. That such a transfer would lead to anything like a new explanation of Ramsauer's puzzling results was not obvious, nor was it already implicit in Kramers' work. Rather, it took the tentative integration of the correspondence argument for scattering, the challenging facts of the Duane-Hunt limit, and Ramsauer's results to formulate Franck's no-interaction interpretation. Taking up Kramers' general idea, Franck thus developed his own take on the correspondence argument for scattering.

Moreover, his and Hund's actual correspondence argument was strikingly different from Kramers'. As Hund realized, making a correspondence argument in the case of scattering was not a trivial thing to do. First of all, it meant describing scattering qualitatively in terms of the state-transition model. Without knowing Kramers' take on it, he, and with him Franck, conceptualized the scattering process within the state-transition model in a new way. This led him to discuss the intricacies involved in associating the Fourier representation of an aperiodic motion with its quantum theoretical radiation spectrum. These intricacies showed that the original formulation of the correspondence principle was inapplicable to scattering, thus forcing him to adapt it. Characteristically, this *adaptive reformulation* left the core idea of the principle intact: The Fourier representation of the radiating system was associated with the frequencies and intensities in the radiation spectrum. Adapting the principle to scattering meant changing the reference system of the principle: the relevant Fourier representation was no longer associated with the initial or the final state of the system; rather, the "frequencies of the classically continued initial and final trajectory" were associated with the deflection in the classical description or, in other words, with the quantum transitions.

5.3 Recognizing Problems: Hund and the Estimate of Transparency

After he had implemented the principle, Hund engaged in the quantitative details of the argument. In the process, as was already pointed out, he recognized much more challenging problems for his application of the correspondence principle. For him, this challenge implied that the principle was inapplicable in the case of scattering and that Franck's explanation of the Ramsauer effect needed to be adapted in a different way.

Hund arrived at this conclusion through a calculation that differed considerably from Kramers'. Where Kramers had Fourier-analyzed the hyperbolic trajectory, Hund made an approximation before taking on such a laborious and complicated task. In this approximation, he considered the frequencies which would "make the biggest contribution" to the Fourier integral. These frequencies, he argued, were associated with the maximal deflection of the electron and thus arose from the electron's "passage in proximity of the perihelion", i.e., the passage at the closest distance from the nucleus.[48]

The advantage of considering only these frequencies was that Hund did not actually need to evaluate the Fourier integral. Rather, it put him in a position to make another approximation: He simply estimated the perihelion frequencies of the hyperbola from the frequencies of an ellipse of comparable extension. These frequencies, Hund argued, would generally be smaller than the frequencies of the hyperbola and therefore provide a lower limit for the energy of the transitions.[49]

To finally obtain the frequencies of the ellipse, Hund considered a "rough model of the atom."[50] On the basis of this model, which was effectively the Bohr model from 1913, with a charged shell representing the eight valence electrons of the argon atom, Hund noted in his diary that the energy $h\nu$ became "considerably larger"[51] than the kinetic energy of the electrons in Ramsauer's experiment. As he put it in his dissertation:

> For most transitions the loss of energy $h\nu$ would be greater than the energy emission in the first quantum jumps; in any case, [the energy $h\nu$] will only have the order of magnitude of the kinetic energy for electrons with a velocity of many volts. Even for velocities of several volts, for which one already knows experiments that indicate a reflection of electrons, the atom would have to be transparent.[52]

Hund extrapolated from his approximation for the dominating frequencies and argued that most of the spectrum would have to be cut off on the basis of Franck's assumptions. Thus the atom would have to be transparent not only for very slow electrons, but also for electrons of higher velocities. For these energies, however, atoms were known to reflect electrons, so that Hund arrived at the drastic conclusion that the "application of the correspondence principle appears not to be possible."[53]

This conclusion was not the final result of Hund's considerations, however. He also considered a way to preserve the basic idea underlying Franck's explanation of the Ramsauer effect. His solution was to calculate the total loss of energy due to radiation classically instead of using the quantum theoretical radiation energy $h\nu$ and the correspondence principle. This classical calculation, which Hund could easily produce on the basis of his earlier work,[54] gave the total energy W radiated

---

[48]Hund (1922, 43–44).

[49]Hund (1922, 44).

[50]Ibid.

[51]Hund's scientific diary, 20 October 1922 (Hund Papers).

[52]Hund (1922, 44). "Bei den meisten Uebergängen wäre der Energieverlust $h\nu$ grösser als die Energieabgabe bei den ersten Quantensprüngen, jedenfalls erst bei Elektronen mit einer Geschwindigkeit von vielen Volt von der Grössenordnung der vorhandenen kinetischen Energie. Noch bei Geschwindigkeiten von mehreren Volt, bei denen man ja genug Versuche kennt, die auf eine Reflexion der Elektronen deuten, müsste das Atom durchlässig sein."

[53]Hund (1922, 44). "Eine solche Anwendung des Korrespondenzprinzips erscheint also nicht möglich."

[54]Hund had the result two days after Franck had presented his new interpretation. In his dissertation this derivation spanned no more than four pages. Hund (1922, 44–47). With it Hund calculated the

by an electron

$$W = c\frac{z^4}{\varrho^5 v^5},$$

depending on the velocity v and the impact parameter ϱ.

This energy, Hund realized, allowed Franck's argument to be made with only a slight modification: For low values of the impact parameter or very slow electron velocity, the electron would lose so much energy that it could no longer leave the atom's force field. This meant that the electron would either have to be bound by the atom or it would crash into the nucleus. Following Franck's reasoning, Hund excluded the first possibility for noble gases and argued that the electron had to cease radiating in order to prevent the crash into the core. According to his "modified Franckian conjecture," this implied that the electrons had to stop interacting with the atom. This made it possible to determine a "cross section in which classical deflection is impossible" depending on the velocity and the impact parameter:[55]

$$\pi\varrho^2 = c\, Z^{8/5}\, v^{-7/5}.$$

In other words, Hund found that there is a range of impact parameters for electrons of every velocity, in which they cease to interact with the atom. For small velocities, the cross section defined in this way becomes larger and larger so that more and more electrons pass through the atom. Considering a field of limited extension, Hund found, the atom eventually becomes transparent below a particular velocity:

> For greater velocities v and greater axial distances ϱ the loss of energy is minor. If ϱ decreases, larger parts of the available energy are drained; finally this energy no longer suffices, and the electron passes through the atom following our assumption. A decrease in velocity has the same result. For electrons of high velocity only a narrow area around the

---

energy w that would be radiated in classical electron theory:

$$w = \frac{4e^6 z^2}{3c^2 m^2}\int_{r_{min}}^{R} \frac{dr}{r^3\sqrt{r^2(v^2 - \frac{2e^2 z}{mR}) + \frac{2e^2 z}{m}r - v^2\varrho^2}}.$$

Considering the special case of a slow electron ($v^2 << \frac{2e^2 z}{mR}$), he arrived at the energy radiated as

$$W = \frac{2e^{12} z^4}{c^3 m^4 v^5 \varrho^5},$$

which has the same dependence on the velocity and impact parameter as the formula given in his diary, cf. Footnote 44.

[55]Hund coined the term "modified Franckian conjecture" in his diary entry given in Footnote 44. For the argument see Hund (1922, 41–43).

> nucleus is freely permeable. With decreasing velocities, this area gets larger until below a certain velocity limit the whole atom is transparent.[56]

Hund checked the feasibility of his "modified Franckian conjecture" and found reasonable results by introducing values for the velocities in Ramsauer's experiments, and comparing the resulting impact parameter with the atomic cross section, i.e., the size of the argon atom.[57]

Hund's modification of Franck's interpretation obviously departed quite drastically from the original conception. Transferring Kramers' correspondence argument to the Ramsauer effect, Franck had assumed that the scattering electron radiated according to quantum theory. In Hund's case, both scattering and the radiation process were described on the basis of classical electron theory and not in terms of the state-transition model. His modification of Franck's hypothesis relied on energy conservation and the contradiction arising from its application in classical electron theory. Making obvious reference to Bohr's 1913 quantum theory of the atom in this way, he recast Franck's hypothesis as a "threshold" where classical interaction ceased to take place.[58]

This alternative conceptualization of the transparency in Ramsauer's experiment led to a different prediction with respect to the threshold at which the atom became transparent. Franck's hypothesis determined this limit from the Duane-Hunt limit, and thereby the kinetic energy of the scattering electron alone. Hund's modification was based on the assumption that the electron could not lose more energy than it required to leave the atomic force field: The threshold of free passage depended on the properties of the scattering electron and the atom.

5.4 Adaptive Reformulation: Franck, Hund and the State-Transition Model for Scattering

While Hund arrived at the conclusion that the correspondence approach had to be abandoned and developed his "modified Franckian conjecture," Franck's reaction to the problems in the correspondence argument was a different one. As Hund noted into his diary, there was a "confrontation with Franck about his conception

---

[56]Hund (1922, 47–48). "Für grösserer Geschwindigkeiten v und grössere Achsenabstände ϱ ist der Energieverlust gering. Sinkt jedoch ϱ, so werden immer grössere Teile der vorhandenen Energie aufgezehrt; schließlich reicht diese nicht mehr aus, und das Elektron geht nach unserer Annahme durch das Atom. Das gleiche Ergebnis hat eine Verringerung der Geschwindigkeit. Für Elektronen grosser Geschwindigkeit ist nur eine nahe Umgebung des Kernes frei durchlässig. Mit abnehmender Geschwindigkeit wird diese Umgebung immer grösser, bis unterhalb einer gewissen Geschwindigkeitsgrenze das ganze Atom durchlässig ist."

[57]Cf. Footnote 44.

[58]Hund (1922, 47–48).

and mine" when Franck learned about Hund's counterargument.[59] We do not know the content of the argument in detail. As a result, however, Hund handed in his dissertation at the university, including his hastily written considerations on Franck's interpretation, his interpretation that the correspondence principle was inapplicable and his modification of Franck's hypothesis.

While approving of this move, Franck did not follow Hund's conclusion. Instead he and Born summarized that Hund's dissertation showed, first of all, that a classical explanation was untenable. As far as a quantum theoretical explanation was concerned, it gave only a "first overview on the possibility" of its formulation.[60] In the following, Franck came up with an adaptation of the argument which—as Hund put it in his diary—was based on "a new, very plausible conception in keeping with the correspondence principle."[61]

Unfortunately, we have very little material to assess the relation between Franck's adaptation of the correspondence argument and the problem of perihelion frequencies encountered by Hund. It is not discussed in Hund's diary, Born's examination report, Franck's letter to Bohr and disappeared entirely in Hund's published paper. Only Hund's explicit remark that Franck's new conception was "in keeping with the correspondence principle" suggests that Franck directly aimed to resolve Hund's counterargument and was successful in doing so in the eyes of his peers in Göttingen.

Analyzing Franck's solution is even more difficult, as Franck's new conception is only partly described in Hund's diary and Born's examination report, so that we need to turn to the conclusion of Hund's published paper for its explicit statement. In this situation, we need to discuss the overall transformation of Franck's and Hund's description of scattering in terms of the state-transition model and analyze the underlying rationale, which suggests how Franck might have arrived at it to save his correspondence argument.

Franck's State-Transition Model for Scattering

As mentioned above, we need to turn to the final result of Franck's and Hund's work to learn about the "new, very plausible conception in keeping with the correspondence principle," which Franck proposed in October 1922. In his letter to Bohr in December 1922, Franck described the basic idea of his correspondence argument in the following way:

> A free electron is on a straight quantum trajectory; a transition to another quantum trajectory can be stimulated through the collision. Thereby deflections occur. For the electrons which

[59] Hund's scientific diary, 24 October 1922 (Hund Papers).

[60] See Born's examination report in "Promotionszulassungen der mathematisch-naturwissenschaftlichen Fakultät" (UAG).

[61] Hund's scientific diary, 29 October 1922 (Hund Papers).

> remain on their quantum trajectory, however, the collision does not exist at all. The electron stays on its straight old quantum trajectory.[62]

This description was similar to the one given by Hund in his dissertation, in which the electron approaching and leaving the atom on a straight line was identified with the initial and final states of the scattering process. Still, Franck's new description was decisively different from Hund's, as is shown in the conclusion of Hund's paper, which Franck probably wrote himself:[63]

> In a collision of an electron with an atom, a *transition probability* is created under the influence of the atomic force field [for a transition] from a straight orbit, on which the electron arrives, to another orbit of lower energy. The new orbit is also a straight line.[64]

The atomic force field no longer deflected the electron on a hyperbolic trajectory as in Hund's dissertation. Rather, it induced a transition probability. This shows that Franck's new conception of scattering in the state-transition model abandoned the hyperbolic trajectories of classical electron theory as a description of the actual motion of the electron. Instead, he seems to have conceptualized the scattering process literally as a transition from one straight line to another.

What was the rationale behind this new description of scattering in the state-transition model, and how could it be related to Hund's counterargument? A first indication for answers to these questions, which I am going to give below, can be found in Born's examination report on Hund's dissertation:

> Maybe one has to disregard the validity of electrostatics for very slow electrons, which would crash into the nucleus according to their radiation damping in the classical calculation, and instead [one has to] introduce quantum theoretical deflection according to probabilistic laws.[65]

When Born wrote the report, the situation was still wide open. On the one hand, he described Hund's argument on the threshold of classical interaction. On the other hand, he thought that this argument called for a description of scattering in terms

---

[62]Franck to Bohr, 23 December 1922 (BSC 2.4). "Ein freies Elektron befindet sich auf einer geradlinigen Quantenbahn; durch einen Zusammenstoss kann ein Übergang in eine andere Quantenbahn angeregt werden. Dabei finden Ablenkungen statt. Für die Elektronen jedoch, die auf ihrer Quantenbahn bleiben, existiert der Zusammenstoss überhaupt nicht. Das Elektron bleibt auf seiner geraden alten Quantenbahn."

[63]In his oral history interview with Kuhn, Hund clarified that he could not possibly have written the concluding remarks of his paper and credited Franck with it. For the transcript see interview of Friedrich Hund by Thomas S. Kuhn on 1963 June 25, Session I (AHQP APS 1419-03_hund-002).

[64]Hund (1923, 262, emphasis in the original). "Beim Zusammenstoß eines Elektrons mit einem Atom entsteht unter dem Einfluß des Atomkraftfeldes eine Übergangswahrscheinlichkeit von der geradlinigen Bahn, auf der das Elektron ankommt, auf eine andere Bahn geringerer Energie. Die neue Bahn ist ebenfalls eine Gerade."

[65]Born's examination report (UAG). "Vielleicht muss man für ganz langsame Elektronen, die auf Grund ihrer Strahlungsbremsung nach klassischer Rechnung in den Kern fallen würden, von der Gültigkeit der Elektrostatik absehen und dafür quantentheoretische Ablenkung nach Wahrscheinlichkeitsgesetzen einführen."

of quantum theory (at least for low-velocity electrons). This quantum theoretical description, Born specified, would be governed by "probabilistic laws."

As becomes clear from Hund's diary and the published paper, this focus on the probabilistic nature of the quantum theoretical description played a central role in Franck's adaptation of the correspondence argument: It appears to be the way in which he sought to resolve Hund's counterargument. To see how his argument might have looked, we need to consider both Hund's diary and the published paper. Hund's diary provides an overview and a detailed chronology of the developments, which allow us to discern a general pattern in Hund's work. This general chronology can be linked to the arguments and conceptions presented in the paper.

Taking only Hund's diary into account, the following pattern in Hund's work emerges from entries between 29 October and 21 December 1922: After the submission of his dissertation, Hund continued working on Franck's approach. He adopted Franck's new conception as a working hypothesis and reworked his correspondence argument by developing at least three different spectra.[66] However, like in his dissertation, Hund remained highly skeptical about the correspondence argument and kept challenging it. These counterarguments were waived in the discussions between Hund, Franck, Born, and Heisenberg, who became involved in the discussions, so that Franck's and Hund's work on the Ramsauer effect came to a close with the new conception of scattering in terms of the state-transition model.

In light of this development, we can approach the analysis of the new correspondence argument and its alteration with respect to Hund's dissertation. On 4 November—a week after the dissertation was finished—Hund noted the general working hypothesis, on which the following work was based:

> The assumptions (in general from Franck):
>
> > every deflected electron radiates with a frequency ν and energy $h\nu$.
> > The multitude of electrons impinging with a certain v, ϱ radiates an intensity for every ν that is not greater than the one classically associated with this ν. It [the intensity] is all the more equal to the classical, the smaller ν gets compared with $\nu_{max} = \frac{mv^2}{2h}$; it is 0, if $\nu \geq \nu_{max}$.[67]

As in Kramers' work and Hund's dissertation, the spectrum radiated by a scattering electron was produced by a multitude of electrons, each radiating within a range of frequencies, while a single electron radiated a definite frequency according to the $h\nu$ relation. The energy lost in this way could be any amount consistent with

---

[66] Hund's scientific diary, 24 October 1922, 29 October 1922 and 4 November 1922 (Hund Papers). Unfortunately, Hund's diary is not very specific and we cannot compare them with each other.

[67] Hund's scientific diary, 4 November 1922 (Hund Papers): "Die Annahmen (i. Wesentlichen nach Franck):

> Jedes abgelenkte Elektron strahlt eine Frequenz v und die Energie $h\nu$.
> Die Gesamtheit aller mit bestimmten (v, ϱ) auftreffenden Elektronen strahlt für jedes ν eine Intensität, die nicht größer ist, als die klassich zu diesem ν gehörige. Sie ist umsomehr [sic!] der klassischen gleich, je kleiner ν gegen $\nu_{max} = \frac{mv^2}{2h}$ ist; sie ist 0, wenn $\nu >= \nu_{max}$ ist."

energy conservation, and the frequency would range from 0 to ν_{max}, i.e., the Duane-Hunt limit. Contrary to his previous work, in which the radiation spectrum up to the Duane-Hunt limit had been identical to the spectrum resulting from classical electron theory, Hund now assumed that the intensity of a particular frequency did not exceed the classical one and used the classical spectrum as an approximation.

With this assumption, Franck and Hund took a first step towards a reevaluation of the meaning attached to the classical spectrum, which appears to have been decisive for Franck's believe that his new conceptualization of the scattering process allowed him to resolve Hund's counterargument. The assumption that the actual intensity of a given frequency was "not bigger than the one classically associated with this ν" and only became identical with the classical one for small ν implies that the classical spectrum only provided an approximation for the actual quantum theoretical spectrum, even for the actual radiation spectrum up to the Duane-Hunt limit.

At least in the final paper, this approximate relation between the classical spectrum and the quantum theoretical spectrum was more than a mere numerical difference. Rather, the two spectra were distinct objects, such that making a correspondence argument meant that the "classical spectrum should now relate to the real emitted spectrum."[68] Franck's new conception took the "real" quantum theoretical spectrum to be primary. The classical spectrum only provided an approximation. In principle, however, it was incorrect for *all* electrons and thus distinct from the quantum theoretical spectrum on a conceptual level.

This new interpretation extended a way of thinking about the correspondence principle and the quantum-classical divide, similar to Bohr's. For Bohr, as we have seen in Chap. 2, the classical description of radiation was conceptually divided from the state-transition model and irrevocably so. This conceptual divide, he stressed time and again, was not overcome by the correspondence principle. Rather, the principle expressed that the relation between radiation and motion in quantum theory was formally analogous to classical radiation theory. Trying to resolve Hund's counterargument, it appears, Franck adopted a similar point of view when he stressed that the observed spectrum was a manifestation of a quantum process and had to be described in quantum terms while the classical spectrum only provided an approximate representation of the observed spectrum.

This understanding meant a subtle change for the correspondence argument. It preserved the idea of mapping the Fourier representation of an aperiodic motion onto the quantum theoretical radiation spectrum. While he continued to argue along this line, Franck decoupled the classical from the quantum theoretical spectrum on a conceptual level. Moreover, he extended the divide. The classical description of scattering, he assumed, could not be regarded as a representation of the actual motion. As we have seen above, Franck now assumed that this motion would be described as a transition process from one straight line to another, so that the atomic

[68]Hund (1923, 254).

force field no longer curved the path of the electron, but created a probability for the transition.

While the change in understanding was minute, it appears to have been the key to resolving Hund's counterargument. One of the most important results of decoupling the classical from the quantum theoretical spectrum was that the classical relation between the radiation spectrum and the electron's deflection no longer carried over to the quantum case. Electrons moved solely according to (yet unknown) laws of quantum theory and lost energy equal to their own kinetic energy according to quantum theory, while the associated frequencies corresponded to rather small deflections in classical theory.

Inversely, this implied that the maximum in the classical spectrum was no longer associated with the electron's actual deflection. It only provided an approximation for determining the probability distribution for electrons deflected in all possible angles, and for those that went through the atom without deflection. From this perspective, Hund's counterargument at best implied that the classical spectrum obtained from the rough estimate on the perihelion frequencies of a hyperbolic trajectory was not a good approximation.

New Problems: The Infrared Divergence

Subscribing to Franck's new hypothesis, Hund resumed his work on the intensity distribution of the continuous spectrum of an aperiodic process. Working out his first distribution—unfortunately, without noting any details in his diary—he encountered a new problem, which again raised questions about the applicability of the quantum theoretical radiation process: he found that the intensity did not vanish for frequencies tending toward zero. Radiating this intensity according to the $h\nu$ relation thus required an infinite number of radiation processes. If infinitely many scattering electrons were needed to radiate with these frequencies, however, how could any of them stop radiating and simply pass the atom according to Franck's hypothesis? Laconically, Hund noted in his diary: "Objection: The transition probability tends towards infinity for decreasing ν."[69]

Before using this problem against the correspondence approach again, Hund developed a second spectral distribution, in which the nonsensical result of infinite transition probabilities did not occur. On 23 November, he had developed a spectrum for which the intensity tended towards zero for vanishing frequencies. However, this distribution yielded values too high for the threshold of free passage. With these two

[69] Hund's scientific diary, 5 November 1922 (Hund Papers). "Einwand: Die Übergangs=Wahrscheinlichkeit geht mit abnehmendem ν gegen unendlich." For the argument, see Hund (1923, 256). Hund encountered the infrared divergence in quantum physics for the first time, a problem which would come to greatly trouble physicists in the development of quantum electrodynamics. For a detailed discussion of the problem and the significance of Hund's discovery, see Blum (2015).

possibilities, Hund turned to Born for help on the theoretical approach. Born decided that the first alternative was the correct one:

> The right distribution of energy over frequencies is determined by the part of the energy absorbed by the respective resonators (following Born). We get the old distribution (5.11.).[70]

According to Born, the continuous spectrum of an aperiodic process should be determined by assigning a harmonic oscillator to each frequency in the spectrum. The intensity associated with the various frequencies would be given by the energy absorbed by the oscillators in a periodic external electric field. With this move, Born and Hund thus determined a continuous spectrum from the classical interaction between oscillators and radiation. This approach finally detached Hund's work from Kramers' approach based on the Fourier analysis of the hyperbolic trajectory of a scattering electron.[71]

Having discarded Hund's counterargument based on the maximum of a continuous spectrum, the problem of infinite transition probabilities for vanishing frequencies still needed to be solved. On 29 November, Hund met with Franck again and discussed the problem with him, Born and his new private assistant Heisenberg. In the end, the problem could be eliminated rather easily, according to Heisenberg. One just had to make the assumption that the electron could emit a multitude of quanta $h\nu$ in one scattering process, in contrast to Hund's and Franck's previous assumption that it only radiated once during the scattering process. Having one electron that could radiate infinitely often if necessary, the infinite transition probability no longer implied that infinitely many scattering electrons were needed to produce the continuous spectrum. Consequently, there were electrons that did not radiate and passed the atom without radiation.[72]

With this solution, the last conceptual difficulty disappeared. Hund elaborated Born's idea of assigning a driven harmonic oscillator to each frequency in the continuous spectrum. Based on this assumption, he first arrived at an integral representation for the continuous spectrum. In this form, Hund could say very little about this spectrum. Without an explicit expression for the spectral distribution, he could only show that the spectrum had a non-vanishing intensity for zero frequencies. As a consequence, his paper gave an "entirely qualitative" curve for the spectrum.[73] The curve $C(\nu)$ shown in Fig. 5.5 incorporated the two results of

[70] Hund's scientific diary, (Hund Papers). "Die richtige Verteilung der Energie auf die Frequenzen ist bestimmt durch den Anteil der Energie, den entspr[echende] Resonatoren aufnehmen (nach Born). Wir erhalten die alte Verteilung (5.11.)."

[71] Instead, Born and Hund now followed a line of thought similar to Planck's in the classical part of his theory of black-body radiation. Hund's paper, however, did not make reference to Planck's *Vorlesung über die Theorie der Wärmestrahlung*, and his diary indicates that Hund developed these considerations on his own, adopting Born's proposal.

[72] Hund's scientific diary, 5 November 1922 (Hund Papers). "Heisenberg lässt Vielfache von $h\nu$ zu, damit Einwand von 5.11. hinfällig."

[73] Hund (1923, 254).

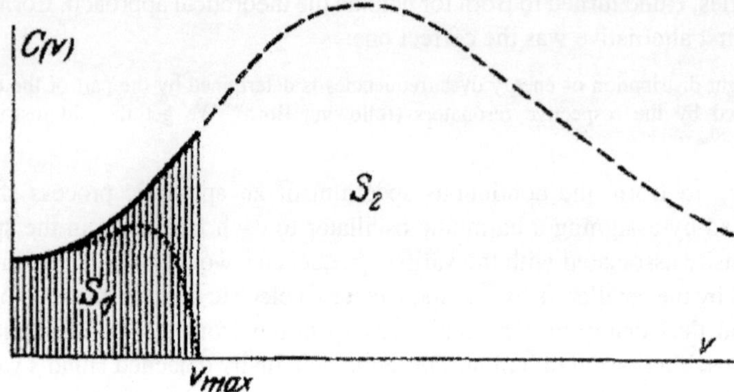

Fig. 5.5 Hund's qualitative representation of the continuous spectrum in Hund (1923, 255)

Hund's work on the continuous spectrum: a non-vanishing intensity for frequencies tending toward zero and a maximum at the former perihelion frequencies, which was beyond the Duane-Hunt limit.

In the next step, Hund knew, the transition probability would have to be determined from the ratio of the intensity distribution in the classical spectrum $C(\nu)$ and the energy $h\nu$ of the respective transition, thus requiring an explicit expression for the classical spectrum. Hund gave such an expression in the form of a first approximation. This approximation, however, was reasonable only for small frequencies and thus did not allow him to calculate the transition probability in the cut-off part.[74]

Without an explicit expression for the classical spectrum in this case, Hund returned to the considerations of his dissertation on the total energy radiated classically by the electron. He argued that the ratio between this energy and the kinetic energy of the electron set the upper limit for the ratio of the classical spectrum and the energy of a quantum transition. As such, it provided an estimate for the probability of free passage, which allowed Hund to determine the cross section Q.[75] Comparing his result with Ramsauer's experimental data, Hund concluded that the theoretical cross section came out too high. This discrepancy, however, could be cast aside on the assumption that the actual quantum theoretical spectrum would have to be different. As such, it did not present a new challenge for the correspondence principle based on the new description of scattering, which Franck

[74]Hund (1923, 256–257). On the assumption that the electron moved with its final velocity throughout the scattering process rather than being accelerated to it, Hund obtained a quadratic expression for $C(\nu)$. This expression worked only for the part of spectrum with increasing intensity and failed to describe the continuous spectrum shown in Fig. 5.5 beyond the maximum (regardless of whether this maximum was beyond the Duane-Hunt limit).

[75]Hund (1923, 257). This ratio, Hund argued, provided an upper limit for the transition probability, as it ensured that the ratio $\frac{C(\nu)}{h\nu}$ was always smaller than one and thus qualifies as a probability at all.

and Hund had arrived at through the implementation of the correspondence principle in scattering. Finally finishing his work, Hund noted in his diary, "the electron paper is put down in writing for Zeitschrift für Physik," before going on to study Hamilton-Jacobi theory and perturbation theory in the next two months to prepare for his work as Born's assistant in the new research program on atomic dynamics.[76]

5.5 Conclusion

Franck and Hund's work on the Ramsauer effect occupies a peripheral place within the history of quantum physics. It did not play a role in later explanations of the Ramsauer effect or even the development of quantum theories of scattering. Nonetheless, as I hope to have shown in this chapter, their work is highly significant for understanding research on quantum physics based on the correspondence principle during the 1920s.

As we have seen, Franck and Hund's work followed the pattern of *transformation through implementation.* Beginning in October 1922, Franck and Hund integrated the principle into their research. While Franck and Hund's work had been focused almost exclusively on a classical explanation of the "argon effect," this integration turned the Ramsauer effect into a quantum phenomenon. Recognizing the correspondence principle as a theoretical resource did not lead to a new explanation of the Ramsauer effect on a straightforward path: First, Franck needed to amalgamate the correspondence argument for scattering developed by Kramers for the continuous X-ray spectrum, the Duane-Hunt threshold and the Ramsauer effect. Second, the subsequent implementation of the correspondence principle was not a mere technicality; it required a conceptualization of scattering in terms of the state-transition model as well as a first adaptation of the correspondence principle.

Even with these shifts, the implementation of the correspondence principle was anything but straightforward. For Hund, a first approximation of the frequencies in the motion of the scattering electron showed that the correspondence principle led to untenable results. Recognizing problems in the correspondence argument, he abandoned the correspondence principle for scattering in his dissertation.

In reaction, Franck tried to adapt the correspondence argument by taking the classical description of scattering as a mere auxiliary for determining the transition probabilities on the basis of the correspondence principle. Thereby he developed a new conception of scattering in terms of the state-transition model, in which the classical description of scattering was abandoned as a representation of the actual motion of the scattering electron. Instead electrons made transitions from one straight line to another. This presents a if not the first conception of scattering

[76] Hund's scientific diary summarized these developments in two entries for the period 14 to 20 December 1922 and 9 January to 4 March 1923 (Hund Papers).

in quantum theory that would become the basis for future quantum theories of scattering.

This reconstruction allows us to probe some of the inner workings of the dissemination of the correspondence principle and its *transformation through implementation*.

Adoption: Motivations for Taking up the Correspondence Principle

First, the case of Franck and Hund presents a prime example for the role of personal communication with Bohr and Kramers in the adaptation of the correspondence principle: A personal encounter with Bohr and Kramers in Copenhagen was crucial for Franck's transfer of the correspondence approach from X-ray spectroscopy to electronic collisions in gases. As such, the case of Franck and Hund shows what exactly was being transferred through personal communication in the case of the correspondence principle. In this respect, as we have seen, Franck picked up little more than the general idea of making a correspondence argument for scattering. This general idea did not lead to the new explanation of the Ramsauer effect, however. Rather than explicating ideas inherent in Kramers' work, Franck tentatively extended Kramers' speculations by combining the known features of the continuous X-ray spectrum, the chemical properties of noble gases, and the Ramsauer effect. Moreover, Franck's transfer of the correspondence argument did not involve detailed information on how to actually make the correspondence argument. Franck and Hund adopted a conception of scattering in the state-transition model that was decisively different from Kramers'. In turn, Hund encountered different problems when implementing the principle and tried to solve them by adapting the correspondence principle. As such, the case of Franck and Hund presents a central insight into the relation between the transfer of the correspondence principle within the quantum network and its integration into a new research field: personal communication between Franck, Bohr, and Kramers was central for motivating the new approach to the Ramsauer effect; however, it had little impact on the integration of the principle into Franck and Hund's research and the development of their argument.

Implementation: Preconditions for Making Correspondence Arguments

Second, the case of Franck and Hund further allows us to probe the conceptual preconditions for making correspondence arguments. As we have seen, Franck and Hund took up the correspondence principle in a situation in which the description of the physical problem was not based on quantum theory. Franck presumably developed his new interpretation of the Ramsauer effect without thinking about scattering in terms of the state-transition model. The general idea of making correspondence arguments was thus independent of the state-transition model—actually making a correspondence argument, however, was not. As soon as Hund tried to implement the correspondence principle, he realized that it was necessary to identify the motion in the stationary states. In this identification, the theoretical framework of Bohr's quantum theory of multiply periodic systems did not play a role (and would have been close to useless); rather, Hund interpreted the purely classical description of scattering in terms of the state-transition model and arrived at a first qualitative description of scattering in quantum theory.

Adaptive Reformulation: Implications for the Formulation of the Correspondence Principle

Finally, the case of Franck and Hund provides another example for what *transformation through implementation* meant for the formulation of the correspondence principle. As we have seen, there were two adaptations of the correspondence principle in Franck and Hund's case. The first was Hund's identification of the corresponding motion, which was associated with the transition rather than with the state. The second was Franck's reinterpretation of the classical spectrum within the correspondence argument, in which the spectrum provided an approximation to the quantum theoretical spectrum, but was otherwise decoupled from the description of scattering in the state-transition model. For the Göttingen community, as we have seen, these conceptual adaptations resolved the main problems encountered in the attempt to implement the principle. These adaptations changed the correspondence principle and went hand in hand with the development of a new conceptualization of scattering in terms of the state-transition model.

At the same time, Franck and Hund's *transformation through implementation* remained confined to a general, conceptual level with respect to both the correspondence principle and the state-transition model for scattering. An adaptation of the calculational techniques did not take place, despite the fact that the quantitative implementation of the correspondence argument never squared well with the experimental data and even led to contradictions. Especially after Franck's adaptation of the description of the scattering process, conflicts between theoretical predictions and empirical results no longer carried much weight. While the classical spectrum had been unambiguously associated with the motion of the scattering electron in Hund's first implementation, this was no longer the case after Franck's adaptation. In the latter, any continuous classical spectrum could be taken as an approximation for the quantum spectrum without making any statements about the actual motion. Making the correspondence relation between radiation and motion more flexible allowed Franck to discard quantitative discrepancies between theory and experiment.

Chapter 6
That I Cannot Conceive of After the Results of Your Dissertation: Fritz Reiche and the F-sum Rule

This chapter discusses the work of Fritz Reiche on the quantum theory of radiation and his application of the correspondence principle in the context of dispersion theory. Following his private correspondence with Kramers in 1923 and 1924, I reconstruct Reiche's attempts to determine transition probabilities on the basis of the correspondence principle and show how they led to the formulation of a relation among transition probabilities, which came to be known as the f-sum rule or the Thomas-Reiche-Kuhn sum rule.[1]

Like the previous case studies, this chapter can be read in two interconnected ways. Within the overall framework of this book, it presents the final case study of the principle's *transformation through implementation* by physicists outside of Copenhagen. It analyzes how Reiche implemented the principle within his work, how he recognized problems, and, finally, how he adapted the principle. At the same time, the chapter reconstructs the emergence of the Thomas-Reiche-Kuhn sum rule and provides a detailed analysis of Reiche's changing approach to the quantum theory of radiation.

As such, the chapter aims to contribute to the extensive scholarship on the history of dispersion theory in the old quantum theory, in which Reiche's considerations leading to the formulation of the f-sum rule have not played a role.[2] This chapter

---

[1]My analysis of Reiche's work with the correspondence principle is based first of all on his correspondence with Hendrik Antoon Kramers (AHQP 8b.9), in which he developed his arguments. It is complemented by Reiche's published articles: Ladenburg and Reiche (1923, 1924), Reiche and Thomas (1925), Reiche (1926a,b, 1929), and Rademacher and Reiche (1927). Focusing on Reiche's work, my reconstruction keeps the discussion of the general development of dispersion theory to a minimum. The quantum theory of dispersion developed by Kramers, Born, and Van Vleck is discussed only as far as it is necessary to understand Reiche's work.

[2]These considerations dropped out of sight as Jammer, Darrigol and Duncan and Janssen analyzed the dispersion theories of Ladenburg and Reiche, Kramers, Born, Van Vleck, and Kramers and Heisenberg in detail and argued that these dispersion theories paved the way to Heisenberg's *Umdeutung*. See Jammer (1966, 181–195), van der Waerden (1968, 10–18), Darrigol (1992, 224–

M. Jähnert, *Practicing the Correspondence Principle in the Old Quantum Theory*, Archimedes 56, https://doi.org/10.1007/978-3-030-13300-9_6

follows Reiche's pathway in research on radiation theory, which he developed within Ladenburg's research program on optical dispersion. The analysis of Reiche's work shows that it centered on a research problem, which was significantly different from the one followed by Kramers and others at the time. Whereas Kramers tried to translate the classical dispersion formula into a quantum dispersion formula, Reiche focused on the determination of the number of dispersion electrons. Working on this problem, he implemented the correspondence principle as a theoretical tool and formulated the f-sum rule as he ran into the pitfalls of integrating the correspondence principle into Ladenburg's approach.

Reiche's work highlights the dynamic interaction between the correspondence principle and the theoretical and empirical domain into which it was integrated. Probing the process of *transformation through implementation*, I will begin in Sects. 6.1 and 6.2 with a discussion of Ladenburg's work on optical dispersion and Reiche's first correspondence argument in 1923, which he developed within this context. As will be shown in Sect. 6.3, Reiche's implementation based on Kramers' *Zwischenbahn* model became problematic by the end of 1923, when Reiche extended the considerations on the number of dispersion electrons and obtained a relation among the transition probabilities associated with a particular state.

Section 6.4 discusses how Reiche eventually dropped the *Zwischenbahn* as a tool for calculation in the spring of 1924 and replaced it by a new method for determining transition probabilities, which emerged from his new relation. This new method, I show, fueled the development of the Thomas-Reiche-Kuhn sum rule as Reiche attempted to establish a connection between his new relation and the correspondence approach to the quantum theory of dispersion.

6.1 Formulating the Problem: Ladenburg, Reiche and the Number of Dispersion Electrons

When Fritz Reiche took the chair for theoretical physics at the *Schlesische Friedrich-Wilhelms-Universität* in Breslau in 1921,[3] he soon joined his old colleague Rudolph Ladenburg in his research program on optical dispersion. This program had just made a breakthrough with Ladenburg's paper "Die

234), Konno (1993), Duncan and Janssen (2007a, 581–597), and Jordi Taltavull (2013, 52–53). In addition, there are a few studies dedicated at least in part to Reiche and his scientific work: Benjamin Bederson and Valentin Wehefritz presented general biographical sketches which focus primarily on Reiche's persecution in Nazi Germany and his late escape to the U.S. in 1941. See Bederson (2005) and Wehefritz (2002). Clayton Gearhart studied Reiche's earlier work on the quantum theory of molecules and contextualized his textbook *Die Quantentheorie. Ihr Ursprung und ihre Entwicklung*. See Gearhart (2010, 2012). At any rate, these works did not discuss Reiche's pathway to the f-sum rule.

[3]For general biographical information, see Bederson (2005) and also Wehefritz (2002).

quantentheoretische Deutung der Zahl der Dispersionselektronen," in which Ladenburg presented his solution to the long-standing problem of the number of dispersion electrons. This problem and Ladenburg's argument provided the background for Reiche's application of the correspondence principle and I will therefore briefly summarize the emergence of this problem and characterize Ladenburg's attempt to solve it.

The number of dispersion electrons had emerged as a problem within classical theories of optical dispersion in the late nineteenth century. These theories, which were based on the classical radiation mechanism linking radiation and motion directly, described optical dispersion as a resonance phenomenon. It resulted from the interaction of matter conceived of as material particles, the ether, ions, or electrons vibrating with some proper frequencies ν_i and an impinging light wave, which caused forced covibrations or *Mitschwingungen* in these particles and led to a relation between the index of refraction n and the frequency of the incoming radiation ν:

$$n(\nu) = 1 + \sum_i \frac{K_i \nu^2}{\nu^2 - \nu_i^2},$$

where ν_i were resonance frequencies coinciding with the proper frequencies of the vibrating particles and K_i was a parameter depending on the properties of matter, which determined the strength of dispersion.[4]

Within this description, the number of dispersion electrons emerged as a problem when the parameter K_i was interpreted as the number of particles that were bound quasi-elastically by the atom. Following the work of Paul Drude, this number should have been identical with the number of valence electrons. In conflict with this assumption, however, several experimental investigations on the values for K_i led to the conclusion that only one in a thousand valence electrons acted as a dispersion electron.[5]

This discrepancy between theoretical predictions and experimental results emerged time and again in research on optical dispersion throughout the late nineteenth and early twentieth century turning the number of dispersion electrons into a major challenge for classical theories of dispersion. While this problem was discussed by physicists working experimentally on optical dispersion within the framework of classical electrodynamics, dispersion provided a challenging problem of a different kind for Bohr and other quantum physicists, who set out to develop the Bohr model after 1913. In connection with the newly emerging quantum theory, the question was first of all whether and how dispersion could be conceptualized in the state-transition model. This was everything but straightforward since the original Bohr model did not describe the interaction of atoms and light with a

[4] Jordi Taltavull (2013, 31–32). For a comprehensive discussion of dispersion in classical radiation theory, see Jordi's forthcoming dissertation.

[5] Jordi Taltavull (2013, 33).

frequency different from the atom's transition frequency, treating only emission and absorption processes. In this situation, Sommerfeld and Debye developed an account which kept the classical model of *Mitschwingungen* in play. They assumed that the impinging light wave still led to covibrations of valence electrons albeit in a different form, namely, by causing covibrations of the entire electronic orbit in the Bohr model. By contrast, Bohr and others believed that such a hybridization of the Bohr model and classical radiation theory was not feasible. Bohr's position, which ultimately came to be accepted, was that electrons in a stationary state would not interact with light waves unless the system made a transition to another state, i.e., when the atom either emitted or absorbed radiation. This meant that the *Mitschwingungen* model could not be integrated into the Bohr model and, more generally, that the Bohr model lacked the conceptual means to describe dispersion.[6]

While Bohr, Sommerfeld and Debye discussed how dispersion could be even conceptualized within the Bohr model, the old problem of the number of dispersion electrons did not play a role. It resurfaced only in 1921 when Ladenburg proposed his quantum theoretical reinterpretation of the number of dispersion electrons.

This reinterpretation was based on an argument, which was of a purely formal character. This formal character is central for understanding Reiche's application of the correspondence principle. It introduced neither a novel conceptualization of the radiation process in quantum theory nor an adaptation of the classical description of *Mitschwingungen* and, overall, left the physical interpretation of the interaction between light and matter aside. Rather, as we will see, Ladenburg developed a metatheoretical comparison of a classical and a quantum theoretical description of the same physical process and ended up with a formal expression mixing quantities of classical and quantum theory.

To develop this comparison, Ladenburg needed to bypass the process of dispersion since it could not be conceptualized in terms of the state-transition model. Instead, he considered the process of absorption and the number of absorbing electrons, which was identical to the number of dispersion electrons in classical radiation theory. Building on this identity, his argument was quite simple. He compared the classical expression for the energy absorbed per second by R electrons with the energy absorbed per second by N quantum systems according to Einstein's quantum theory of radiation. Equating these two expressions, he expressed the number of classical absorption (and hence dispersion) electrons in terms of Einstein's quantum theory.[7]

On the classical side of this comparison, Ladenburg departed from Planck's *Vorlesung zur Wärmestrahlung*,[8] in which Planck had given the mean energy J_{el}

[6]Jordi Taltavull (2013, 34–48).

[7]Note that although they discussed the number of dispersion electrons and the dispersion constant, Ladenburg's and Reiche's considerations focused entirely on this comparison in the case of absorption until March 1924. Their argument did not involve Ladenburg's central assertion that the number of absorbing systems was equal to the number of dispersing systems, and dispersion played no constructive role until they received Kramers' quantum theory of dispersion.

[8]Planck (1921).

emitted per second by R electrons performing damped oscillations with the proper frequency ν_0 (expressed in Gaussian units):

$$J_{el} = \frac{\bar{U}R}{\tau}(\tau = \frac{3mc^3}{8\pi^2 e^2 \nu_0^2}),$$

where τ is the attenuation time of electrons performing damped oscillations and $\bar{U}$ is their mean energy. Using a relation between $\bar{U}$ and the (spectral) radiation density u_0 for oscillators in equilibrium with surrounding radiation derived by Planck:

$$\bar{U} = \frac{3c^3}{8\pi \nu_0^2} u_0,$$

Ladenburg then obtained a classical expression for the emitted energy per second:

$$J_{el} = \frac{\pi e^2}{m} R u_0.$$

In thermal equilibrium, he argued, this expression also gave the energy absorbed by the electrons.[9]

The discussion of the quantum theoretical side of the problem was based on Einstein's quantum theory of emission and absorption.[10] In it, Einstein considered N systems in the states k and i and introduced the transition probabilities a_{ki} for spontaneous emission, as well as the transition probabilities b_{ki} and b_{ik} for emission and absorption induced by a radiation field with the radiation density u_{ik}.[11] In order to secure equilibrium between radiation and matter, Einstein argued, the total number of transitions $k \Leftrightarrow i$ needed to balance each other in thermal equilibrium. Thus the number of induced transitions from a lower to an upper state needed to be equal to the number of spontaneous and induced transitions from an upper to a lower state:

$$N_k a_{ki} + N_k b_{ki} u_{ik} = N_i b_{ik} u_{ik}.$$

Taking up Einstein's equilibrium condition, Ladenburg identified the left-hand side of the equation with the process of emission and the right-hand side with the process of absorption. The energy absorbed per second by a quantum system, he assumed, could thus be obtained by multiplying the right-hand side with energy $h\nu_{ik}$:

$$A_Q = h\nu_{ik} N_i b_{ik} u_{ik}.$$

[9]Ladenburg (1921, 452).

[10]Ladenburg (1921, 453) and Einstein (1916).

[11]Note that I am following Ladenburg's notation here. Einstein had assumed that the radiation density was classical and had not associated it with a particular transition. This difference did not play a role, as Ladenburg also assumed that his u_{ki} was equivalent to the radiation density u_0 of the classical radiation field so that it cancelled out in his comparison.

Following Einstein's theory further, Ladenburg then expressed this equation in terms of the transition probability a_{ki} and equated it with Planck's classical expression for the absorbed energy. Assuming that the radiation densities u_0 and u_{ik} and the radiation frequencies ν_0 and ν_{ik} were identical, he obtained the expression:

$$R = \frac{mc^3}{8\pi e^2 \nu_{ik}^2} N_i \frac{g_k}{g_i} a_{ki},$$

where g_i and g_k are the statistical weights of the respective states. As will be discussed in detail in Sect. 6.3, this argument was anything but unproblematic and became subject to important shifts throughout Ladenburg and Reiche's work. In 1921, however, Ladenburg did not problematize the assumptions underlying his argument and focused primarily on his new expression for the number of dispersion electrons.

Exploring the new expression, Ladenburg faced questions of different kinds. On the one hand, the question was whether the new relation would actually account for the low number of dispersion electrons found in experiment. On the other hand, due to the formal nature of Ladenburg's argument, it remained unclear how the expression for the number of dispersion electrons was to be interpreted in terms of a physical process.

These two aspects and their relative importance for Ladenburg are central for understanding the point of departure for Reiche's correspondence argument. Whereas the secondary literature has put an emphasis on the question of physical interpretation as a starting point for developments leading to the concept of the substitute oscillator, Kramers' dispersion theory, and the Bohr-Kramers-Slater theory, we will see that Reiche's correspondence argument first of all contributed to the quantitative exploration of Ladenburg's relation and the attempts to extend and revise Ladenburg's formal argument.

As such, it followed the main line of inquiry of Ladenburg's 1921 paper, in which the question of physical interpretation did not play a significant role. Developing his purely formal argument, Ladenburg refrained from discussing the conceptual implications or the physical interpretation of the new relation for the number of dispersion electrons. He made little more than side remarks in this direction, which remained purely formal, just like his argument. For example, he observed that transition probabilities took "the place of the dispersion constant R" or that "the number of dispersion electrons R [had] the meaning of [a] product in quantum theory."[12] In light of these formal comments, the physical interpretation

[12]For the first quote, see Ladenburg (1921, 451). "Infolgedessen ist es [...] diese W[ahrscheinlichkeit] der spontanen Rückgänge (multipliziert mit dem in jener Einsteinschen Beziehung auftretenden Verhältnisse der statistischen Gewichte der Quantenzustände), die an die Stelle jener Dispersionskonstante R tritt [...]" For the second quote, see Ladenburg (1921, 468). "Die [...] Zahl R der Dispersionselektronen hat nach der Quantentheorie die Bedeutung des Produktes $N_i a_{ki} \frac{g_k}{g_i} \frac{mc^3}{8\pi e^2 \nu_{ik}^2}$."

of the number of dispersion electrons remained deeply unclear and Ladenburg acknowledged that they did not have "a definite meaning in quantum theory."[13]

This conceptual vacuum and uncertainty remained inconsequential, however, as Ladenburg refrained from problematizing the conceptual implications of his argument or from speculating about a physical interpretation of his relation. Instead he focused on the quantitative determination of the number of dispersion electrons. Exploring whether his new relation could account for the low number of dispersion electrons found in experiment, he realized that it provided a formal link between the dispersion constant on the one hand and the state-transition model on the other. This link presented a means to translate experimental values for the dispersion constant R into values for the statistical weights and transition probabilities. Ladenburg then compared these values obtained from experiments on dispersion and magneto-rotation with values obtained in different empirical domains, like the intensity of spectral lines or the intensity decay along canal rays,[14] and found these values to be consistent with each other. Consequently, Ladenburg concluded that this relation between the transition probabilities and the dispersion constant yielded meaningful results.[15]

6.2 Implementing the Correspondence Principle: Reiche, Ladenburg and the Determination of Transition Probabilities

The previous assessment of Ladenburg's approach to the number of dispersion electrons provides the starting point for the reconstruction of Reiche's correspondence argument. As we have seen, Ladenburg's argument was purely formal and left conceptual clarification and physical interpretation aside for the time being. It aimed primarily at the quantitative determination of the number of dispersion electrons and provided a possible solution to this long-standing problem of dispersion theory.

Joining Ladenburg's research program, as we will see in the following, Reiche extended this highly formal approach. Rather than attempting to solve the problem of physical interpretation, he used the correspondence principle to calculate

[13]Ladenburg (1921, 454). "Indem wir die Gleichung (2) als die Definition für die experimentell bestimmbare Größe (2) ansehen, die natürlich in der Quantentheorie keine bestimmte Bedeutung hat, erhalten wir durch Gleichsetzen der beiden Gleichungen (2) und (6) die quantentheoretische Deutung der Dispersionskonstante R [. . .]."

[14]Examples considered by Ladenburg were (a) the transition probabilities for excited hydrogen, which were obtained from the measurements on anomalous dispersion, and the decay experiments of Stark and Wien. See Stark (1916) and Wien (1919, 1921). Moreover, Ladenburg considered the transition probabilities for the sodium doublet obtained from intensity measurements and measurements on dispersion. See Ladenburg (1921, 456–464).

[15]Ladenburg (1921, 468). Note that Ladenburg's consistency check was limited to showing that the different values for transition probabilities were of the same order of magnitude.

transition probabilities. Thereby he sought to provide a theoretical counterpart to Ladenburg's consistency check based on experimental data obtained from various phenomena.

The main source for this implementation is Reiche and Ladenburg's paper "Absorption, Zerstreuung und Dispersion in der Bohrschen Atomtheorie," which they wrote as a contribution to the *Bohrheft* of the *Naturwissenschaften* in 1923.[16] The issue was intended as a collection of overview articles on the development of quantum theory related to the Bohr model in celebration of its ten-year anniversary. Ladenburg and Reiche's paper, however, went beyond a presentation of the state of the art as they revised and extended Ladenburg's initial argument considerably.

This put Ladenburg and Reiche in a peculiar position. They started from general expositions of Einstein's theory of emission and absorption and the correspondence principle and then revised and extended Ladenburg's initial argument for the number of dispersion electrons as they shifted towards more specific questions.

In this manner, Reiche and Ladenburg first of all reworked Ladenburg's original argument. As we have seen in the previous section, Ladenburg had initially equated the classical and quantum theoretical expressions for the energy absorbed by a radiating system per second:

$$\frac{\pi e^2}{m} R u_0 = h\nu_{ik} N_i b_{ik} u_{ik}.$$

This equation connected the classical expression for the absorption energy of R harmonic oscillators with the energy of quantum transitions from a lower to a higher energy level. In other words, Ladenburg assumed that the classical and quantum expressions described a process in which the energy of the radiating system increased due to the interaction between the system and the external field.

Focusing on the resulting expression for the number of dispersion electrons, as we have seen, Ladenburg took this interpretation of absorption for granted without further discussion. This changed in Ladenburg and Reiche's 1923 paper. Rather than looking at processes in which the energy increased, they considered the reaction of a radiating system to an external radiation field in general:

> According to Einstein's assumptions, the effect of external radiation on a quantum atom corresponds to that which a classical oscillator receives from an impinging wave. If its frequency is only a little or no different from the proper frequency of the oscillators, the reaction of the oscillator consists in an increase or decrease of its energy, depending on the phase difference between the external wave and the motion of the oscillator. In analogy, Einstein assumes that the atom in state i possesses a probability, characterized by the factor b_{ik}, to go over to the higher state k under the absorption of the energy $h\nu$ from the impinging wave ("positive irradiation"), and that the atom in state k possesses another probability (b_{ki}) to return to state i under the influence of the external wave ("negative irradiation").[17]

---

[16]Ladenburg and Reiche (1923). For the following correspondence argument, see Ladenburg and Reiche (1923, 586–587).

[17]Ladenburg and Reiche (1923, 586). "Ebenso entspricht nach Einsteins Annahmen die Wirkung einer äußeren Strahlung auf ein Quantenatom derjenigen, die ein klassischer Oszillator durch eine

This textbook presentation had important consequences for Ladenburg's argument. Contrary to the initial assumption, Ladenburg and Reiche realized that Planck had calculated the average absorption of a harmonic oscillator, which included both the increase and decrease in energy corresponding to different phase relations. In quantum theory, one thus also needed to consider all transitions which were induced by the radiation field. In addition to the transitions from the lower state i to the higher state k, one thus also needed to take the inverse transition from k to i into account. Following this assessment, Ladenburg and Reiche thus arrived at a revision of Ladenburg's initial expression:

$$\frac{\pi e^2}{m} R u_0 = h\nu_{ik}(N_i b_{ik} - N_k b_{ki}) u_{ik}.$$

Including the term $(N_k b_{ki}) u_{ik}$, which had been absent from Ladenburg's initial argument, Ladenburg and Reiche's revision thus differed from the initial argument both with respect to the conception of the absorption process and the resulting formula. This difference remained inconsequential for their following exposition, however, since they found a way to retrieve Ladenburg's initial expression. Using the relations of Einstein's radiation theory as before, Reiche and Ladenburg rewrote the expression as:

$$R\frac{\pi e^2}{m} = N_i h\nu_{ik} b_{ik} \frac{1}{1 + \frac{c^3 u_{ik}}{8\pi h\nu_{ik}}},$$

and argued that, under certain conditions, the factor $\frac{1}{1+\frac{c^3 u_{ik}}{8\pi h\nu_{ik}}}$ "can be practically set to 1" so that the expression reduces to Ladenburg's initial formula.[18]

auffallende Welle erfährt. Wenn deren Frequenz sich von der Eigenfrequenz des Oszillators nur wenig oder gar nicht unterscheidet, besteht die Reaktion des Oszillators in einer Vermehrung oder einer Verminderung seiner Energie, je nach Phasenunterschied zwischen der äußeren Welle und der Bewegung des Oszillators. In Analogie hierzu nimmt Einstein an, daß das Atom im Zustand i eine durch den Faktor b_{ik} charakteristische Wahrscheinlichkeit besitzt, unter Aufnahme der Energie $h\nu$ aus der auffallenden Welle in den höheren Zustand k überzugehen ('positive Einstrahlung'), und daß ein Atom im Zustand k eine andere Wahrscheinlichkeit (b_{ki}) besitzt, unter dem Einfluss der äußeren Welle in den Zustand i zurückzukehren ('negative Einstrahlung')."

[18]Ladenburg and Reiche (1923, 588). "Der Faktor [...] kann bei nicht zu hoher Strahlungsdichte am Ort der Atome und bei mäßiger Temperatur der Gasschicht praktisch gleich 1 gesetzt werden." The only discussion of problems with Ladenburg's formula is given by Duncan and Janssen (2007a, 586–587). They argue that Ladenburg and Reiche "derived a result for emission consistent with the correspondence principle (i.e., merging with the classical result in the limit of high quantum numbers), but their attempts to derive similar results for absorption and dispersion were unconvincing." They diagnosed that the reason for this failure was that they "did not limit their 'correspondence' arguments to the regime of high quantum numbers [...]. These problems invalidate many of the results purportedly derived from the correspondence principle in their paper." The approximation responsible for Ladenburg and Reiche's result, however, is not a result

In addition to revising the initial argument, Reiche and Ladenburg went further in the quantitative exploration of Ladenburg's relation. In this direction, Ladenburg had initially used his relation to translate experimental values for the dispersion constant into statements about transition probabilities and then compared them with values for the transition probabilities obtained in different experimental domains. Reiche and Ladenburg now engaged in the calculation of transition probabilities and turned to the correspondence principle for the first time.[19]

This calculation presented a theoretical counterpart to Ladenburg's comparison of experimental data and thus provided another consistency check for Ladenburg's initial argument. Developing their argument as part of their overview article, Reiche and Ladenburg began by summarizing Bohr's correspondence argument:

> The correspondence principle builds on the fact that in the region of large quantum numbers n [...] the frequency emitted in a transition $n' \rightarrow n''$ coincides with the harmonic component $(n' - n'')\omega$ in the Fourier representation of the motion of the electron, where ω is the orbital frequency. [...] For large frequencies, that is, small quantum numbers, such a coincidence is of course impossible because the orbiting frequencies in the neighboring orbits are totally different, while the frequency ν always depends on the energy difference of both orbits. It is possible to show, however, that in this case, too, ν can be represented as a certain *mean value* over the respective harmonic component $(n' - n'')\omega$ of the initial and the final orbit as well as a continuous series of imagined *Zwischenbahnen*. We say with *Bohr* that the frequency ν emitted in the transition $n' \rightarrow n''$ "corresponds" to the harmonic component $(n' - n'')\omega$.[20]

---

of an unwarranted extrapolation beyond the high quantum number limit. As will be discussed in the following, it was in conflict with expectations on this limit, and it was this conflict that became the source for the revision of the argument.

[19]That Reiche was the driving force behind the theoretical calculations may be seen from a letter by Reiche to Kramers, 9 May 1923 (AHQP 8b.9). In it, Reiche discussed the details of the calculations. Ladenburg had been interested in Kramers' dissertation before and asked for a separate copy of the dissertation, which was not available in Breslau. He had already mentioned that the transition probabilities could be determined from the correspondence principle in his 1921 paper. Actual calculations based on Kramers' *Zwischenbahn*, however, did not appear in his work and are not part of the correspondence between Ladenburg and Kramers. For Ladenburg's interest in the correspondence principle, see Ladenburg to Kramers, 26 June 1920 (AHQP 8a.9) and Ladenburg (1921, 455).

[20]Ladenburg and Reiche (1923, 586). "Das Korrespondenzprinzip knüpft an die Tatsache an, daß im Gebiet großer Quantenzahlen n [...] die bei einem Übergang $n' \rightarrow n''$ ausgestrahlte Schwingungszahl übereinstimmt mit der harmonischen Komponente $(n' - n'')\ \omega$ in der Fourierzerlegung der Bewegung des Elektrons, wobei ω die Umlaufzahl ist. [...] Bei großen Schwingungszahlen, d.h. kleinen Quantenzahlen ist eine solche Übereinstimmung natürlich nicht möglich, da dort die Umlaufzahlen in benachbarten Bahnen ganz verschieden sind, während die Schwingungszahl ν stets von der Energiedifferenz beider Bahnen abhängt. Es lässt sich jedoch leicht zeigen, daß auch in diesem Falle ν als ein bestimmter *Mittelwert* über die entsprechenden harmonischen Komponenten $(n' - n'')\ \omega$ der Anfangs- und Endbahn sowie einer kontinuierlichen Reihe gedachter Zwischenbahnen darstellbar ist. Wir sagen mit *Bohr*, daß die beim Übergang $n' \rightarrow n''$ ausgesandte Schwingungszahl ν mit der harmonischen Komponente $(n' - n'')\ \omega$ in der Bewegung des Elektrons 'korrespondiert'."

This paraphrase of Bohr's correspondence argument took the reader from the equality of mechanical frequencies and radiation frequencies in the limit of high quantum numbers to the frequency correspondence for all quantum numbers and Kramers' *Zwischenbahn* model as a solution to the initial-final-state problem.

In the following, Reiche and Ladenburg extended the correspondence idea from frequencies to intensities and transition probabilities and introduced the idea of determining transition probabilities from the principle. Using Kramers' *Zwischenbahn* model, they argued, one compared the classical and the quantum theoretical description of the energy emitted per second. A quantum system with frequency ν_0 would radiate with the energy $S_q = a_{ki} h\nu_0$. This energy had to be equal to the energy radiated according to classical radiation theory $S_{kl} = \frac{16\pi^4 e^2 \nu^{*4}}{3c^3} C^2$, where ν^* and C are the frequency and Fourier coefficient of "an electron [performing a] purely harmonic oscillation." As the radiation frequency ν_0 was identical to the frequency ν^* of the *Zwischenbahn*, this comparison gave the transition probability as:

$$a_{ki} = \frac{16\pi^4 e^2 \nu_0^3}{3c^3 h} \bar{C^2},$$

where the Fourier coefficient of the *Zwischenbahn* $\bar{C^2}$ is the average taken over the initial and the final state of the transition.[21]

Up to this point, Ladenburg and Reiche thus again recapitulated the correspondence principle on a textbook level and presented the general procedure for determining the transition probabilities on its basis. In the following, they introduced values for the transition probabilities calculated from the principle into Ladenburg's relation for the number of dispersion electrons and thus connected the correspondence principle with Ladenburg's initial argument. This meant to integrate the correspondence principle into Ladenburg's initial comparison of classical and quantum theory or, conversely, to show that Ladenburg's argument was a "consequence of the *Bohrian correspondence principle*."[22]

In so doing, Reiche and Ladenburg considered a specific type of quantum system: the linear harmonic oscillator and the spatial harmonic oscillator, i.e., the harmonic oscillator with one or three degrees of freedom. Focusing on the harmonic oscillator first of all meant to narrow the scope of the argument considerably. Whereas the comparison of classical and quantum theory in the initial and the revised version of the argument referred to radiating systems in general, the harmonic oscillator presented a system with a set of unique features. These features turned the harmonic oscillator into a prime testing case for Reiche's considerations and allowed him to investigate the implications of his implementation of the correspondence principle and its relation to Ladenburg's argument.

---

[21] Ladenburg and Reiche (1923, 586–587).

[22] Ladenburg and Reiche (1923, 586). "Wir sehen in dieser Analogie der klassischen und quantenmäßigen Emissions- und Absorptionsgesetze die Folge des *Bohrschen Korrespondenzprinzips*."

The harmonic oscillator could function in this capacity first of all because it was a harmonic system. Its transitions from a particular state were thus limited to adjacent states. Moreover, unlike other harmonic systems, the transitions frequencies of the harmonic oscillator are the same for all transitions and are equal to the mechanical frequency. Finally and most importantly, its transition probabilities could be expressed explicitly in terms of quantum numbers in a digital and strikingly simple form. As Reiche and Ladenburg argued in their paper, the Fourier coefficient C was nothing but the oscillator's displacement from its equilibrium position; it could be determined from the energy of the oscillator ($E = nh\omega = 2\pi^2 mC^2\omega^2$):

$$C^2 = \frac{h}{2\pi^2 m\omega} n.$$

Ladenburg and Reiche then introduced this Fourier coefficient into the *Zwischenbahn* formula discussed above. Using the equality of the mechanical frequency ω and the radiation frequency ν_0 again, they obtained the transition probability:

$$a_{n+1,n} = \frac{8\pi^2 e^2 \nu_0^2}{3mc^3} \bar{n} = \frac{1}{3} \frac{\bar{n}}{\tau},$$

where the factor $\frac{8\pi^2 e^2 \nu_0^2}{mc^3}$ is the reciprocal of the attenuation time of a classical harmonic oscillator τ, which had already played a role in Ladenburg's considerations in 1921. $\bar{n}$ is the average quantum number of the *Zwischenbahn*, for which Reiche and Ladenburg considered the arithmetic mean value $n + \frac{1}{2}$ to be the simplest possible average.[23]

These features alone turned the harmonic oscillator into a prime testing case for Reiche's further considerations on the relation between the correspondence principle and Ladenburg's argument and became extremely important for the subsequent development of his work. In addition, the harmonic oscillator became even more important for Reiche since his teacher Max Planck had determined its transition probabilities on the basis of a different argument and arrived at a similar, yet different result. In his *Vorlesung über Wärmestrahlung* Planck had compared a number of systems in a quantum state n, which made spontaneous transitions in a certain period of time, with the same number of systems in this quantum state, which emitted radiation in the same time according to classical radiation theory.[24]

[23]Reiche and Ladenburg preferred the arithmetic mean value $n + \frac{1}{2}$ for its simplicity, although they knew that Kramers' dissertation had advocated the more complicated logarithmic mean value $n = \frac{1}{e}\frac{(n+1)^{n+1}}{n^n}$. See Ladenburg and Reiche (1923, 589) and Reiche to Kramers, 9 May 1923 (AHQP 8b.9).

[24]Ladenburg and Reiche cited Planck (1921, 179). Planck had compared the "number of those molecules N_n lying in the elementary region n, which perform an act of emission in a certain time τ:"

$$N_n A_n \tau$$

In this manner, Planck found the transition probability of the harmonic oscillator radiating with its proper frequency ν:

$$A_n = \frac{8\pi^2}{3}\frac{e^2\nu^2}{mc^3}n.$$

While differing in the specifics of their arguments,[25] Ladenburg and Reiche understood that Planck's approach was closely connected to their argument. Both compared the classical description of the radiation process and the corresponding quantum transition. The only difference was whether one considered the quantum system in the initial state or in the *Zwischenbahn* as emitting energy according to classical radiation theory. From this perspective, it was not difficult to recast Planck's argument in terms of the correspondence argument. Planck's argument suggested considering the initial excited state of the transition $n+1$ rather than the average value of the *Zwischenbahn*, otherwise the argument remained the same: the Fourier coefficient was determined from the energy of the harmonic oscillator just as before and Planck's result for the transition probability of the harmonic oscillator was recovered from the correspondence relation:

$$a_{n+1,n} = \frac{1}{3}\frac{n+1}{\tau}.$$

Planck's calculation implied a different solution to the initial-final-state problem. Taken seriously from a conceptual point of view, this argument might have suggested several questions or even appeared in flat contradiction with the state-transition model. For example, one might ask whether it makes sense to consider the classical radiation energy of a non-radiating stationary state. In addition, one might ask how it is possible to associate this energy with an actual transition probability. Considering these questions, one might then be pressed to assume that the transition

---

with the number of molecules whose "state variables g decreases by r_a in the time τ:"

$$NW(g)r_a.$$

In this equation, N is the total number of molecules and $W(g)$ is the "Verteilungsdichte" [distribution density]. r_a is the energy emitted per frequency in the time τ according to classical radiation theory. Planck had calculated its value for a quantized harmonic oscillator as:

$$r_a = \frac{8\pi^2}{3}\frac{e^2\nu^2 hn}{mc^3}\tau.$$

[25]Planck gave his arguments in terms of the numbers of systems that made a transition or radiated classically, Ladenburg and Reiche's argument was based on considerations of the radiation energy of an individual system.

probabilities are properties of a particular initial state instead of depending on the initial and the final states.

It is a fact worth noting that, just like Ladenburg in his 1921 paper, Reiche and Ladenburg did not discuss these conceptual issues. Rather they turned to Ladenburg's expression for the number of dispersion electrons and compared the predictions resulting for the transition probabilities $a_{ki,\mathrm{Zw.}}$ calculated from the *Zwischenbahn* and its counterpart $a_{ki,\mathrm{ini}}$ determined from the initial state. This comparison would become crucial in the following, when Reiche was forced to decide which of the two values for transition probabilities was correct. In their 1923 paper, however, Ladenburg and Reiche did not feel the need to make such a decision and remained uncommitted as they developed the consequences of the two alternatives.

Ladenburg and Reiche made this comparison for both the linear and the spatial harmonic oscillator. While the transition probabilities were the same for both, these two systems were markedly different in another respect. For the linear harmonic oscillator, the number of mechanical frequencies was equal to its degrees of freedom so that it presented a non-degenerate system. The spatial harmonic oscillator, by contrast, was a degenerate system, because its number of degrees of freedom was greater than the number of mechanical frequencies.

As a consequence, the statistical weights were identical for each state for the linear harmonic oscillator, whereas the statistical weights of the spatial harmonic oscillator were different. Ladenburg's relation for the number of dispersion electrons thus led to different results in both cases: For the linear oscillator, only the transition probabilities had to be taken into account, so that the number of dispersion electrons obtained for the transition probabilities $a_{ki,\mathrm{Zw.}}$ differed only slightly from the number obtained when considering the transition probability $a_{ki,\mathrm{ini}}$:

$$R = N_i \tau a_{ki,Zw.} = \frac{1}{3}(n + \frac{1}{2})N_i$$

$$R = N_i \tau a_{ki,ini} = \frac{1}{3}(n + 1)N_i .$$

For the spatial harmonic oscillator, the situation was different. Since the statistical weights depend on the respective states, according to the formula $g_n = \frac{1}{2}(n + 2)(n + 1)$, the transition probabilities determined from the initial state and the *Zwischenbahn* yielded rather different results: When the quantum number $n + 1$ of the initial state is used, it cancels out against the denominator of the ratio of the statistical weights:

$$\begin{aligned} R &= \tfrac{1}{3}\tfrac{(n+3)(n+2)}{(n+2)(n+1)}(n + 1)\ N_i \\ &= \tfrac{1}{3}(n + 3)\ N_i . \end{aligned}$$

For the *Zwischenbahnen* this is not the case, so that Reiche and Ladenburg instead obtained the relation:

$$\begin{aligned} R &= \tfrac{1}{3}\tfrac{(n+3)(n+2)}{(n+2)(n+1)}\left(n+\tfrac{1}{2}\right) N_i \\ &= \tfrac{1}{3}\tfrac{n+3}{n+1}\left(n+\tfrac{1}{2}\right) \quad N_i\,. \end{aligned}$$

As we will see in the next section, these different expressions for the spatial harmonic oscillator became crucial for Reiche's ongoing considerations. In writing their joint paper, however, Reiche and Ladenburg considered their formulas only for the special case of transitions between the first excited and the ground state and therefore took n equal to zero.[26] In this much simplified form, their relation between the classical number of dispersion electrons and the number of atoms reduced to:

	Linear harmonic oscillator	Spatial harmonic oscillator
$a_{ki,ini} = \frac{1}{\tau}$	$R = \frac{1}{3}N_i$	$R = N_i$
$a_{ki,Zw.} = \frac{1}{2\tau}$	$R = \frac{1}{6}N_i$	$R = \frac{1}{2}N_i$

In other words, in each case the classical number of dispersion electrons R associated with the spectral frequencies ν_{ik} was proportional to N_i quantum systems in the lower state of the transition.[27] In this form, the difference between the various predictions remained insignificant and led to the implicit conclusion that the determination of transition probabilities based on the correspondence principle led to acceptable results in any case.

Ersatzoszillatoren *and the Interpretation of Ladenburg's Relation*

Apart from the revision of Ladenburg's formal argument and the incorporation of theoretical values for the transition probabilities obtained from the correspondence principle, Ladenburg and Reiche also considered the physical interpretation of Ladenburg's relation. This consideration, as mentioned in the previous section, led to the introduction of the *Ersatzoszillator* concept, which plays a central role in the secondary literature.

[26] Ladenburg and Reiche (1923, 589).

[27] In the specific case, this means that the respective classical number of dispersion electrons is associated with the number of atoms in the ground state. This interpretation follows from the definitions of the transition process. While Ladenburg and Reiche did not dwell on this when making their calculations, they identified the number of quantum systems $N_{i=0}$ with "a singular spatial quantum oscillator *in the lowest quantum state*" in a later part of their paper. (Ladenburg and Reiche 1923, 591, my emphasis).

Going beyond Ladenburg's initial grappling with the question, Reiche and Ladenburg interpreted Ladenburg's equation as a relation between a number of classical oscillators and the number of quantum systems. As they argued, the relation implied that "R oscillators have to be distributed equally among N points in space."[28] This distribution, Ladenburg and Reiche suggested, could also be understood as the assignment of a classical oscillator with a specific electric moment to each of the N quantum atoms:

> We can express this distribution [...] formally in such a way that we imagine a classical oscillator of charge xe and mass xm in each of the N points in space, whereas, like above $x = \frac{R}{N}$, or in other words: each of these N '*Ersatzoszillatoren*' should have an electric moment under the influence of an external wave, whose amplitude is x times as large as that of a classical oscillator of charge e and mass m.[29]

Ladenburg and Reiche thus assigned a classical oscillator to each transition of a quantum system and thereby offered a first interpretation of Ladenburg's expression for the number of dispersion electrons in terms of a spatiotemporal model. The *Ersatzoszillatoren* could be thought of as dispersing light according to classical radiation theory, only their oscillator strength was set by the transition probabilities and the correspondence principle.

On a structural level, this model interpretation was similar to Kramers' *Zwischenbahn*: in both cases, radiation processes—dispersion or emission and absorption—were described in terms of a classical model, while the actual radiation process was supposed to be governed by the state-transition model. This similarity, however, did not play a role in Reiche and Ladenburg's argument. For them, the two models existed separately and were used to describe different aspects of the radiation process: The *Ersatzoszillator* allowed them to attach a classical model to the quantum theoretical determination of the number of dispersion electrons. The *Zwischenbahn* was the basis for calculating the transition probabilities.

In this way, Reiche and Ladenburg had come to a position from which Ladenburg's initial argument on the number of dispersion electrons, and with it the process of dispersion itself, no longer appeared to be separate from the quantum theory of radiation: Just as the emission and absorption of spectral lines were the result of transition processes in the atom, dispersion appeared to be the result of the quantum theoretical radiation process, which was described on the basis of the correspondence principle:

> [...] *the probability of the possible quantum transitions* [...] is a measure not only for the contribution of the quantum absorption but also *for the contribution of scattering and*

---

[28]Ladenburg and Reiche (1923, 590). "[Es ist zu beachten, daß] die R Oszillatoren auf die N Raumpunkte gleichsam aufgeteilt werden müssen."

[29]Ladenburg and Reiche (1923, 590). "Wir können diese Aufteilung [...] formal so ausdrücken, daß an jedem der N Raumpunkte ein klassischer Oszillator von der Ladung xe und der Masse xm zu denken ist, wobei, wie oben $x = \frac{R}{N}$ ist, oder mit anderen Worten: jeder dieser N 'Ersatzoszillatoren' soll unter dem Einfluß der äußeren Welle ein elektrisches Moment annehmen, dessen Amplitude x mal so groß ist wie die eines klassischen Oszillators von der Ladung e und der Masse m."

dispersion [...]. We do not assume that the transitions actually come about under the influence of the wave ν. Rather we have to imagine, following the correspondence principle [...], that this probability for the realization of the quantum transitions is determined by the amplitude of the harmonic component of the motion corresponding to ν_0, that is, determined by the configuration of the atom, and it is conceivable that this amplitude not only regulates the rate of the actual quantum transitions [...] but also the reaction of the atom to waves of arbitrary frequency.[30]

Interim Conclusion

This section showed how Reiche implemented the correspondence principle in the context of Ladenburg's work on the number of dispersion electrons. Seeking to provide a theoretical counterpart to Ladenburg's original consistency check based on a comparison of experimental data, Reiche used the correspondence principle as a tool for determining transition probabilities and thereby integrated the principle into Ladenburg's initial argument without adapting either of them. Rather he used the principle in its original formulation and followed Kramers' model of the *Zwischenbahn*. While the details of this approach were still uncertain, it was applicable in principle to all kinds of quantum systems and led to consistent results for the harmonic oscillator as a particularly simple case.

6.3 Recognizing Problems: Reiche and the Comparison of Absorption in Classical and Quantum Theory

In the course of 1923/1924, Reiche's perspective on Ladenburg's argument in relation to the correspondence principle underwent a considerable transformation. As I will show in this section, Reiche's integration of the correspondence principle into Ladenburg's approach led to discrepancies between theory and experiment as well as to theoretical inconsistencies. In light of these challenges, he revised Ladenburg's argument again and arrived at a new relation between transition probabilities, which allowed him to remove these inconsistencies. At the same time, the new relation presented a challenge for Reiche's implementation of the

[30]Ladenburg and Reiche (1923, 591, emphasis in the original). "[...] *die Wahrscheinlichkeit der möglichen Quantenübergänge* [...] ist ein Maß nicht nur für den Betrag der Quantenabsorption sondern auch *für den Betrag der Zerstreuung und Dispersion* [...]. Dabei nehmen wir nicht etwa an, daß die Übergänge unter Einfluß der Welle ν wirklich zustande kommen. Vielmehr müssen wir uns nach dem Korrespondenzprinzip vorstellen [...], dass diese Wahrscheinlichkeit für das Zustandekommen der Quantenübergänge durch die Amplitude der mit ν_0 korrespondierenden harmonischen Komponente der Bewegung, also durch die Konfiguration des Atoms bestimmt ist, und es ist begreiflich, dass diese Amplitude nicht nur die Häufigkeit der wirklichen Quantenübergänge [...] sondern auch die Reaktion des Atoms auf die Wellen beliebiger Schwingungszahl regelt."

correspondence principle. Reconstructing this development, I will analyze the conditions in which epistemic challenges emerged in Reiche's work, and discuss the dynamic interaction of Ladenburg's comparison of classical and quantum theory and the correspondence principle.

Discrepancies and Challenges

Reiche encountered discrepancies between his implementation of the correspondence principle and experimental data early on, when he worked on his and Ladenburg's paper in 1923. Having found the transition probabilities of the harmonic oscillator, he turned to the hydrogen atom and with it to Kramers' consideration of the Kepler problem. Here, he realized, the transition probabilities determined from the *Zwischenbahn*-model did not match the available experimental data. Seeking an explanation for this discrepancy, Reiche began to study Kramers' dissertation in detail. Eventually he wrote a letter to Kramers, asking for clarification of "additional problems" he and Ladenburg had discovered:

> If one considers for example an electron orbiting on a Kepler orbit [...] to calculate the a_{ki} for a certain transition, does one not have to identify $a_{ki}h\nu$ with the radiation emitted according to classical electrodynamics by an electron on a corresponding harmonic *circular motion*? [...] I am asking this question because it seems to follow from your dissertation p. 46 that the $a_{ki}h\nu$ have to be compared with the radiation of a *linearly* oscillating electron [...] (You are there talking about an electron "performing a *simple harmonic* vibration [...]").[31]

In his answer, Kramers agreed with Reiche and admitted that "the discussion of the respective points was regrettably kept rather short in my dissertation."[32] Following this response, which did not help much to improve agreement with experiment, Reiche came to argue that the discrepancies were due to the complicated

[31]Reiche to Kramers, 9 May 1923 (AHQP 8b.9, emphasis in the original). "Wenn es sich z. B. um ein in der x-y=Ebene, etwa in einer Keplerbahn umlaufendes Elektron handelt (wie z.B. beim ungestörten H=Atom), [...] muss man dann nicht, um a_{ki} für einen bestimmten Übergang zu berechnen $a_{ki}h\nu$ gleichsetzen derjenigen sekundlichen Strahlung, die ein Elektron bei der korrespondierenden harmonischen *Kreisbewegung* [...] nach der klassischen Elektrodyanmik ausstrahlt? [...] Ich stelle diese Frage deshalb, weil aus Ihrer Dissertation S. 46 [...] hervorzugehen scheint, dass man $a_{ki}h\nu$ vergleichen müsse mit der Strahlung eines *linear* schwingenden Elektrons, *die ja nur halb so gross ist.* (Sie sprechen dort von einem Elektron 'performing a *simple harmonic* vibration [...]')."

[32]Kramers to Reiche, 15 May 1923 (BSC 15.1). "Mit ihren Ansichten, die gestellten Fragen betreffend, bin ich im grossen Ganzen ganz einig; an den betreffenden Punkten ist die Diskussion in meiner Dissertation leider etwas knapp gefasst."

experimental situation and the theoretical insecurities connected with the averaging procedure in the correspondence approach.[33]

From this position, Kramers' *Zwischenbahn* was kept in place as the standard operationalization of the correspondence principle and provided the basis for the doctoral dissertation of Reiche's student Willy Thomas. In it, Thomas determined the transition probabilities of the *sp* transition for the sodium atom, making extensive numerical calculations based on the central field approximation developed by Fues.[34] As in the case of hydrogen, the transition probabilities came out too high in comparison with the values obtained from experiment. Moreover, Thomas considered several types of averages without finding significant differences among them, so that his negative results appeared to be independent of his choice of averaging procedure.[35]

Since these transitions had played a crucial role in Ladenburg's research, one might expect that Thomas' negative results presented a major blow against the attempt to calculate transition probabilities on the basis of the correspondence principle. Reiche, however, reacted to the discrepancy between theory and experiment in the same way he had done before. Without the possibility of blaming the problem on the averaging procedure, he argued that the discrepancy was due to the fact that magnetic effects, which played a role in the respective experiments, were not taken into account in Thomas' calculations.[36] The discrepancies that had arisen between theoretical predictions and empirical results might thus disappear by incorporating these effects.

These two examples indicate that discrepancies between theoretical predictions and experimental data did not constitute a problem which led Reiche to question the validity of the correspondence approach. The situation changed by the end of 1923 when Reiche and Ladenburg were confronted with a challenge of a different kind. This challenge was put forward by Kramers, who had studied the 1923 paper and argued that Ladenburg and Reiche's determination of the number of dispersion electrons led to untenable results. This assessment was based neither on a discrepancy between theoretical predictions and experimental data nor on a critique of the basic assumptions underlying Reiche and Ladenburg's argument. Instead, Kramers took issue with the specific formulas for the number of dispersion electrons

---

[33]Ladenburg and Reiche (1923, 594). As Reiche argued, Kramers' way of estimating the values for different transitions and the experimental conditions in the Stark effect were responsible for the discrepancies.

[34]Thomas (1924) and Fues (1922a,b). One might ask here how Thomas' work related to Sommerfeld and Heisenberg's work on multiplets, discussed in Chap. 4. Thomas referenced their work and knew that there actually was a precessional motion of the electronic orbit around the axis of the total angular momentum of the atom. Due to the central field approximation, he argued, his method was "not sufficient" to deal with Sommerfeld and Heisenberg's precessional motion and its underlying dynamics. See Thomas (1924, 186).

[35]Thomas (1924, 195).

[36]For Reiche's assessment see footnote 1 in Thomas (1924, 195).

in the case of the harmonic oscillator discussed in the previous section and argued that they were not consistent with general physical expectations.

For the harmonic oscillator, as we have seen, Reiche and Ladenburg had arrived at expressions in which the number of dispersion electrons was proportional to the number of quantum systems N and the quantum number n. This implied that less quantum systems would function as classical dispersion electrons as n increased. To the contrary, Kramers asserted, one should expect "an identity of the number of classical oscillators R and the number of quantum systems N, at least for high quantum numbers n." Consequently, the number of *Ersatzoszillatoren* should not depend on the quantum number n in the limit of high quantum numbers.[37]

This argument did not involve an approximate description of some complex physical system. Rather, Kramers considered the harmonic oscillator as an idealized physical system free from approximations of any kind and discussed its transition probabilities, which, as we have seen in the previous section, were given in a simple, digital form. He then argued that these expressions led to a conflict with respect to a general physical intuition about the limit of high quantum numbers. This argument presented a challenge, which Reiche could not cast aside in the previous manner. Rather, he accepted the harmonic oscillator as an exact representation of a quantum system and acknowledged the validity of Kramers' challenge.

Reacting to Kramers' challenge, Reiche formulated a solution to it in a long letter on 28 December 1923, which represents at least two temporally distinct stages in the development of his thoughts on the problem. In it, Reiche developed the new perspective on the quantum theory of radiation mentioned in the beginning of this section and established a new relation among the transition probabilities. As we will see, this relation turned into the main object of study in his subsequent work and ultimately turned into the f-sum rule.

Developing this perspective, Reiche also realized that the new relation was not consistent with the determination of transition probabilities based on the *Zwischenbahn* model and thus problematized his integration of the correspondence principle into his and Ladenburg's approach. These two aspects were tightly connected and were disentangled gradually as Reiche developed his argument. In the following, I will discuss them separately only for the sake of isolating those issues connected with the development of the f-sum rule and those connected to the implementation of the correspondence principle.

[37]Reiche to Kramers, 28 December 1923 (AHQP 8b.9). "Auf dem Wege über Herrn Pauli und Herrn Minkowski erfuhren Herr Ladenburg und ich vor wenigen Tagen von einem Einwand, den Sie gegen die Formeln 7 und 8 unserer gemeinsamen Arbeit im Bohrheft der Naturwissenschaften geäussert haben und der sich darauf bezog, dass man, zum mindesten für grossen Quantenzahlen n, ein Übereinstimmen der Zahl R der klassischen Oszillatoren mit der Zahl N der Quantenoszillatoren erwarten müsse. Mit diesem Einwand haben Sie vollkommen Recht."

The Emergence of a New Relation

Reiche's new relation emerged when he came to realize that Kramers' challenge was a necessary consequence of Ladenburg's initial expression for the number of dispersion electrons and his implementation of the correspondence principle: Whenever the transition probability was directly proportional to the quantum number n, the number of dispersion electrons would not be equal to the number of quantum systems in the limit of high quantum numbers.

This conclusion, Reiche found, could be avoid by amending the original argument on absorption in classical and quantum theory and otherwise leaving the correspondence argument untouched. In their 1923 paper, as we have seen, Reiche and Ladenburg had introduced an approximation in order to retrieve Ladenburg's relation for the number of dispersion electrons and thereby effectively dropped the term $N_k b_{ki}$ from the exact formula:

$$\frac{\pi e^2}{m} R u_0 = h\nu_{ik}(N_i b_{ik} - N_k b_{ki}) u_{ik}.$$

Trying to resolve Kramers' challenge, Reiche kept the exact formula. Considering the harmonic oscillator again, he argued:

> The issue is resolved, as we think, in the following manner: if one considers especially spatial harmonic quantum oscillators as "quantum atoms," which one compares with the R classical (harmonic, spatial) oscillators, one has to write a sum over *all* quantum states in our equation 4^a; because of the strict harmonicity all quantum oscillators respond to ν_0, no matter on which energy level they might be.[38]

The central assertion of this argument was that the harmonic oscillator absorbs radiation of frequency ν_0 in all of its states so that the external radiation triggers transitions in oscillators in all possible states. As a consequence, the total energy absorbed by R classical oscillators was not equal to a pair of inverse transitions between two particular states; rather, one had to consider pairs of transitions between all possible states with the same energy $h\nu_0$ and arrived at the expression:

$$R\frac{\pi e^2}{m} = h\nu_0 \left(\overbrace{N_0 b_{01} - N_1 b_{10}}^{\text{inverse transitions}} + \overbrace{N_1 b_{12} - N_2 b_{21}}^{\text{inverse transitions}} + \overbrace{N_2 b_{23} - N_3 b_{32}}^{\text{inverse transitions}} \ldots \right).$$

[38]Reiche to Kramers, 28 December 1923 (AHQP 8b.9). "Die Sache klärt sich, wie wir glauben, folgendermassen auf: wählt man als 'Quantenatome', die man mit den R klassischen (harmonischen, räumlichen) Oszillatoren vergleicht, speziell räumliche, harmonische Quantenoszillatoren, so muss man in Gleichung 4^a unserer Arbeit rechts eine Summe über *alle* Quantenzustände schreiben; denn wegen der strengen Harmonizität sprechen alle Quantenoszillatoren, auf welcher Energiestufe sie sich auch befinden, auf ν_0 an."

Reiche thus extended the initial consideration based on Einstein's condition of thermal equilibrium. Just as in the initial case, he equated the net absorption of a number of classical resonators with transitions of a number of quantum harmonic oscillators in a radiation field. The only difference was that, instead of a single pair of transitions between two states i and k, the new expression now involved the sum of these pairs for all possible states i and k.

While this extension was purely formal and did not introduce a new physical assumption, it allowed Reiche to consider the radiation process in a different way and thereby to resolve Kramers' challenge. Instead of looking at inverse transitions between different states, he realized, one could also consider transitions associated with one and the same state:

$$R\frac{\pi e^2}{m} = h\nu_0 \left(N_0 b_{01} \underbrace{-N_1 b_{10} + N_1 b_{12}}_{\text{transitions from state 1}} \underbrace{-N_2 b_{21} + N_2 b_{23}}_{\text{transitions from state 2}} - N_3 b_{32} \ldots \right).$$

Making this move away from inverse processes to transitions involving a particular state, Reiche considered expressions of the form:

$$N_n(b_{n,n+1} - b_{n,n-1}).$$

Evaluating these expressions individually, Kramers' challenge disappeared if the difference between the transition probability to an upper state and the transition probability to a lower state was constant. In this case, the number of classical oscillators was proportional to the number of quantum harmonic oscillators and no longer depended on the quantum number of the particular system. Kramers' challenge could thus be interpreted as a constraint on the transition probabilities.

Developing this solution, Reiche did not write down these difference expressions explicitly nor did he interpret the new pairs of transitions in connection to a particular physical process or a physical quantity. Instead he identified the difference expressions as crucial for his argument as he imposed Kramers' requirement on his equation for the number of dispersion electrons. The new relation among transition probabilities thus did not emerge as an independent theoretical entity but rather as part of a larger argument.

Repercussions for the Zwischenbahn

Logically, this argument relied entirely on a revision of Ladenburg's comparison of Einstein's radiation theory and classical radiation theory and not on the implementation of the correspondence principle. As mentioned in the beginning, this was not the case for Reiche. As his letter to Kramers shows, he relied on the explicit expressions for the transition probabilities, which he had obtained for the harmonic oscillator, to resolve Kramers' challenge.

Looking at these explicit expressions for the spatial harmonic oscillator, Reiche considered the transition probabilities determined from the initial state $n+1$ and the one for the *Zwischenbahn* $n + \frac{1}{2}$. For the former, the transition probability $b_{n,n+1}$ for the transition from a lower to an upper state was obtained from the expressions for the statistical weights $g_n = \frac{1}{2}(n + 1)(n + 2)$ and the transition probability $a_{n+1,n,ini} = \frac{8\pi^2 e^2 \nu_o^2}{3mc^3}(n + 1)$. This led to the transition probability $b_{n,n+1}$:

$$
\begin{aligned}
b_{n,n+1} &= \frac{g_{n+1}}{g_n}\frac{c^3}{8\pi h\nu_0^3}a_{n+1,n} \\
&= \frac{c^3}{8\pi h\nu_0^3}\frac{8\pi^2 e^2\nu_o^2}{3mc^3}\frac{(n+3)(n+2)}{(n+2)(n+1)}(n+1) \\
&= \frac{\pi e^2}{3mh\nu_0}(n+3).
\end{aligned}
$$

In addition, Reiche now determined the remaining transition probability $b_{n,n-1}$ for the induced transition from the state n to the lower state $n - 1$. This transition probability was not given immediately by the relations of Einstein's radiation theory as the corresponding transition did not play a role in considerations on inverse transitions between n and $n + 1$. It could be obtained, however, from the expression for $b_{n+1,n}$ by considering the argument $n - 1$ instead of n:

$$
b_{n,n-1} = \frac{\pi e^2}{3mh\nu_0}n.
$$
[39]

Using these two expressions, each difference associated with a particular number of quantum systems N_n became independent of n:

$$
b_{n,n+1} - b_{n,n-1} = \frac{\pi e^2}{mh\nu_0}\frac{1}{3}((n+3) - n) = \frac{\pi e^2}{mh\nu_0}.
$$

In accordance with Kramers' demand, the relation between the number of classical resonators and the total number of quantum systems in different states became identical.

This realization not only resolved Kramers' challenge. As already mentioned, it also turned into a challenge for the implementation of the correspondence principle and problematized its integration into Ladenburg and Reiche's approach. As Reiche

[39]Reiche to Kramers, 28 December 1923 (AHQP 8b.9). Note that Reiche obtained this expression in a somewhat pedestrian but essentially equivalent way. Introducing the same values for the statistical weights, he obtained the transition probability $b_{n+1,n}$ from the transition probability $b_{n,n+1}$ using the relations of Einstein's radiation theory. The transition probability $b_{n,n-1}$ then followed in the same way from the expression for $b_{n+1,n}$ by considering the argument $n-1$ instead of n.

realized, his solution relied crucially on the expression for the transition probability $a_{n+1,n}$. It worked only "for the transition probability, which is calculated *without averaging* over the different intermediate states lying between the initial state $n+1$ and the final state n, but in which only the initial state is present."[40] In the case of the *Zwischenbahn*, the factor arising in the transition probability $n+\frac{1}{2}$ did not cancel out against the ones of the statistical weight $(n+3)(n+2)/(n+2)(n+1)$. In contrast to the transition probabilities determined from the initial state, the quantum number n did not vanish when evaluating the difference between the transition probabilities.

Inverting this assessment, Reiche thus concluded:

> If one thus elevates the relation
>
> $$R = \sum_{i=0}^{\infty} N_i$$
>
> into a *requirement*, it seems to me that it follows directly that one has to accept the above formula for the harmonic oscillator for $a_{n+1,n}$ for arbitrary quantum numbers and thus to refrain from averaging in this special case.[41]

Situating this argument within the framework of *transformation through implementation*, Reiche thus began to problematize the connection between Ladenburg's initial argument for the number of dispersion electrons and the correspondence principle. Previously he had calculated transition probabilities on the basis of the intensity correspondence and introduced the resulting expression into Ladenburg's fundamental equation. Considering both Kramers' *Zwischenbahn* and the initial state as possible corresponding motions, he had concluded that both yielded acceptable results. Following Kramers' challenge, Reiche now recognized that Kramers' *Zwischenbahn* did not fit into his revised argument.

This realization depended on two aspects. First, Kramers' challenge demanded an exact identity between the classical number of dispersion electrons and the number of quantum systems and thus formulated a robust constraint on the transition probabilities. Second, Reiche considered explicit expressions for the transition

[40] Reiche to Kramers, 28 December 1923 (AHQP 8b.9). "Man gelangt aber offensichtlich zu diesem Resultat *nur dann*, wenn man, wie es oben geschehen ist, für die Übergangswahrscheinlichkeit den Wert (vergl. Gleichung 2^a):

$$a_{n+1,n} = \frac{n+1}{\tau} \left(\text{wo } \frac{1}{\tau} = \frac{8\pi^2 e^2 \nu_o^2}{3mc^3} \right)$$

benutzt, der *ohne Mittelwertbildung* über die verschiedenen zwischen Anfangszustand $(n+1)$ und Endzustand (n) liegenden Zwischenzustände, berechnet ist, in welchem vielmehr nur der Anfangszustand auftritt."

[41] Reiche to Kramers, 28 December 1923 (AHQP 8b.9). "Erhebt man also die Beziehung $R = \sum_{i=0}^{\infty} N_i$ zur *Forderung*, so scheint mir daraus direkt zu folgen, dass man beim harmonischen Oszillator für $a_{n+1,n}$ die obige Formel für beliebige Quantenzahlen anzunehmen, also in diesem speziellen Fall von einer Mittelung abzusehen hat."

probabilities in terms of quantum numbers. These two aspects were independent of considerations on the accuracy of experiment, the vagueness of theoretical techniques, and the complexity of physical effects involved. Instead, the question was whether the implementation of the correspondence principle was consistent with the new relation between the transition probabilities associated with a particular state.

While Reiche recognized that there was a problem in this respect, its implications were of limited scope. For him, the argument did not suggest that the initial state had to be identified as the corresponding motion, but rather that one had to "refrain from averaging." In addition, this methodological rather than conceptual implication was limited to the harmonic oscillator. As such, the argument did not imply that the method of the *Zwischenbahn* needed to be abandoned in general, let alone that the correspondence principle had become problematic on a conceptual level.

Saving the Zwischenbahn

This conclusion marked the end of a first stage in the development of Reiche's thinking, which is expressed in his letter to Kramers. In the second part of his letter, Reiche took another swing at the problem and searched for a way to resolve the problem within the method of the *Zwischenbahn*. In this direction, he pointed out that "the correspondence-like element, i.e. the limit of high quantum numbers in the above *requirement*,"[42] was missing from the argument and reconsidered his comparison between classical and quantum theory again in order to incorporate this limiting case:

> By the way, the following possibility just came to mind: if one from the beginning considers *only quantum oscillators in the* n^{th} *state* and compares them with R classical oscillators *of a certain energy*, only the induced transitions $n \rightarrow n+1$ and $n \rightarrow n-1$ would presumably have to be considered. Then one sees:
>
> $$R\frac{\pi e^2}{m} = h\nu_0(N_n b_{n,n+1} - N_n b_{n,n-1}).$$
>
> Using the (*unaveraged*) values given above [...] it follows:
>
> $$R = N_n.$$[43]

[42]Reiche to Kramers, 28 December 1923 (AHQP 8b.9). "Jedoch vermisse ich selbst in der obigen *Forderung* das korrespondenzmässige Element, nämlich den Grenzübergang zu hohen Quantenzahlen. Wie denken Sie darüber?"

[43]Reiche to Kramers, 28 December 1923 (AHQP 8b.9). "Es fällt mir übrigens eben noch folgende Möglichkeit ein: wenn man von vornherein *nur Quantenoszillatoren im* n^{ten} *Zustand* betrachtet und sie mit R klassischen Oszillatoren *von einer ganz bestimmten Energie* vergleicht, so würden wohl nur die erzwungenen Übergänge $n \rightarrow n+1$ und $n \rightarrow n-1$ in Betracht kommen. Dann erblickt man:

With respect to the question as to whether the *Zwischenbahn* or the initial state should be used to determine the transition probabilities, this argument meant a subtle change: Just as before, the "unaveraged" transition probabilities gave the desired result, while the "averaged" transition probabilities did not. In contrast to the previous approach, however, the quantum number n no longer assumed all of its values throughout a summation but rather turned into a free variable, which Reiche could consider in the high quantum number limit. From this perspective, the averaged transition probability of the *Zwischenbahn* found its place: By introducing the transition probability $a_{n,n+1}$ in the general form $a_{n,n+1} = \frac{1}{\tau} f(n)$, the relation for the number of dispersion electrons became:

$$R = \frac{1}{3} N_n \left(\frac{n+3}{n+1} f(n) - f(n-1) \right).$$

Considering the limit of high quantum numbers n, Reiche reconciled his new relation with the values predicted by the *Zwischenbahn*:

> [...] the *correspondence requirement* would say:
>
> $$\lim_{n=\infty} \left(\frac{n+3}{n+1} f(n) - f(n-1) \right) = 3.$$
>
> This is obviously fulfilled for *every* $f(n)$, which take the form $f(n) \sim n$ for large n, thus e.g. for the *arithmetic* [...] or for the *logarithmic* mean value [...].[44]

The "correspondence requirement" in the limit of high quantum numbers, Reiche thus observed, was not only fulfilled by the "unaveraged" transition probabilities. It was also consistent with every transition probability, which was proportional to n in this limit. The *Zwischenbahn* model could thus be reconciled with Kramers' initial requirement of an identity between classical and quantum theory in this case.

This attempt to save the *Zwischenbahnen* was a peculiar one. While he had already found that an alternative approach led to a satisfactory result for all quantum

---

$$R\frac{\pi e^2}{m} = h\nu_0(N_n b_{n,n+1} - N_n b_{n,n-1}).$$

Unter Benutzung der oben angebenen (*ungemittelten*) Werte [...] folgt:

$$R = N_n."$$

[44]Reiche to Kramers, 28 December 1923 (AHQP 8b.9). "die *Korrespondenzforderung* würde lauten:

$$\lim_{n=\infty} \left(\frac{n+3}{n+1} f(n) - f(n-1) \right) = 3.$$

Dies ist offenbar für *alle* $f(n)$ erfüllt, die, für grosse n, die Form $f(n) \sim n$ annehmen, also z.B. für den *arithmetischen* Mittelwert $f(n) = n + \frac{1}{2}$, oder für den *logarithmischen*: $f(n) = \frac{1}{e}\frac{(n+1)^{n+1}}{n^n}$."

numbers, Reiche believed that the *Zwischenbahnen* needed to be kept in play, even if this meant retreating to the high quantum number limit for the time being. Keeping in mind that the model was introduced as a means to extend the correspondence principle beyond this limit, this shows how much weight he attached to the model.[45]

Focusing on the implications for the correspondence principle, Reiche tacitly developed his argument for the new relation among transition probabilities in a new direction and moved away from the comparison between classical and quantum theory as it was conceived in the 1923 paper and his first attempt to resolve Kramers' challenge. In the first part of his letter, he had considered the response of an ensemble of oscillators to radiation of a particular frequency. Following Einstein's theory of emission and absorption, this meant to look at inverse transitions between different initial and final states, which needed to balance each other to assure thermal equilibrium. Taking off from this perspective, Reiche had assumed that the harmonic oscillator responded to incident radiation in all of its states and had been able to reconsider his expression by effectively looking at oscillators in a particular state and their response to incoming radiation.

Through his attempt to save the *Zwischenbahn*, he now realized that his argument did not depend on this initial assumption and moreover that inverse transitions and considerations of thermal equilibrium were not essential. While these considerations had led him to a new relation between the transition probabilities, the same type of argument could also be made by considering transitions of oscillators in a certain state from the beginning (Fig. 6.1).

[45]In light of this attempt to preserve as much of Kramers' approach as possible, it is worth noting that there was a different line of defense, which Reiche did not consider. For the spatial harmonic oscillator, as we have seen, the equality between the number of classical oscillators and the total number of quantum oscillators did not hinge on the transition probabilities alone but rather on the cancellation of the factors arising from the transition probability and the statistical weights. This was not the case for the linear oscillator. Here, the statistical weights did not play a role so that difference expressions in Reiche's relation only depended on the transition probabilities calculated from the correspondence principle so that the difference of the transition probabilities in two adjacent states was independent of n for both the initial state $n+1$ and the *Zwischenbahn* $n+\frac{1}{2}$. Considering the spatial and the linear harmonic oscillator, one could thus have concluded that the correspondence approach failed for degenerate systems but remained valid for nondegenerate ones or that the calculation of the statistical weights needed to be amended. This was done by Edwin Kemble in his work on band spectra at the same time; Kemble (1924, 1925a,b). For Kemble, the solution to the problem was to construct an average value for the statistical weight of the degenerate system by considering it as the sum of nondegenerate systems. While he took the linear harmonic oscillator into consideration, Reiche did not discuss this point but rather developed his own attempt to save the *Zwischenbahn* model, which was certainly not the only possible response to the inconsistency he had encountered. See Reiche to Kramers, 28 December 1923 (AHQP 8b.9).

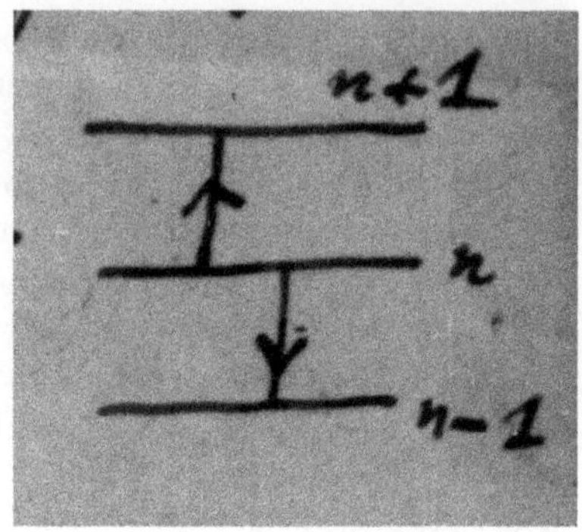

Fig. 6.1 Reiche's term diagram of transitions departing from the state n. Reiche to Kramers, 28 December 1923 (AHQP 8b.9)

Interim Conclusion

As we have seen in this section, Reiche encountered discrepancies between his implementation of the correspondence principle and experimental data early on. These discrepancies did not coalesce into an epistemic challenge, however, as they could be explained away on account of theoretical complexity and uncertainties in his calculation techniques. This changed with Kramers' challenge, which showed that the determination of transition probabilities was inconsistent with Reiche and Ladenburg's determination of the number of dispersion electrons.

Reiche solved the problem by considering transitions associated with a particular state and developed a new relation among the transition probabilities. This relation in turn became the source of a first problematization of the implementation of the correspondence principle: As Reiche realized, it was inconsistent with the transition probabilities obtained from the *Zwischenbahn*. At the same time, the new relation was obeyed in general by transition probabilities determined from the initial state.

The implications for the correspondence principle, which Reiche saw arising from this result, were of limited scope. For him, the problems with the *Zwischenbahn* model occurred only in the special case of the spatial harmonic oscillator. This particularity affected the determination of transition probabilities on an operational level: It implied that a different technique had to be used. In this situation, Reiche tried to keep the *Zwischenbahn* model in general and took a first step in this direction by rehabilitating it in the high quantum number limit.

6.4 Adaptive Reformulation: Reiche, Thomas and the F-sum Rule

Reiche's perspective on the determination of transition probabilities entered a new stage in April 1924 when he received his next letter from Copenhagen. In it, Kramers communicated his ideas for a new quantum theory of dispersion and, as I will show in this final section, led Reiche to reconsider the relation among transition probabilities again.[46] In reaction, Reiche dropped the *Zwischenbahn* model and adopted a new method for determining transition probabilities based on the relation among transition probabilities.

Finally, he reconsidered the connection between his relation for the transition probabilities departing from a particular state and the correspondence principle. Together with his assistant Willy Thomas, he used the techniques of Van Vleck's and Kramers' dispersion theory to interpret his relation as a permanence relation for the transition probabilities, which he called the f-sum rule.

Reiche's Determination of Transition Probabilities and Kramers' Dispersion Formula

Kramers' dispersion theory, which has been discussed extensively in the secondary literature, was based on the idea of translating a classical dispersion formula into a quantum theoretical counterpart. To do so, Kramers relied on the action-angle formalism of analytical mechanics, perturbation theory and a new technique for translating differential quotients into difference quotients, which he developed by going back to the correspondence argument of his dissertation.

In it, as we have seen in Chap. 2.4, he had shown that the frequency of a quantum transition was equal to the average mechanical frequency of an electron moving on a state lying in between the initial and the final state and could be determined by an integral over all these intermediate states. In 1919, this relation led to the introduction the model of the *Zwischenbahn* as a means to calculate transition probabilities. In the context of his new dispersion theory, Kramers returned to his correspondence relation and developed a different approach. Rather than considering it as an expression linking the quantum theoretical radiation process to the classical motion of the *Zwischenbahn*, he interpreted it as a relation between a quantum description and a classical description and used it to translate one into the other. To make this translation, Kramers considered the correspondence relation in a differential form. In the case of the frequency correspondence, the mechanical frequency of the system was given by Bohr's golden rule:

$$\omega = \left(\tau_1 \frac{\partial}{\partial J_1} + \ldots + \tau_u \frac{\partial}{\partial J_u}\right) E.$$

It corresponded to the transition frequency ν:

$$\nu = \frac{\Delta}{h} E.$$

[46]This reception can be reconstructed from a letter from Reiche to Kramers on 9 April 1924 (AHQP 8b.9).

Kramers' new interpretation of the correspondence relation implied that one needed to replaced a classical differential quotient of the form $\tau \frac{\partial}{\partial J}$ by a quantum theoretical difference quotient of the form $\frac{1}{h}\Delta$.

Implementing this procedure within his dispersion formula, Kramers considered a multiply periodic system, whose initial unperturbed motion was characterized by the Fourier coefficient of the τth overtone B_τ and the sum of the overtone frequencies $\tau_1\omega_1 + \ldots + \tau_f\omega_f = \omega$.[47] Using perturbation techniques borrowed from celestial mechanics, he established a classical expression for the polarization P of this system under the influence of an external field E varying with the frequency ν in the form:

$$P = \frac{1}{2}E \sum_{\tau_1 \ldots \tau_f} \left(\tau_1 \frac{\partial}{\partial J_{\tau_1}} + \ldots + \tau_f \frac{\partial}{\partial J_{\tau_f}} \right) \left[\frac{B^2_{\tau_1 \ldots \tau_f}\omega}{(\omega^2 - \nu^2)} \right].$$

In a second step, he then applied his new technique and translated the classical dispersion formula into a quantum theoretical one. To do so, he resolved the differential in the classical dispersion formula into a difference with two terms associated with neighboring states. Next, he replaced the Fourier coefficients B_τ with the corresponding transition probabilities A_i^a and A_i^e, and the mechanical frequencies ω with the respective transition frequencies ν_i^a and ν_i^e, arriving at the quantum expression for the polarization:[48]

$$P = E \sum_i A_i^a \tau_i^a \frac{e^2}{m} \frac{1}{4\pi^2(\nu_i^{a2} - \nu^2)} - E \sum_j A_j^e \tau_j^e \frac{e^2}{m} \frac{1}{4\pi^2(\nu_j^{e2} - \nu^2)}.$$

Kramers communicated his new theory to Reiche in a letter, which is not available. Responding to this letter on 9 April, Reiche informed Kramers that he had been able to retrace the two-step argument and agreed with it:[49]

[47] Reiche to Kramers, 9 April 1924 (AHQP 8b.9) Reiche only gave the Fourier series for a multiply periodic system and the classical dispersion formula. For reconstructions of this classical derivation see Darrigol (1992, 225–228) and Duncan and Janssen (2007b, 646–652), as well as Konno (1993, 122–123) for a shorter exposition of Kramers' derivation.

[48] For the details of this translation procedure see Duncan and Janssen (2007b, 635–637). A discussion of the problems encountered by Kramers in developing the formula is given in Konno (1993).

[49] This point allows an interesting side remark on the reconstruction of Kramers' work. As is well known, Kramers did not mention the replacements of differentials by differences in his first note to *Nature* and only briefly mentioned it in his second note. See Kramers (1924a,b). As Slater and with him Dresden and Konno have stressed, Kramers already had the central idea of translating differentials into differences in January 1924. See Slater (1975, 15), Dresden (1987, 155) and Konno (1993, 125). The self-evidence of Reiche's recapitulation of the argument strengthens this point and suggests that Kramers had communicated a short description of his derivation based on the replacement of differentials by differences in his letter to Reiche.

> Following Epstein's paper and using the Born-Pauli method, I have easily derived the classical expression for P, *which you gave in your letter,* and I was also able to realize the correspondence-like transition to the quantum formula without further ado.[50]

In the following, Reiche focused on the implications of the new dispersion theory for his and Ladenburg's work. Among other things, he considered a remark in Kramers' letter, which connected the new dispersion formula to his thinking about the relation among transition probabilities.[51] In this remark, Kramers had pointed out that it was possible to determine the transition probabilities by considering "the limiting case ν, i.e., of the case of a constant electric field."[52]

Coming to terms with this suggestion, which he "unfortunately had not understood up to this point,"[53] Reiche compared the polarization of a system in a static external field with the polarization of the system for vanishing radiation frequencies ($\nu \rightarrow 0$). To do so, he turned to the harmonic oscillator as his prime test case again and obtained the polarization P in a static external electric field E by considering the perturbation energy W_2:

$$P = -2\frac{W_2}{E} = \frac{e^2 E}{4\pi^2 m \nu_0^2},$$

where ν_0 is the proper frequency of the oscillator. He then considered Kramers' dispersion formula for the harmonic oscillator in the case of vanishing external frequencies $\nu \rightarrow 0$. Due to the fact that a harmonic oscillator only makes transitions to adjacent states, the dispersion formula has only two terms, corresponding to a transition between a particular state and an upper and a lower state. In addition, both frequencies ν_a and ν_e of these transitions are equal and identical to the proper

---

[50]Reiche to Kramers, 9 April 1924 (AHQP 8b.9, my emphasis). "Ich habe mir, im Anschluss an Epsteins Arbeit, unter Benutzung der Born-Paulischen Methode, den klassischen Ausdruck für P, den Sie in Ihrem Brief angeben, leicht abgeleitet und mir auch den korrespondenzmässigen Übergang zu der Quantenformel ohne weiteres klar machen können."

[51]Reiche to Kramers, 9 April 1924 (AHQP 8b.9). Reiche also discussed questions concerning Reiche and Ladenburg's arguments on the equality of the total scattering energy [*Gesamtstreuung*] and absorption, which Kramers felt was not valid. Defending his and Ladenburg's former considerations, he considered the polarization to be complex and incorporated a damping term into the classical expression for the polarization. He then considered the imaginary part associated with absorption according to classical radiation theory, and made the transition to the corresponding quantum formula to reproduce his former results within Kramers' approach. For the present reconstruction, this argument did not play a role, as Reiche did not connect it with the determination of transition probabilities at the time.

[52]Reiche to Kramers, 9 April 1924 (AHQP 8b.9). "Die Bemerkung in Ihrem Briefe, dass man aus dem Grenzfall $\nu = 0$, also eines konstanten elektrischen Feldes, Aufschluss über die Werte der A gewinnt, habe ich leider bisher nicht ganz verstanden."

[53]Ibid.

frequency ν_0. The dispersion formula could therefore be factorized in the form:

$$P = \frac{e^2 E}{4\pi^2 m \nu_0^2}(A^a \tau^a - A^e \tau^e).$$

A comparison of these two expressions for the polarization implied that the difference expression in the dispersion formula needed to be equal to 1:

$$A^a \tau^a - A^e \tau^e = 1.$$

Obtaining this relation for spontaneous transition probabilities from the dispersion formula, Reiche then considered the ground state ($n = 0$). Since transitions to lower states do not exist, he argued, the transition probability A^e was zero so that the probability A^a for the transition from the first excited state to the ground state could be determined directly. Feeding this result back into the difference relation for the first excited state gave the probability for the transition $2 \rightarrow 1$. In general, the transition probability in the nth state could thus be obtained recursively as:

$$
\begin{aligned}
A^0_{-1} &= 0 \\
A^1_0 &= \frac{1}{\tau} \\
A^2_1 &= \frac{2}{\tau} \\
&\dots \\
A^{n+1}_n &= \frac{n+1}{\tau},
\end{aligned}
$$

where $\tau = \tau^a = \tau^e$ as a result of the equality of the spectral frequencies.

Making this argument, Reiche thus obtained an expression for transition probabilities associated with a particular state, according to which the difference between spontaneous transition probability leading to a particular state and the spontaneous transition probability departing from this state was equal to one. This relation was similar to the one he had already obtained in the case of absorption, in which the difference between induced transition probabilities associated with transitions from a particular initial state needed to be a constant independent of the quantum number n.

While the relations between transition probabilities were structurally similar in the two cases, Reiche's consideration had shifted in important ways.[54] Previously he had compared expressions for the absorption energy, which were obtained using two conceptually incompatible frameworks for describing the radiation process. In

[54] Moreover, due to Kramers' derivation of the dispersion formula, the central model in Reiche's consideration was no longer the degenerate spatial harmonic oscillator but the nondegenerate linear oscillator. Thereby the statistical weights, which had been central in Reiche's previous considerations, came to play a secondary role.

the case of dispersion, this was no longer the case since Reiche now compared expressions for the polarization of a mechanical system for static and vanishing external fields, which were obtained within the same theoretical framework.

Most importantly, however, Reiche interpreted the relation among the transition probabilities in a new way. Following Kramers, he no longer used it as a consistency check for the determination of transition probabilities according to a physical model, like the *Zwischenbahn* or the stationary state of the system. Instead, the relation itself became the starting point for the determination of the transition probabilities. In this way, Reiche was able to determine the transition probability for the harmonic oscillator uniquely without making any assumptions about the correspondence between the underlying motion and the transition probability, i.e., whether the transition probabilities were associated with the Fourier coefficients of the *Zwischenbahn* or the initial state. As such, the new approach implied at the very least that the initial state led to the correct value or even that the determination of transition probabilities from an underlying motion was superfluous in general. In any case the *Zwischenbahn* model needed to be abandoned.[55]

This abandonment of the *Zwischenbahn* model was costly. As Reiche realized immediately, his new method depended crucially on the singular properties of the harmonic oscillator. Unlike other harmonic, multiply periodic or anharmonic systems, the harmonic oscillator presented the only quantum system making transitions to its adjacent upper or lower states with the same radiation frequencies. These two features were necessary to obtain the difference expression among the transition probabilities from the dispersion formula and to obtain the transition probabilities recursively on the basis of the ground-state argument. Consequently, the new method worked only in a singular case and was not able to compensate for the loss of a spatiotemporal model for the correspondence relations, which was applicable at least in principle to all quantum systems.

This is where things stood in April 1924. While Reiche prolonged his investigations making calculations for hydrogen,[56] his attempts came to an end when Reiche, Ladenburg and Kramers agreed that the Copenhagen community deserved priority on the subject. In the following, Reiche and Ladenburg sought to remain in contact

[55]In principle, this ground-state argument would have been possible already on the basis of Reiche's earlier considerations on absorption. Since he conceived the relation between transition probabilities as a consistency check for transition probabilities, this possibility did not occur to him.

[56]Reiche's calculations are mentioned but not discussed in a letter from Ladenburg to Kramers, 31 May 1924 (AHQP 8.9). Otherwise, the AHQP does not contain letters from Reiche to Kramers after April 1924 or any other material pertaining to his subsequent work. The Reiche papers at the AIP contain little material dating from the 1920s. Reiche's dismissal from Breslau in 1933 and his escape from Germany in 1941, which has been described by Bederson (2005), make it likely that his notebooks and correspondence are no longer extant. Willy Thomas, as reported by Reiche in his oral history interview with Kuhn, died young of tuberculosis. His papers, if they existed, could not be traced. See interview with Fritz Reiche by Thomas S. Kuhn on 9 May 1962, Niels Bohr Library & Archives, American Institute of Physics, College Park, MD USA, www.aip.org/history-programs/niels-bohr-library/oral-histories/4841-3. [Accessed on 21 March 2019: 11:38]

with the Copenhagen community and invited Kramers to Breslau from 22 to 27 June 1924 to give a talk on the new quantum theory of radiation. After Kramers' visit, however, the intense exchange between Reiche and Kramers broke off and Reiche's investigation appears to have been put on hold.[57]

Reiche, Thomas, and the F-sum Rule

More than a year after Kramers' visit to Breslau, Reiche eventually returned to the relation among transition probabilities. Together with his former doctoral student Willy Thomas, they transformed his considerations into the f-sum rule and published the results in a preliminary note by Thomas and a paper by Reiche and Thomas.[58] With these two publications, Reiche's thinking about the relation among transition probabilities reached its conclusion. Developing a new argument, they isolated Reiche's relation among transition probabilities from concrete physical problems like the absorption process or Kramers' dispersion formula and turned it into the f-sum rule, which presented a permanence relation for transition probabilities in its own right.

The reconstruction of this final stage of Reiche's work is difficult. On the one hand, there is little information on the details of the genesis of their argument in the two papers, which provide the only available sources. On the other hand, as we will see, Reiche and Thomas relied crucially on the work of John H. Van Vleck when formulating their argument. In fact, the connection between the two was so close that Van Vleck later suggested that he had already anticipated if not formulated the f-sum rule.

To reconstruct the genesis of the f-sum rule in Reiche's work and discuss Van Vleck's claim of priority, I will briefly consider, as far as the available sources permit, how Reiche and Thomas might have become aware of Van Vleck's work. Then I will briefly present Van Vleck's considerations as far as they are relevant to the formulation of the f-sum rule and discuss to which extent he had indeed

[57]For the negotiations between Copenhagen and Breslau, see Ladenburg to Kramers, 31 May 1924, Kramers to Ladenburg, 3 June 1924, Kramers to Ladenburg, 5 June 1924 and Ladenburg to Kramers, 8 June 1924 (all given in AHQP 8b.9). Pointing out that Reiche had begun to work on hydrogen, Ladenburg assured Kramers that they would cease to pursue the issue if Kramers or Bohr wanted to work on the problem simultaneously. Kramers took up the offer and asserted that it was his and Bohr's "intention to think this point [...] through ourselves, or to give it to some of the gents at the institute." Kramers to Ladenburg, 3 June 1924 (AHQP 8b.9). Kramers eventually did not consider the problem himself prior to the advent of matrix mechanics and offered the problem to Werner Kuhn, who arrived in Copenhagen from Zurich in the spring 1925. Kuhn's approach to the problem differed remarkably from the one of Reiche and Thomas. Reiche and Thomas arrived at the f-sum rule following a derivation from the general framework of multiply periodic systems and the new correspondence techniques. Kuhn, on the other hand, argued solely on the basis of the dispersion formula in the limit of high frequencies of the external radiation.

[58]Thomas (1925) and Reiche and Thomas (1925).

anticipated the f-sum rule. Finally, I will discuss what put Reiche and Thomas in a position to see Van Vleck's work as the key to formulate their argument and the way, in which it led them to transform Reiche's previous considerations into the f-sum rule.

Within the historiography on quantum physics, Van Vleck is widely known as one of the most talented theoretical physicists of a generation, which included John C. Slater, Ralph de Laer Kronig, David Dennison, Franck C. Hoyt and others. As discussed in detail by Alexi Assmus as well as by Anthony Duncan and Michel Janssen, these physicists formed a cohort of theoretical physicists trained in the U.S. and engaged in active discussions on quantum physics with their European peers by exchanging letters or visiting important research centers during a post-doctoral fellowship.[59] Somewhat exceptionally, Van Vleck got a permanent position at the university of Minnesota right after his PhD. Nonetheless he established contacts with Bohr, Born and other European physicists, who he had met during a short, privately funded tour in 1923, and exchanged letters with them on a regular basis.[60]

While Van Vleck was connected to the European centers of the quantum network, there is no evidence indicating that he communicated with Reiche or Thomas at the time. This makes it highly unlikely that direct communication played a role in the formulation of the f-sum rule. Rather, Reiche and Thomas appear to have found the key to their argument in Van Vleck's two-part paper "The Absorption of Radiation by Multiply Periodic Orbits, and its Relation to the Correspondence Principle and the Rayleigh-Jeans Law."[61]

In this paper, Van Vleck presented a comprehensive discussion of the problem of absorption in classical radiation theory and its connection to quantum theory. Treating a subject central to Reiche and Ladenburg's work, he did so using different techniques and focusing on questions which were significantly different from Reiche's. In particular, Van Vleck did not focus on a relation for transition probabilities similar to the one considered by Reiche. Rather, he paid close attention to the classical description of absorption, which had not played a role in Ladenburg and Reiche's work. Whereas they had simply taken up the resulting expressions from Planck's *Theorie der Wärmestrahlung*, Van Vleck developed a new derivation for the energy absorbed by a multiply periodic system based on the action-angle formalism and perturbation theory.

[59]See Assmus (1993) for a more detailed discussion of post-doctoral education in the U.S. before World War II.

[60]See Duncan and Janssen (2007a, 561).

[61]This is also suggested by Van Vleck's own comments on the f-sum rule in his 1926 *Bulletin for the National Research Council* "Quantum Principles and Line Spectra," in which he claimed to have anticipated the f-sum rule in his 1924 paper. Here, Van Vleck did not mention any personal communication with Reiche and Thomas and acknowledged that their work had developed an argument that was well beyond the scope of his own considerations on the subject. See Van Vleck (1926).

Using these methods, he arrived at the expression for the energy ΔW absorbed by a multiply periodic system in a radiation field in the time t:

$$\Delta W = \frac{2}{3}\pi^3 e^2 t \sum_\sigma \rho(\omega_\sigma)\sigma \frac{\partial}{\partial J_\sigma} C_\sigma^2 \omega_\sigma,$$

where C_σ is the Fourier coefficient associated with the frequency ω_σ of the overtone σ and the spectral density $\rho(\omega_\sigma)$, which does not depend on the radiation frequency.[62]

Discussing this expression, Van Vleck developed a corollary to his argument, which is particularly important for the formulation of the f-sum rule by Reiche and Thomas. In it, Van Vleck discussed the special case of a radiation density independent of ω and evaluated the term $\sigma \frac{\partial}{\partial J_\sigma} C_\sigma^2 \omega_\sigma$ in his general expression. Using the relation between the action variable J_k and the mean kinetic energy $\overline{T_k}$:

$$J_k = \frac{\delta \overline{T_k}}{\delta \omega_k},$$

he obtained:

$$\begin{aligned} J_k &= \frac{\delta}{\delta \omega_k} \pi^2 m \sum_{\sigma_1 \ldots \sigma_s} C_{\sigma_1 \ldots \sigma_s}^2 (\sigma_1 \omega_1 + \ldots + \sigma_s \omega_s)^2 \\ &= 2\pi^2 m \sigma C_\sigma^2 \omega_\sigma. \end{aligned}$$

Using this result, the absorbed energy became:

$$\begin{aligned} \Delta W &= \frac{1}{3} \frac{2\pi^3 e^2 t}{2\pi^2 m} \sum_\sigma^3 \rho(\omega_\sigma)\sigma \frac{\partial}{\partial J_\sigma} \frac{J_k}{\sigma} \\ &= \frac{\pi e^2 t}{m} \rho. \end{aligned}$$

As we will see below, the evaluation of this classical expression was crucial for Reiche and Thomas' formulation of the f-sum rule. In order to understand the position, which allowed them to develop it, it is helpful to consider Van Vleck's own thoughts on the subject, which he and, following him, Duncan and Janssen later identified as already containing a formulation of the f-sum rule.

For Van Vleck, his corollary showed that the energy absorbed classically by a multiply periodic system was independent of the action variable and thus of the shape or the energy of the orbits. This result corresponded to Planck's expression

[62] Van Vleck (1924b). See Duncan and Janssen (2007b, 640–643) for a detailed reconstruction of Van Vleck's calculations leading to this result.

for the average energy absorbed by a harmonic oscillator, which was used by Ladenburg and Reiche, and presented an extension of Planck's calculation from the linear harmonic oscillator to multiply periodic systems.[63] In a footnote, Van Vleck considered the quantum theoretical side of the absorption process and discussed the possibility of formulating a relation similar to the classical expression in terms of Einstein's quantum theory of emission and absorption and the correspondence principle. As he argued, such a "quantum analogue" was indeed possible. The quantum theoretical counterpart to the total absorption ΔW in classical radiation theory was the total absorption ΔF, which was given by the equation:

$$\Delta F = h\nu_{rs}\rho(\nu_{rs})B_{rs} - h\nu_{st}\rho(\nu_{st})B_{st}.$$

This expression described "all the different kinds of differential absorption of various frequency which may commence at a given orbit" and therefore corresponds to the central expression used by Ladenburg and Reiche in their 1923 paper.[64]

In his next step, Van Vleck used the relations of Einstein's quantum theory of emission and absorption to express his quantum analogue in terms of transition probabilities for spontaneous emission. Unlike Reiche and Ladenburg, however, he introduced a generalized version of the *Zwischenbahn* formula and argued that one could obtain a relation equivalent to his classical expression for the absorbed energy by adjusting the parameters of this generalized *Zwischenbahn* formula.[65]

Assessing the possibility of such an adjustment, Van Vleck ultimately discarded his "quantum analogue." First, he argued that the adaptation of the *Zwischenbahn* expression for the transition probabilities was in conflict with the work of Franck C. Hoyt on X-ray lines. Second, he asserted that the sum of transitions associated with a particular state "would in general lead to transitions from positive to negative quantum numbers, which can scarcely correspond to any physical reality." Finally, he doubted whether the conditions, which he had to assume to obtain the classical expression for absorbed energy in the first place, would be of "much real physical significance."[66]

Overall, Van Vleck thus formulated the idea of a "quantum analogue" with the same features as the classical expression, i.e., that it was independent of the shape or energy of the orbits. He considered whether such a quantum analogue could be derived from the transition probabilities determined from an adapted *Zwischenbahn* model and ultimately rejected this possibility.

This approach was quite different from the one of Reiche and Ladenburg. Rather than deriving a "quantum analogue" to the classical calculation of absorption, as we have seen, they had compared Einstein's quantum theory of emission and absorption

[63]Note that Van Vleck considered an individual system, whereas Planck discussed the absorption of R systems.

[64]Van Vleck (1924b, 359).

[65]Van Vleck (1924b, 359).

[66]Van Vleck (1924b, 359–360).

and classical radiation theory to obtain an expression for the number of dispersion electrons. Following Kramers' challenge, Reiche had then considered the harmonic oscillator and isolated a relation between the transition probabilities within this expression. This relation became an independent relation used for checking the *Zwischenbahn* model and later on for determining transition probabilities. Van Vleck, by contrast, considered multiply periodic systems in general and tried to derive his quantum analogue from the *Zwischenbahn* model as the accepted solution for the initial-final-state problem. A major consequence of this approach was that Van Vleck did not isolate a relation between transition probabilities or discuss its implications.

Returning to Reiche and Thomas' work on the f-sum rule, it is thus clear that they did not simply take over an argument for the f-sum rule already given in Van Vleck's work. Rather they developed the f-sum rule by incorporating Van Vleck's classical calculation into their existing considerations on the relation between the transition probabilities.

From this perspective, as already pointed out, Van Vleck's classical calculation provided the central ingredient for their argument. In Reiche's previous attempts, the classical description of absorption had entered as a well-established result, which was simply equated with the quantum theoretical expression. This changed with Van Vleck's derivation based on the action-angle formalism. Through it, the classical expression for the absorbed energy was connected at once to the new correspondence techniques for translating classical expressions into a quantum theoretical version. This offered a new way to relate classical and quantum theory. Rather than *equating* classical and quantum theoretical expressions of a particular quantity, Reiche and Thomas realized, they could *translate* the classical equation into a quantum theoretical counterpart.

Following this approach, Thomas and Reiche considered the classical expression:

$$\sigma \frac{\partial}{\partial J} 2\pi^2 m C_\sigma^2 \omega_\sigma ,$$

given in Van Vleck's work, and Van Vleck's evaluation of it, according to which the classical expression was equivalent to the degree of periodicity s of the radiating system:

$$\sum_{k=1}^{s} \sum_{\sigma_k} \sigma_k \frac{\partial}{\partial J_k} 2\pi^2 m C_{\sigma_1 \dots \sigma_s}^2 \omega_{\sigma_1 \dots \sigma_s} = \frac{\partial}{\partial J_k} J_k = s.$$

Applying the correspondence techniques to this classical equation as in the theory of dispersion, Thomas and Reiche translated the differential quotient on the left-hand side into the quantum theoretical difference and replaced the Fourier coefficients C by the respective transition probabilities A, obtaining τ from the remaining constants. The right-hand side, by contrast, remained unchanged, because the degree of periodicity of a multiply periodic system was the same in classical and quantum theory:

$$\sum A_i^a \tau^a - \sum A_i^e \tau^e = s,$$

or—introducing the oscillator strength $f = A\tau$ from Ladenburg and Reiche's work:

$$\sum f^a - \sum f^e = s.$$

This derivation of the f-sum rule was different from Van Vleck's as well as Reiche's and Kramers' earlier considerations. The former—as we have seen throughout this chapter—had focused on physical processes like net absorption or the dispersion of light by a physical system. In their work in 1925, Thomas and Reiche did not develop their argument by considering a physical process. Rather, they considered the relation between the transition probabilities itself and isolated it from concrete physical problems.

This new take had one major advantage: within the previous arguments on absorption and dispersion, Reiche always needed to isolate the relation among transition probabilities within a given expression, which also involved transition frequencies, and therefore relied on the harmonic oscillator, whose transition frequencies were identical. With the establishment of the f-sum rule, this was no longer the case and Reiche could thus tackle other harmonic systems, like the rotator or the hydrogen atom, whose transition frequencies differed from each other. The new relation thus not only led to the results already obtained for the harmonic oscillator, but also allowed Reiche and Thomas to go beyond this special case.[67]

Moreover, Reiche and Thomas extended their f-sum rule to degenerate systems. As they realized, this extension was possible by treating degenerate systems as a sum of nondegenerate systems. With respect to the right-hand side of the f-sum rule, this meant little change. The degree of periodicity remained the same. The transition probabilities of the degenerate system, however, had to be calculated by summing over the transition probabilities of the nondegenerate system.[68] Following

[67] Reiche and Thomas (1925, 520–521).

[68] Reiche and Thomas (1925, 514–515). The energy emitted by a number of N_a degenerate systems in the transition from an upper state to a lower state: $S = N_a \bar{a}_a h\nu_a$ had to be identical to the energy S^* emitted by the same number of nondegenerate systems making transitions from a g_a-fold upper level to the g-fold lower level. To establish the total energy, Reiche and Thomas first considered the transitions from the g_a-fold upper levels to one of the lower levels given by $S^* = \frac{N_a}{g_a} \sum\limits_m a_a h\nu_a$ and summed up these transitions according to the "Ornstein-Burger-Dorgelo-sum rules," which implied that the energy radiated in a transition to one lower state was identical for all g-sublevels. In the transition from the f-sum rule for nondegenerate systems to the f-sum for degenerate system one thus had to take the sum of all the transition probability a_a of the nondegenerate system and replace them by: $\sum\limits_m a_a = \frac{g_a}{g} \bar{a}_a$ For the inverse transition, on the other hand, the situation was different. Instead of $\frac{N_a}{g_a}$-systems making the transition from the upper state, one had now to

this line of argument, Reiche and Thomas obtained the f-sum rule for degenerate systems:

$$\sum \frac{g_a}{g} \bar{f}_a - \sum \bar{f}_e = s,$$

where g_a and g are the statistical weights of the upper level and the lower level, respectively. This expression was equivalent to the one Reiche had obtained in his letter to Kramers. It allowed Reiche and Thomas to reproduce the initial expression for the net absorption in Ladenburg and Reiche's 1923 paper. Considering the Einstein relations for induced transitions b_{ik} and b_{ij}, they rewrote them in terms of the f values:

$$b_{ik} h\nu_{ik} = \frac{c^3}{8\pi\nu_{ki}^2} \frac{g_k}{g_i} a_{ki} = \frac{\pi e^2}{3m} f_{ki} \frac{g_k}{g_i}$$

$$b_{ij} h\nu_{ij} = \frac{c^3}{8\pi\nu_{ij}^2} a_{ij} = \frac{\pi e^2}{3m} f_{ij}.$$

They then reexpressed the f-sum rule for degenerate systems in terms of the transition probabilities for induced transitions:

$$\sum_k b_{ik} h\nu_{ki} - \sum_j b_{ij} h\nu_{ij} = \frac{\pi e^2}{3m} s.$$

This expression, they realized, only needed to be multiplied with the energy density u to be identical with Ladenburg and Reiche's formula given in 1923 and thus unified Reiche's considerations on the spatial harmonic oscillator in early 1923 with his considerations on the linear harmonic oscillator in the case of dispersion.

Having come to this position, Reiche and Thomas presented a physical interpretation of the relation Reiche had considered from various perspectives since 1923. For them, the f-sum rule was:

> a proposition for the totality of all f (and therefore the transition probabilities) associated with a stationary state, which does not entail mechanical symbols and which, if it avails itself in this form, could be called a permanence law for the transition probabilities.[69]

---

consider $\frac{N}{g}$-systems making the transition from one of the lower states thus radiating the energy $S^* = \frac{N}{g} \sum_m a_e h\nu_e$ so that taking the Ornstein-Burger-Dorgelo-sum rule into account the transition probability of the degenerate system was identical to the sum of the transition probabilities of nondegenerate systems. $\sum_m a_e = \bar{a}_e$ Combining these two results, Reiche and Thomas obtained their f-sum rule for degenerate systems.

[69]Reiche and Thomas (1925, 511). "Wir wollen im folgenden für die Gesamtheit der f (und damit der Ü[bergangs]w[ahrscheinlichkeiten]), die einem stationären Zustand zugeordnet sind, einen

The first part of this interpretation mirrors Kramers' interpretation of the dispersion formula and his rhetoric of observable nonmechanical quantities. Claiming that mechanical symbols were absent from their new relation, Reiche and Thomas did not take issue with the fact that the degree of freedom is a central property of any mechanical system. The second part of their interpretation was much more to the point. As they argued, the constant difference between the transition probabilities implied that there was a "permanence law for the transition probabilities" associated with a particular state.

With the establishment of the f-sum rule, Reiche's correspondence arguments came to an end. The relation he had first considered in 1923 had found a new interpretation within the context of the new quantum theory of dispersion. In the form of the f-sum rule, the relation, initially understood as conflicting with the transition probabilities determined from the *Zwischenbahn*, became the starting point for the determination of the transition probabilities themselves.

This new relation replaced the original model representation of the correspondence principle. For Reiche, however, abandoning the *Zwischenbahn* in favor of the f-sum rule was a costly move. Just like in his letter to Kramers in April 1924, the determination of transition probabilities from the f-sum rule remained severely limited since the new method was only applicable to simple harmonic systems like the harmonic oscillator, the spatial harmonic oscillator or the rotator. The possibilities to extend it were limited and Reiche published only one follow-up paper on the Zeeman effect.[70] Without a model representation for the correspondence relations, the developing matrix mechanics, in which the relation among the transition probabilities played a central role as a quantum condition, did not appeal to Reiche: He only returned to the f-sum rule within the framework of Schrödinger's wave mechanics and used Schrödinger's wave functions to determine the transition probabilities. For Reiche, only wave mechanics reestablished a physical model of the radiating system and replaced the *Zwischenbahn*.[71]

6.5 Conclusion

In this chapter, I presented a detailed analysis of Fritz Reiche's work on the determination of transition probabilities resulting in the formulation of the Thomas-Reiche-Kuhn-sum rule. Such an analysis has been absent from the existing literature, as the main narrative has focused on the development of Kramers' dispersion formula and its impact on Heisenberg's *Umdeutung*.

Satz ableiten, der keine mechanischen Symbole mehr enthält und der, wenn er sich in dieser Form bewährt, ein Permanenzgesetz der Ü.W. genannt werden könnte."

[70]Reiche (1926b).

[71]Reiche (1926a, 1929) and Rademacher and Reiche (1927).

This absence might be baffling at first sight. After all, the f-sum rule provided a pillar for this narrative as it became the quantum condition for Heisenberg's *Umdeutung* and emerged as a "corollary" to Kramers' dispersion theory. Yet, the formulation of the f-sum rule appeared as little more than a footnote in the history of quantum mechanics. Reiche's pathway to it was not analyzed let alone compared with Heisenberg's thinking on it.

After the present reconstruction, the value of such a comparison might become evident. It presents the opportunity to contrast two historically alternative approaches that led to similar results while offering different prospects for further research—one leading to a dead end, the other to matrix mechanics. This chapter has provided the first half on which such a comparison can be based. As will be shown in Chap. 7, it offers key insights into Heisenberg's work on *Umdeutung* and the role of dispersion theory within it.[72]

Even the present reconstruction should suffice, however, to show that Reiche's work would be misconstrued if taken as an example for a physicist who discovered a central ingredient of matrix mechanics but did not take the next, seemingly logical step. Reiche's work was not directed at a new theory of quantum mechanics. Rather, he aspired to no more, but also no less, than solving a specific problem in the quantum theory of radiation. To this end, he took up the correspondence principle like many other physicists in the 1920s and used it as a research tool.

From this perspective Reiche's work followed the pattern of *transformation through implementation*. He implemented the correspondence principle in its original form and used it to determine transition probabilities within his research based on Einstein's radiation theory. Integrating the correspondence principle into this approach, Reiche initially believed, this tool fit right into his considerations.

This situation changed with Kramers' challenge of Reiche and Ladenburg's argument. In reaction to it, Reiche adapted the argument, began to study transitions associated with a particular state and established a new relation among the respective transition probabilities. This new relation then turned into a challenge for Reiche's initial implementation of the correspondence principle and eventually led him to abandon the *Zwischenbahn* model. At the same time, it led to a new method for determining transition probabilities and became the starting point for Reiche's adaptive reformulation. In it, Reiche developed a new correspondence argument based on the translation techniques of the quantum theory of dispersion and interpreted his relation as a permanence relation between transition probabilities.

[72]Such a comparison was made by Duncan and Janssen in the case of John H. Van Vleck. As they concluded, Van Vleck was "on the verge of Umdeutung" but did not take the next step. Asking why he did not, they argue that Van Vleck "was too wedded to the orbits of the Bohr-Sommerfeld theory to discard them." (Duncan and Janssen 2007b, 665). Following their assessment, the absence of any reference to a program of finding a new quantum mechanics, and Van Vleck's own recollections that he would have to have been a lot more "perceptive" to come up with such a new theory, I am inclined to draw a more radical conclusion. Like other physicists discussed in this book, there was no next step for Van Vleck to take as he did not think about dispersion as the key for a new quantum mechanics, but as a particular physical problem to be solved within a given framework.

Like the other case studies, Reiche's work thus exhibits the central characteristics of *transformation through implementation*. At the same time, his work again allows us to engage in the analysis of this process in detail.

Adoption: Motivations for Taking Up the Correspondence Principle

First, we can probe the motivations for making correspondence arguments and the importance of the input coming from Copenhagen. As we have seen, Reiche's work was motivated by the turn to transition probabilities. They became the central theoretical entity as Ladenburg aimed to describe dispersion within the state-transition model, and the correspondence principle was the central tool for calculating them.

The development of Reiche's work clearly depended on input from Copenhagen. Communicated via third parties and personal letters, Kramers' challenges acted as a catalyst for the development of Reiche's work. Yet, these challenges did not predetermine his adaptations or the conclusions he drew from them. Rather, as we have seen, Reiche developed his own, rather conservative solutions within his approach to the quantum theory of radiation and thereby formulated a new perspective on the state-transition model that was strikingly different from the one held by Kramers.

Implementation: Preconditions for Making Correspondence Arguments

Second, Reiche's case allows us to analyze the preconditions for making correspondence arguments. As we have seen, Reiche's work relied first of all on a description of the radiation process in terms of the state-transition model. This description was already in place when he began his work. It was embedded in Ladenburg's comparison of Einstein's quantum theory and Planck's classical theory of radiation and—by that time—did not include a specification of the quantum system. While such a description was sufficient to make Ladenburg's argument, Reiche's implementation of the correspondence principle further required him to work with a particular type of quantum system like the harmonic oscillator, the rotator or a multiply periodic system. Initially, Reiche also needed to be able to quantize this system; yet this precondition disappeared in Reiche's adaptive reformulation, when his correspondence argument itself provided the quantized values for the transition probabilities.

More importantly, Reiche's work points to another central precondition for implementing the correspondence principle. As we have seen, Reiche eventually abandoned the *Zwischenbahn* model and thus no longer determined transition probabilities directly from the Fourier coefficients of a corresponding motion. Rather his determination based on the f-sum rule essentially relied on a relationship between the transition probabilities themselves and was based on a reading of the correspondence principle as a symbolic translation, in which the Fourier coefficients of the motion were replaced by transition probabilities while preserving the relationship among them.

This adaptation, however, was severely limited, as the new method for determining transition probabilities was applicable only to simple harmonic systems. This limitation was rooted deeply in Reiche's adaptation, which was not able to replace the physical model for the correspondence relation. He would not find such a description of the motion until the advent of Schrödinger's wave mechanics. In other words, Reiche needed a physical model to implement the correspondence principle.

Recognizing Problems: How Challenges Arise

Third, Reiche's case provides key insights into the way contradictions arose from the implementation of the correspondence principle. As we have seen, Reiche's initial implementation of the correspondence principle within the quantum theory of radiation led to results that did not match experimental results. They appeared to result from the complexity of the physical situation or from the theoretical uncertainties involved in approximate calculations. In this manner, they were disregarded rather than taken as a sign for a conflict on a more fundamental level.

The situation was different for simple harmonic systems, where the implementation of the correspondence principle also led to problematic results. It was the simplicity and exactness of the description for the harmonic oscillator that made these results more robust. Rather than being pushed aside like the previous approximate results, they were in contradiction with general expectations on the relation between classical and quantum theory and became the source for Reiche's adaptive reformulation.

Adaptive Reformulation: Implications for the Formulation of the Correspondence Principle

Finally, Reiche's case provides insights into the implications of *transformation through implementation* for the correspondence principle as a tool, and the dynamic interplay between the state-transition model and the correspondence principle. As we have seen, Reiche's implementation of the correspondence principle eventually called for an adaptation of Reiche and Ladenburg's initial argument. The new conception of the state-transition model resulting from this adaptation, in turn, became a challenge for *Zwischenbahn* model and led to Reiche's adaptive reformulation in the form of the f-sum rule.

While dynamically transforming both the state-transition model and the correspondence principle, Reiche's adaptive reformulation remained incomplete. Arriving at a partial solution to his problem, he did not find a way to turn his method for determining transition probabilities into a general scheme or to develop a new conception of the physical core of the correspondence principle. This is how far his approach based on Einstein's and Planck's radiation theory came to turning the correspondence principle from a qualitative into a quantitative description of the radiation process within the state-transition model.

Chapter 7
Copenhagen Reactions: The Intensity Problem in Copenhagen, 1924–1925

After the discussion of the applications of the correspondence principle in Munich, Breslau and Göttingen, the final chapter of this book returns to Copenhagen. It studies how physicists around Niels Bohr approached the multiplet intensity problem and its relation to the correspondence principle. Based on a set of letters by Bohr, Kramers, Heisenberg, Pauli, and Kronig, this chapter analyzes how the intensity problem turned from one aspect of the *patchwork of problems* into a challenge for the conceptual development of quantum theory which played a central role in Heisenberg's work leading to his seminal paper "Über die quantentheoretische Umdeutung kinematischer und mechanischer Beziehungen" (henceforth *Umdeutung*).[1]

This discussion in Copenhagen is a pivotal phenomenon with respect to dissemination and adaptation of the correspondence principle: It presents a reaction to the applications of the correspondence principle outside of Copenhagen and can be interpreted as a feedback of the *transformation through implementation* of the principle into the conceptual framework of quantum theory.[2] This chapter analyzes

[1]Heisenberg (1925a). Following up on the reconstruction presented in Chap. 4 and Sects. 7.1, 7.2, and 7.3 of this chapter, Alexander Blum, Christoph Lehner, Jürgen Renn and myself have analyzed Heisenberg's work and the subsequent emergence of matrix mechanics and presented a reinterpretation of his pathway to *Umdeutung*. See Blum et al. (2017). Section 7.4 discusses our joint reconstruction in the context of the present chapter.

[2]Focusing on the multiplet intensity problem, the analysis of the Copenhagen reaction is restricted to a particular case, extending the discussion of Sommerfeld's approach in Chap. 4. I will not devote a separate chapter to the reaction to Franck and Hund's or Reiche's work mainly for pragmatic reasons. In both cases it is not possible to probe the Copenhagen reaction extensively. In the case of Franck and Hund, as we saw in Chap. 5, the reaction of the Copenhagen community was rather brief and had limited repercussions on the formulation of the correspondence principle. In Reiche's case, one can anticipate that his work had a profound influence on Kramers' dispersion theory (see Chap. 6), yet unfortunately, we lack Kramers' letters to Reiche as well as his actual calculations. As a result, we can only speculate about the specific impact of Reiche's work, arriving at results that corroborate Konno's reconstruction of Kramers' dispersion theory; see Konno (1993).

M. Jähnert, *Practicing the Correspondence Principle in the Old Quantum Theory*, Archimedes 56, https://doi.org/10.1007/978-3-030-13300-9_7

this process and highlights the different roles of the correspondence principle in the work done on the *patchwork of problems* vis-a-vis the emergence of matrix mechanics.

Thus concluding the overarching analysis of *transformation through implementation*, the present chapter contributes to the understanding of the emergence of Heisenberg's *Umdeutung* and the transition from the so-called "old quantum theory" to quantum mechanics. The reconstruction and interpretation presented here depart considerably from standard accounts of the history of quantum physics. In these accounts, it is generally argued that the development leading to Heisenberg's *Umdeutung* was driven by the study of the interaction of light and matter, especially dispersion theory or the Bohr-Kramers-Slater theory (BKS), and that they were based on a new interpretation of the correspondence principle as a "symbolic translation" of classical into quantum theory.[3] Following the Copenhagen reaction to the intensity problem from 1924 up to June 1925, this chapter advocates a reassessment of Heisenberg's *Umdeutung*. I will show, first, how Heisenberg developed his *Umdeutung* within the context of the discussions of the intensity problem for multiplets rather than dispersion. Second, I will argue that he did not adopt a new interpretation of the principle as a metatheoretical statement about classical and quantum theory. Rather, the new quantum kinematics of his *Umdeutung* emerged from an attempt to adapt the principle's core idea of a connection between radiation and motion.

This assessment does not merely replace one crucial problem with another. Reinterpreting the experiential basis of Heisenberg's work offers a new perspective on the transformation that came with it: The emergence of a quantum theoretical concept of motion in *Umdeutung* presents an integration of spectroscopic classifications, encoded in intensity schemes and the state-transition model, into the description of motion underlying classical mechanics.

My analysis is divided into five parts. Section 7.1 reconstructs how the intensity problem was received in Copenhagen in the fall of 1924. It analyzes how Bohr, Kramers, and Heisenberg tried to defend the correspondence principle against the "attack" presented by Sommerfeld's *Gesetzmäßigkeiten* approach. Section 7.2 discusses the formal solutions to the intensity problem developed by Ralph Kronig in January 1925. It reconstructs how the perspective on the intensity problem shifted from earlier defensive positions to an attempt to solve the intensity problem.

[3]See MacKinnon (1977, 1982) and Darrigol (1992, 235–246 and 260–276). McKinnon and Darrigol have analyzed some of the letters and papers discussed in this chapter. For them, the Copenhagen reaction to Sommerfeld's attack on the correspondence principle played a role in the discussion of Heisenberg's idea of sharpening the correspondence principle and his paper on the polarization of fluorscence radiation. They did not follow the prolonged discussions of the intensity problem in Copenhagen, however. As a result, the multiplet intensity problem ultimately remained peripheral in their reconstructions. Darrigol mentioned the discussions of Pauli, Kronig, and Heisenberg in April 1925 in his reconstruction of Heisenberg's pathway to *Umdeutung*, without, however, realizing a genetic connection between them. For him, Heisenberg and Pauli had simply "guessed" the intensity formula for the anharmonic oscillator that Heisenberg obtained when developing the central idea of his *Umdeutung*. See Darrigol (1992, 264 and 267).

Moreover, it shows how this attempt turned into a challenge for the interpretation of the correspondence principle as a statement about the motion of a quantum system. Section 7.3 reconstructs the reaction to this challenge by Pauli, Kronig, and Heisenberg in the spring of 1925. Through this reaction, the intensity problem was associated with the formulation of a new "quantum kinematics" and was thus conceived as a problem for the conceptual development of quantum theory. Section 7.4 discusses Heisenberg's formulation of this new quantum kinematic description and how it related to the intensity problem. The conclusion returns to the overarching questions addressed above and compares the developments discussed in the present chapter with the previous case studies.

7.1 Defending the Correspondence Principle: The Sum Rules in Copenhagen

The multiplet intensity problem reached Bohr and the physicists in Copenhagen in the summer of 1924. To understand their take on the sum rules, it is necessary to consider them within the developments in Munich and Utrecht. By the time the discussion of multiplet intensities arrived in Copenhagen, the work of Sommerfeld, Dorgelo, Burger and Ornstein had yielded its first, promising results. As we saw in Chap. 4, Sommerfeld had understood that the intensity ratio of simple multiplets was governed by the inner quantum number. Dorgelo and Burger had extended this idea to complex multiplets, formulating the Utrecht sum rules. These new rules had then been used to set up intensity schemes and allowed the determination of intensities in a number of simple cases. This procedure was closely tied to the idea that statistical weights were central for determining the intensity ratios. Within this approach, the correspondence principle did not play a constructive role, as the transition probabilities were assumed to be equal for the transitions making up the multiplet.

In this situation, Bohr first heard about the new approach to multiplet intensities from Ornstein in the summer of 1924 and reacted to it positively:

> At the moment we are here very interested in the question of the intensity of spectral lines, from which it seems possible to draw conclusions on the character and fixation of the electron orbits. In this respect such simple *Gesetzmäßigkeiten* as the ones found in your institute [. . .] should of course be of fundamental importance.[4]

Welcoming the new developments, Bohr immediately connected the sum rules with the correspondence principle and the attempts to use it "to draw conclusions on the

[4]Bohr to Ornstein, 5 July 1924 (BSC 14.2). "Gegenwärtig sind wir hier sehr interessiert an der Frage von der Intensität von Spektrallinien, von der es möglich scheint Schlüsse über den Charakter und Festlegung von Elektronenbahnen zu schliessen. In dieser Beziehung dürften natürlich solche einfachen Gesetzmäßigkeiten wie Sie in Ihrem Institute gefunden sind, und die Sie in Ihrer Abhandlung analysieren, von grundsätzlicher Bedeutung sein."

character and fixation of the electron orbits" from the intensity of spectral lines. In this respect, Bohr augured, the "simple *Gesetzmäßigkeiten*" for multiplets would be of "fundamental importance" for future developments. At the same time, Bohr did not expand on the details of such a development. He discussed neither the problems of intensity schemes nor how they could be interpreted in light of the physical core of the correspondence principle.

Bohr's focus on the principle's general significance rather than specific problems is not surprising, considering the problems that the Copenhagen community had focused on since 1922: Bohr had been working on his general treatises on quantum theory and on X-ray spectra with Coster, the Bohr-Kramers-Slater theory and the problem of the polarization of fluorescence radiation. Kramers' work focused on the absorption of X-rays, the continuous X-ray spectrum of scattering electrons, optical dispersion and the BKS theory.[5] The intensity problem for multiplets and the Zeeman effect had not played a role in these cases and Bohr and Kramers had only begun to consider its intricacies.

This is emphasized as well by the reaction of Bohr and Kramers to Sommerfeld's critique of the correspondence principle in the fall of 1924. In it, as we saw in Chap. 4, Sommerfeld claimed that the principle was conceptually flawed and introduced "foreign elements" into quantum theory. More importantly, he diagnosed that it was ill-suited to account for the new experimental data on multiplet intensities. Overall, Bohr and Kramers reacted evasively and tried first of all to dispel Sommerfeld's programmatic critique. Responding to Sommerfeld in Bohr's diplomatic style, Kramers argued:

> [...] it is not Bohr's intention to regard the correspondence principle as a foundation of an axiomatic presentation of quantum theory. Bohr's formulation of the principle is always cautious and tentative, and at least it would be too soon to want to conclude that the beautiful Utrechtian intensity measurements indicate a "failure" or "inappropriateness" of the correspondence principle.[6]

Stressing the unfinished and open character of the correspondence principle, Kramers (and Bohr) argued that any claims about the incompatibility of the principle and the sum rules were premature. Instead of discarding the principle, they would seek to incorporate the sum rules into the correspondence approach.

In this respect, Bohr and Kramers' response to the second part of Sommerfeld's challenge is telling. Bohr and Kramers did not present a detailed argument to show that the correspondence principle was able to account for the intensity measurements and the Utrecht sum rules. Instead they argued that no such account

[5]Bohr (1923b, 1924), Bohr and Coster (1923), and Bohr et al. (1924b,a) as well as Kramers (1923, 1924a,b).

[6]Kramers to Sommerfeld, 6 September 1924 in Sommerfeld (2004, 165–166). "Es liegt Bohr fern, das Korrespondenzprinzip als eine Grundlage einer axiomatischen Darstellung der Quantentheorie anzusehen. Bohrs Formulierung des Prinzips ist ja überall tastend und vorsichtig, und es wäre mindestens verfrüht aus den schönen Utrechter Intensitätsmessungen auf ein 'Versagen' oder 'Unzweckmäßigkeit' des Korrespondenzprinzips schließen zu wollen."

could be given until a more definite model had been developed for multi-electron atoms:

> Rather it is the case that we know so little about electronic coupling in the atom that the correspondence principle, as far as we have now uncovered it, does not enable us to favor any assumption on multiplet intensities; but we are dealing with a problem here in which the present form of the theory of the fixation of the stationary states essentially fails; one might hope, however, that guided by experiment [an der Hand der Experimente] the general correspondence point of view will give a *Fingerzeig* for eliminating these difficulties.[7]

For Kramers and Bohr, the incompatibility of the sum rules and the correspondence approach claimed by Sommerfeld was thus first of all a result of the failure of the mechanical framework of quantum theory for multi-electron atoms and the resulting incomplete knowledge about electronic coupling, i.e., about the underlying dynamics of the multiplet atom. In light of the uncertainties concerning atomic motion, Kramers argued, one could not expect to deduce any concrete statements about intensities from the correspondence principle and hence the correspondence principle could not fail. Invoking the strategy of using the correspondence principle inductively, he argued that the correspondence principle presented a way to interpret the empirical regularities governing the spectra in terms of a model. These regularities might hence give hints (*Fingerzeige*) for describing multiplet atoms.

While this perspective seemed viable for Bohr and Kramers at the time, it was beside the point for Sommerfeld or Pauli. For them, as we saw in Chap. 4, the incompatibility of the sum rules and the correspondence principle was not rooted in the specifics of the underlying physical model. Rather, it was inherent in the *Zwischenbahn* model and thereby in the correspondence principle itself. This issue was not addressed by Kramers and, as we will see below, came to light only in subsequent discussions between Pauli and Heisenberg in October 1924. By the time Bohr and Kramers responded to Sommerfeld's critique, they had thus only begun to consider the problem of multiplet intensities after working on different problems in the previous years. Their first tentative attempts to capture the relation between the correspondence principle and Sommerfeld's *theory of intensities* remained general and programmatic without addressing the problems that had been discussed in the context of the Sommerfeld-Heisenberg paper in 1922.

[7] Kramers to Sommerfeld, 6 September 1924 in Sommerfeld (2004, 165–166). "Es liegt vielmehr so, dass wir von der Elektronenkoppelung im Atom noch so wenig wissen, dass das Korrespondenzprinzip, soweit wir es bis jetzt erkannt haben, uns nicht im Stande stellt, irgend eine Vermutung über Multiplettintensitäten zu bevorzugen; wir begegnen aber hier einem Problem, wie die bisherige Form der Theorie der Festlegung der stationären Zustände wesentlich versagt; man darf aber vielleicht hoffen, dass der allgemeine Korrespondenzgesichtspunkt an der Hand der Experimente einen Fingerzeig zur Beseitigung dieser Schwierigkeiten geben wird."

Heisenberg, Pauli and the Sharpening *of the Correspondence Principle*

The discussion of the multiplet intensity problem within the Copenhagen community was prolonged with Werner Heisenberg's arrival on 17 September 1924.[8] As we saw in Chap. 4, Heisenberg was familiar with the topic of multiplet intensities. He knew the challenge that the initial-final-state problem held in store for the correspondence principle when it came to the total intensity of a multiplet. Moreover, he had diagnosed a "flaw" in the correspondence approach in connection with this problem as early as 1921. From this position, Heisenberg discussed the problem with Bohr in September 1924 and convinced Bohr that the problem of the sum rules was not one of the underlying dynamics of the multiplet atom alone, but also posed a challenge for the correspondence principle as a general kinematic relation.

As Heisenberg explained to Pauli in a letter on 30 September 1924, he and Bohr had found a solution in which the sum rules appeared as an "inescapable consequence of the correspondence principle."[9] This conclusion, Heisenberg claimed, "totally disprove[d] the attacks on the correspondence principle" by Sommerfeld, and he urged Pauli to write a short note making this point.[10] Pauli declined to write such a note and Heisenberg eventually published it himself under the title "Über eine Anwendung des Korrespondenzprinzips auf die Frage nach der Polarisation des Fluoreszenzlichtes."[11] In the paper, he famously called for a "sharpening of the correspondence principle" that made it possible to draw unambiguous conclusions from it.[12]

In this context, Heisenberg developed his main argument on the sum rules in two letters to Pauli in September and October 1924 and debated its validity with

---

[8]Heisenberg's arguments and his discussions with Pauli have also been analyzed by Darrigol in a somewhat different manner. See Darrigol (1992, 237–246). His analysis is framed in terms of the development of BKS theory and discusses the relation between a use of the principle as a formal analogy and the virtual oscillators of BKS. The difference in perspective between the present analysis and Darrigol's will be discussed below.

[9]Heisenberg to Pauli, 30 September 1924 in Pauli (1979, 162). "Mit Bohr hab' ich mir die Frage nochmal genau überlegt und wir sind zu dem Schluß gekommen, daß die Summenregeln nicht etwa—wie Sommerfeld sagt—, durch's Korrespondenzprinzip nicht verstanden werden können, sondern daß sie eine *zwangsläufige* Folge des Korrespondenzprinzips sind und eigentlich das allerschönste Beispiel dafür daß das Korrespondenzprinzip manchmal eindeutige Schlüsse zulässt."

[10]Heisenberg to Pauli, 30 September 1924 in Pauli (1979, 163).

[11]Heisenberg (1925b).

[12]Heisenberg (1925b, 617). The full argument on the polarization of fluorescence radiation is not essential for understanding the discussion of the sum rules in Copenhagen. For a more detailed reconstruction of Heisenberg's argument, which relied mainly on the sum and polarization rules, see MacKinnon (1977) and Darrigol (1992). MacKinnon's analysis of the significance of Heisenberg's paper for his pathway to *Umdeutung* is discussed in Blum et al. (2017).

Pauli. His defense against Pauli provides important insight into the interpretation of the correspondence principle and the position towards the sum rules in Copenhagen. Heisenberg first formulated this position in his letter to Pauli on 30 September 1924:

> In the *classical* theory the sum rules merely imply that (for example, in the case of *md-np*) the total intensity of the image that is the sum of the three components $J_{+1} + J_0 + J_{-1}$, has to be independent of the angle ϑ, [...] because, indeed, we always have the *same* electronic orbit. Only the *distribution* of the total intensity among J_{+1}, J_0 and J_{-1}, of course, depends on ϑ. *Likewise in quantum theory*. If we have, for example, three *d*- and three *p*-levels, the *total intensity* has to be the same for each of the three *d*-levels, because they have the *same* electronic orbit, but different angles.

	d_1	d_2	d_3		j
$j =$	3	2	1		
	x	x	x	p_1	2
		x	x	p_2	1
			x	p_3	0

> The *distribution* of the total intensity on the *different* oscillators is, of course, dependent on ϑ.[13]

Heisenberg's argument, first of all, points to the shift in perspective towards the sum rules: Bohr and Heisenberg no longer argued that the problem of the sum rules was associated with the hitherto unknown coupling mechanism, i.e., the specific dynamics of the multiplet atom. Instead, they accepted that any description of the multiplet atom would involve some kind of precessional motion around a fixed axis and therefore would always yield the intensity ratios given in the Sommerfeld-Heisenberg paper. From this perspective, the dynamics of the system thus governed only how the total intensity was distributed among the individual components. By contrast, the sum rule, i.e., the idea that the total intensity of the multiplet components was equal to the intensity of the unresolved line, was independent of the particular model. They had to be understandable from the Sommerfeld-Heisenberg formulas and thus became a challenging problem for the correspondence principle itself.

Second, Bohr and Heisenberg's attempt to resolve this challenge shows a central difference in the interpretation of the sum rules by Sommerfeld on the one hand, and Heisenberg and Bohr on the other. For Sommerfeld, as we have seen, the sum

[13] Heisenberg to Pauli, 30 September 1924 in Pauli (1979, 162, emphasis in the original). "Die Summenregel in der *klassischen* Theorie bedeutet nämlich einfach, daß (z.B. im Falle md-np) die Gesamtintensität des Bildes, d.h. die Summe der drei Komponenten $J_{+1} + J_0 + J_{-1}$ unabhängig sein muß vom Winkel ϑ, (vgl. Zeitschrift für Physik **11**,142 (12)), deswegen, weil wir ja immer *dieselbe* Elektronenbahn haben. Nur die *Verteilung* der Gesamtintensität in J_{+1}, J_0 und J_{-1} hängt natürlich von ϑ ab. *Ebenso in der Quantentheorie*. Haben wir etwa drei d- und drei p-Niveaus, so muß die *Gesamtintensität* für jedes der drei d-Niveaus dieselbe sein, weil sie ja *dieselbe* Elektronenbahn haben, nur verschiedene Winkel. [Figure of the intensity scheme] Die *Verteilung* der Gesamtintensitäten auf die *verschiedenen* Oszillatoren ist natürlich von ϑ abhängig."

rules essentially presented a proposition about the statistical weights of a system: they stated that the intensity ratios were determined by the statistical weights of the respective states. Further, he assumed that the statistical weight was distributed among the different components in integral units (*Gewichtseinheiten*) in accordance with the sum rules.

For Heisenberg, by contrast, the proportionality of the total intensity to the statistical weights was a secondary property, introduced "after the fact, because the number of atoms in the states d_1, d_2, d_3 is given by the weights 3, 2, 1." The primary problem was to show that the total transition probability of the individual components was identical with the transition probability of the unresolved line. On the basis of the correspondence principle and the original Sommerfeld-Heisenberg paper, this meant that the total intensity of the different components was "independent of the angle" describing the splitting. As such, the sum rules were primarily a statement about the effect of a superimposed precessional motion on the motion of a multiply periodic system.

Finally, Heisenberg's argument indicates a decisive shift in the physical interpretation of the correspondence principle in Copenhagen. At first glance, Bohr and Heisenberg's argument is based on an extrapolation from classical to quantum theory: since classical radiation theory yields the aforementioned result that the total intensity is independent of the angle, they appear to argue, the same has to be true in quantum theory. Based on this literal interpretation, Mehra and Rechenberg as well as Darrigol have argued that Heisenberg interpreted the correspondence principle as a statement about the formal analogy between classical and quantum theory, and that his work was essentially a search for a "symbolic translation" between the two.[14] A closer reading of the above argument suggests, however, that this interpretation is unsatisfactory. To begin with, it is not clear that Heisenberg actually presented an extrapolation from the classical to the quantum case. His argument can be read equally well as a parallelism of two explanations. Moreover, Heisenberg, as will be discussed in detail below, insisted that his argument was based on a physical assumption rather than on a formal analogy.

This latter point suggests that Heisenberg's argument should be interpreted in a different way. Whether emerging from an extrapolation from classical to quantum theory or presenting a parallelism of a classical and quantum theoretical explanation, the argument is carried by a new interpretation of the core idea underlying the correspondence principle. As Heisenberg stresses, the key for understanding the sum rules is that each multiplet component is associated with "the same electronic orbit." In the classical case, this is a direct consequence of the radiation mechanism. In quantum theory, however, it is the case only if the transition probabilities are associated with the electronic orbit of the initial state.

This assumption solved the initial-final-state problem in a new way. As we have seen throughout this book, transition probabilities were thought of in analogy to the frequency condition as depending on both the initial and the final state, or

[14]Mehra and Rechenberg (1982a, 157–160) and Darrigol (1992, 242–254).

a *Zwischenbahn*. Connecting the transition probabilities to the initial state was pondered in Copenhagen for some time. As we saw in Chap. 6, it was adopted by Kramers in the context of the quantum theory of dispersion as early as March 1924. Moreover, Bohr, Kramers, and Slater had introduced it to the BKS theory and Bohr continued to see it as a particularly important feature of the theory, even after its demise in January 1925:

> Especially, I felt it was far more harmonious from the point of view of the correspondence principle to connect the spontaneous radiation with the stationary states themselves and not with the transitions.[15]

When they accounted for the sum rules in the fall of 1924, Bohr and Heisenberg used this assumption that the transition probabilities were associated with the initial state. They thus discussed the intensity problem with respect to the correspondence principle as a statement about radiation and motion in quantum theory. Thereby, they prolonged Bohr's original take on the principle as a law of quantum theory, instead of adopting a new interpretation of the principle as a metatheoretical statement.

As mentioned above, this point came to the fore in the subsequent discussions when Pauli reacted critically to Bohr and Heisenberg's argument and refused to write the note in defense of the correspondence principle. For Pauli, the explanation of the sum rules was based on a purely "formal" analogy between classical and quantum theory and therefore departed considerably from the hitherto accepted interpretation of the correspondence principle. Confronted with this critique, Heisenberg identified the different understandings of the correspondence principle as the source of the problem:

> If one, like you, equates the correspondence principle with the wrong assumption that one could get from the classical intensities to the quantum theoretical ones via averaging, then *you* are right that one does *not* get to the Ornstein rule; if one, however, considers it as an analogous logical connection to the classical theory, then *I* am right.[16]

For Pauli, transition probabilities depended on both the initial and the final states of the system, just as transition frequencies did. The angles ϑ in the Sommerfeld-Heisenberg formulas were thus associated with the *Zwischenbahnen* of the respective transitions and involved the same initial, but different final states of the multiplet. As such, there were three angles, prohibiting the summation of the cosine and sine terms taken for granted in Heisenberg's argument. From this perspective, Pauli was not convinced by Heisenberg's argument and criticized the analogy argument as being purely "formal."[17]

---

[15]Bohr to Slater, 10 January 1925 (BSC 16.2).

[16]Heisenberg to Pauli, 8 October 1924 in Pauli (1979, 167–168). "Wenn man, wie Sie, unter Korrespondenzprinzip die falsche Behauptung versteht, man könne durch Mittelung der klassischen Intensität zur quantentheoretischen kommen, so haben *Sie* recht, daß man durch's Korrespondenzprinzip *nicht* zur Ornsteinregel kommen kann; wenn man aber darunter meint: sinngemäßen logischen Anschluß an die klassische Theorie, so habe *ich* recht."

[17]See Heisenberg to Pauli, 8 October 1924 in Pauli (1979, 168). In his letter discussed below Heisenberg responded to Pauli's criticism.

In his reaction, Heisenberg acknowledged the validity of Pauli's (and Sommerfeld's) point that the intensities associated with different angles ϑ could not be trivially summed up on the basis of the *Zwischenbahn* approach. Just as for Kramers in his discussion with Reiche, however, this meant only that the *Zwischenbahn* approach had to be abandoned. As Heisenberg saw it, this did not imply that the correspondence principle and its core physical assumption became problematic. On the contrary, he still maintained that the sum rule was a direct consequence of the correspondence principle, if one understood the correspondence principle as a "*sinngemäße*, logical connection to the classical theory."

This interpretation, Heisenberg argued explicitly, was not—as Pauli believed—purely formal. Rather it was "physical" in the sense of being based on the core idea of the correspondence principle:

> If one now goes over to quantum theory, it is totally within the meaning of every radiation theory to say: the total quantum theoretical radiation probability is unambiguously determined by the (virtual) electronic trajectory and its energy radiated virtually per sec.; if one does not say that every definition of lifetime etc. becomes meaningless. If one assumes it, however [...] (I could not think of a theory without this assumption), that is, the correspondence principle, one has the Ornstein rule. Only the *distribution* of the various *possible* jumps is given by the coupling.[18]

In this reiteration of his previous argument, Heisenberg stressed that the sum rules were an expression of the core idea behind the correspondence principle: transition probabilities are determined by the electronic trajectory. As long as an external field merely imposes a precession on this motion without otherwise affecting the motion, the total transition probability remains the same and the summation of the individual components has to yield the transition probability of the undisturbed system. This explanation, Heisenberg stressed explicitly, was not—as Pauli had criticized—a "formal" but rather a "physical" one.

Giving up the physical core, he argued further, would have tremendous consequences:

> From your letter, I take it that you have not entirely understood what I intended with my letter and that you had not understood which misfortune occurs and how much one departs from the classical theory if one abandons the unambiguous relation between the electron trajectory and the transition probability. My arguments are thus indeed *physical*, and not formal ones that also classically

[18]Heisenberg to Pauli, 8 October 1924 in Pauli (1979, 167). "Wenn man nun zur Quantentheorie übergeht, so liegt es vollkommen im Sinn jeder Strahlungstheorie, zu sagen: die quantentheoretische gesamte Strahlungswahrscheinlichkeit ist eindeutig durch die (virtuelle) Elektronenbahn und deren pro sec. gestrahlte (virt[uelle]) Energie gegeben; wenn man das nicht sagt, so hört jede Definition von Lebensdauer u.s.w. auf. Wenn man es aber annimmt—und diese Annahme ist ja vielleicht kein absolut zwingendes, aber doch sehr naheliegendes Analogon zur klassischen Theorie (ich könnte mir keine Theorie ohne die Annahme vorstellen) d.h. Korrespondenzprinzip so hat man die Ornsteinsche Regel. Nur die *Verteilung* auf die verschiedenen *möglichen* Sprünge wird durch die Kopplung gegeben."

$$\cos^4 \frac{\vartheta}{2} + 2\cos^2 \frac{\vartheta}{2} + \sin^4 \frac{\vartheta}{2} = 1.^{19}$$

Heisenberg thus clearly perceived the potential abandonment of the physical core of the correspondence principle as a major threat, which had implications well beyond the special case of the sum rules.

Heisenberg's emotional warning against abandoning the physical core of the correspondence principle shows how thoroughly the Copenhagen community was committed to the interpretation of the correspondence principle as a "physical" statement about the motion of a quantum system. Holding on to this core idea, Bohr, Kramers, and Heisenberg clearly distinguished it from its operationalization. Pauli (and with him many other physicists like Reiche) understood the *Zwischenbahn* model as a tentative but integral part of the correspondence principle. As such, Pauli designated Heisenberg's argument, which bypassed the problems of the *Zwischenbahn*, as purely formal, in opposition to what he understood to be the physical content of the principle. In contrast, Bohr, Kramers, and Heisenberg abandoned the *Zwischenbahn* as an operationalization of the principle rather easily, adopting the virtual oscillator model as an alternative. This new operationalization embodied the association of the transition probabilities with the motion in a particular initial state and thus rendered the qualitative adaptation of the principle's core idea explicit.

This assessment has further reaching implications for understanding the overall development of the correspondence principle up to Heisenberg's *Umdeutung* in May and June 1925. In his reconstruction, Darrigol has asserted that the correspondence principle became a means to symbolically translate classical into quantum theory, and consequently framed his analysis of Heisenberg's *Umdeutung* as an elaboration of this "symbolic translation."[20] Thereby he introduced a subtle shift in the interpretation of the principle. For him the principle was no longer primarily a statement about the motion of a quantum system, but became a metatheoretical statement about classical and quantum theory. By contrast, the present reconstruction asserts that the Copenhagen community continued to interpret the correspondence principle as a statement about the connection between radiation and motion. From this perspective, as will be shown below, the continued discussions of the intensity problem and Heisenberg's *Umdeutung* focused on the adaptation of the concept of motion.

[19]Heisenberg to Pauli, 8 October 1924 in Pauli (1979, 168, emphasis in the original). "Aus ihrem Brief, glaube ich, daß Sie nicht ganz verstanden hatten, was ich wollte mit meinem Brief und sich nicht klar gemacht hatten, welches Unglück passiert und wie weit man sich von der klassischen Theorie entfernt, wenn man die eindeutige Beziehung zwischen Elektronenbahn und Sprungwahrscheinlichkeit aufgibt. Meine Argumente sind also durchaus *physikalische*, nicht etwa die formalen, daß auch klassisch $\cos^4 \frac{\vartheta}{2} + 2\cos^2 \frac{\vartheta}{2} + \sin^4 \frac{\vartheta}{2} = 1$ ist."

[20]For Darrigol's reconstruction of Heisenberg's *Umdeutung*, see Darrigol (1992, 261–267, especially 275–276 for his emphasis on symbolic translation).

7.2 Reformulating the Intensity Problem: The *Vanishing At the Edges* Argument in Copenhagen

The qualitative adaptation proposed by Bohr and Heisenberg first of all saved the correspondence approach from the gravest conceptual problems encountered in the case of multiplet intensities. They provided a qualitative argument for the consistency of the sum rules and the principle. At the same time, Bohr, Kramers, and Heisenberg left the problem of actually determining the intensities untouched and thus did not contribute to a quantitative solution of the intensity problem.

The Copenhagen community came to attack this problem in early 1925 when the American postdocs Ralph de Laer Kronig and David Dennison as well as Cambridge professor Ralph Fowler arrived in Copenhagen. Kronig, Fowler, and Dennison were working to establish explicit expressions for the intensities in various cases.[21] Kronig, in particular, focused on the problem of multiplets and the Zeeman effect, establishing a connection between the original correspondence formulas for multiplet intensities and Sommerfeld's intensity schemes.

Tellingly, Kronig's approach did not follow up on the qualitative argument of Bohr and Heisenberg. Rather than trying to arrive at quantitative predictions from a physical model, his approach closely resembled Sommerfeld and Hönl's. He employed the same formal methods and ideas and adapted the Sommerfeld-Heisenberg formulas to the intensity schemes. Thus obtaining virtually the same results, he put the formal solutions to the intensity problem on the agenda of the Copenhagen community. As the adaptation of the correspondence formulas was discussed in detail in Chap. 4, I will summarize the main line of Kronig's argument rather briefly, highlighting those aspects that are different from Sommerfeld and Hönl's. Moreover, I will discuss the implication for the correspondence principle that Kronig saw arising from his intensity formulas.

Kronig's Solution to the Intensity Problem

Kronig began his work on the Zeeman effect in late 1924, working as a postdoc in Paul Ehrenfest's group in Leiden. The central insight upon which Kronig's work was built was the observation that the Zeeman intensities given by the intensity schemes of Ornstein and Burger could be described by quadratic functions of the magnetic quantum number.[22] Following this observation, Kronig wrote down

[21] See Goudsmit and Kronig (1925a,b), Kronig (1925a,b), Fowler (1925a,b), and Dennison (1926, 1928). Kronig worked on the intensities of the Zeeman effect and multiplets for atoms. The intensities of band spectra of diatomic molecules were tackled by Fowler on the basis of the sum rules, and by Dennison on the basis of matrix mechanics.

[22] Kronig had taken up this general insight from Ehrenfest's student Samuel Goudsmit. See Kronig to Goudsmit, 15 December 1924 (AHQP 60.3). It is clear, however, that Kronig had developed this

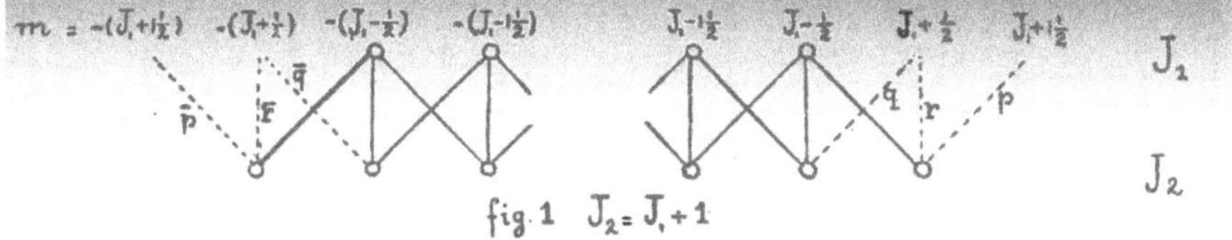

Fig. 7.1 Kronig's term diagram for Zeeman splittings with $\Delta j = \pm 1$ in Goudsmit and Kronig (1925a, 420)

quadratic equations of the inner quantum number J and magnetic quantum number m, which both referred to the initial state, in the most general form:

$$
\begin{aligned}
I_{+1} &= A(J + m_1 + a_1)(J + m_1 + a_2) = A(m_1^2 + \alpha_1 m_1 + \alpha_2)\\
I_0 &= 4A(J + m_1 + b_1)(J + m_1 + b_2) = 4A(m_1^2 + \beta_1 m_1 + \beta_2)\\
I_{-1} &= A(J - m_1 + c_1)(J - m_1 + c_2) = A(m_1^2 + \gamma_1 m_1 + \gamma_2).
\end{aligned}
$$

He then determined the coefficients A, α, β, and γ.

In this determination,[23] Kronig considered the sum of the intensity of a particular triplet and demanded that this sum be independent of the magnetic quantum number, as Heisenberg had done in his qualitative argument. This demand did not suffice to determine the coefficients completely so that Kronig imposed additional constraints on his formulas: Just like Sommerfeld and Hönl, he assumed that the sum and polarization rules held at the "edges of the scheme," i.e., that the quadratic equations also applied to the "virtual" transitions beyond the scheme. As in Sommerfeld and Hönl's case, these transitions, represented by dotted lines in his scheme (Fig. 7.1), remained unobserved, as the respective states in the lower level did not exist. Therefore, their intensity formulas needed to "vanish" for the "virtual" components and thus provided the zeros for Kronig's quadratic equations. This, Kronig observed, gave "just enough relations for the determination of the unknown [coefficients]."[24]

solution without Goudsmit: In February Kronig wrote another letter with a detailed explanation of his calculations to help Goudsmit understand the argument. Giving credit to Goudsmit for the initial observation, they published a first paper together (Goudsmit and Kronig 1925a) in the proceedings of the *Koninklijke Akademie van Wetenschappen* in Amsterdam and presented the results in a short note (Goudsmit and Kronig 1925b) for *Die Naturwissenschaften*.

[23] Kronig presented his determination in detail in a letter to Goudsmit and in their joint paper for the proceedings of the *Koninklijke Akademie van Wetenschappen* in Amsterdam, while leaving it out of a short note for *Die Naturwissenschaften* and his later papers on multiplet intensities. Only the latter papers, however, were more widely received. They are the only ones mentioned in the historiographical literature, see Darrigol (1992, 236).

[24] For this argument see Kronig to Goudsmit, 21 February 1925 (AHQP 60.4) and Goudsmit and Kronig (1925a).

Thus solving the intensity problem in the case of the Zeeman effect, Kronig found the following intensity formulas:

$$
\begin{aligned}
i_{+1} &= A[m_1^2 + 2(J+1)m_1 + (J_1 + \tfrac{1}{2})(J_1 + 1\tfrac{1}{2})] \\
i_o &= 4A[m_1^2 - (J_1 + \tfrac{1}{2})^2] \\
i_{-1} &= A[m_1^2 - 2(J+1)m_1 + (J_1 + \tfrac{1}{2})(J_1 + 1\tfrac{1}{2})].
\end{aligned}
$$

One might note here that Kronig arrived at this result from a representation of the intensity schemes in a diagrammatic form. This representation is somewhat different from Sommerfeld and Hönl's tables. It does not introduce the statistical weights explicitly, and the relevant sums are not as visually accessible as in the tabular representation. These differences, however, did not play out in the *vanishing at the edges* argument, which Kronig and Sommerfeld and Hönl developed essentially in the same form.

The two arguments are different in one major aspect. Up to this point, Kronig had followed a line of argument that was based entirely on the intensity scheme. Unlike Sommerfeld and Hönl, he did not rely on the original Sommerfeld-Heisenberg formulas within this argument. As such, Kronig still had to establish a connection between his intensity formulas and the correspondence formulas. To show the resemblance to the binomial equations of the original Sommerfeld-Heisenberg formulas, he factorized the new intensity formulas and found:

$$
\begin{aligned}
i_{+1} &= A(J_1 + m_1 + \tfrac{1}{2})(J_1 + m_1 + 1\tfrac{1}{2}) \\
i_0 &= 4A[(J_1 + \tfrac{1}{2})^2 - m_1^2] \\
i_{-1} &= A(J_1 - m_1 + \tfrac{1}{2})(J_1 - m_1 + 1\tfrac{1}{2}).
\end{aligned}
$$

With this factorization, Kronig arrived at formulas that were analogous to Hönl's.[25] As he understood, these equations had the form of products of two different factors, instead of the square terms of the Sommerfeld-Heisenberg formulas.[26] For Kronig, this difference appeared to be minor: overall both expressions were remarkably similar.

To Kronig, this suggested that the original correspondence formulas, obtained from classical kinematic considerations, prescribed general mathematical expres-

[25] The only difference being that Kronig assumed half-integer values for the magnetic and inner quantum number, whereas Hönl had formulated his formulas for integer values (and observed the possibility of taking half-integer values as well).

[26] Note that this formulation appeared only in the short note in *Die Naturwissenschaften* and was omitted as an intermediate step in the paper for the proceedings. He thought, the two terms had the apparently significant feature that they could be written symmetrically in the quantum numbers of the initial and the final state. This interpretation, however, had no immediate consequences and I will therefore not discuss it further.

sions for the intensity. Having come to the same position as Sommerfeld and Hönl, he took the original correspondence formulas as a point of departure to determine the intensity formulas in more complicated cases of spectral multiplets. In these cases, he did not start from the existing intensity schemes, but "proceed[ed] half empirically and half with the help of the correspondence principle," as his friend Enrico Fermi characterized their mutual position.[27] Kronig, like Sommerfeld and Hönl, integrated the original correspondence formulas into the intensity schemes and formally adapted them.

After this brief summary of his argument, we can turn to the discussion of the implications for the correspondence principle that Kronig saw arising from his work. As we have seen, Sommerfeld and Hönl had been content with solving the intensity problem on this formal level and had left all theoretical implications aside. Kronig was not. He thought that the formal approach to the intensity problem led to "interesting conclusions on the sharpening of the correspondence principle."[28]

What Kronig had in mind specifically is hard to tell as he did not develop his conclusions in detail. A letter from him to Landé, however, gives some indication as to the kind of conclusion he was thinking of. In it, Kronig considered a specific class of transitions—known as intercombinations of primed and unprimed terms. In this case, the intensity formulas implied that "[t]he atom apparently behaves as if it had a component in the motion perpendicular to the electronic orbit."[29] Kronig's thinking about the implications of the intensity formulas was thus directed towards

[27]Fermi to Kronig, 10 January 1925 (AHQP 16.4). "Ich glaube auch, dass der richtige Weg um Gesetzmäßigkeiten über die Intensitäten zu finden, wohl der ist, dass man halb empirisch, und halb mit Hilfe des Korrespondenzprinzips vorschreitet [sic!]." Fermi and Kronig met during their stay in Leiden in 1924 and remained in close contact afterward. Fermi had also worked on the intensity problem on his own in Leiden and calculated the transition probabilities for multiplets on the basis of Sommerfeld and Heisenberg's correspondence formulas. Considering the problem as one of matching quantitative predictions, Fermi did not discuss the conceptual problems of the *Zwischenbahn* approach at all, concluding instead that the correspondence principle was in agreement with the empirical data within the margin of error if one calculated the multiplet intensities from the sum of their Zeeman intensities. See Fermi (1924a, 1925). Fermi, who had not shared his work with Kronig until February 1925, did not perceive Kronig's work to be fundamentally different from his own and informed Kronig that the transition probabilities, which he had calculated in this way, "only differ from yours as the intensities of the perpendicular components come out a quarter bigger for the case of the bathtub [most likely the case of $\Delta j = 0$ MJ]."

[28]Before realizing that he needed to explain the approach to his coauthor, Kronig made these comments with respect to the extension of his work to multiplet intensities. See Kronig to Goudsmit, 9 February 1925 (AHQP 60.4). "Aus den kleinen Anfängen hat sich nun während der letzten 10 Tage eine größere Arbeit mit langen Formeln für die Mehrfachlinien ohne, im schwachen, im starken, sowie teilweise auch im mittleren Magnetfeld und mit einem Permanenzprinzip entwickelt, welche interessante Schlüsse über die Verschärfung des Korrespondenzprinzips zulässt."

[29]Kronig to Landé, 11 February 1925 (AHQP 4.17). "Was besonders interessiert [sic!] ist, dass ich auch Formeln gefunden habe, welche bei Kombinationen zwischen gestrichenen und ungestrichenen Termen die Verhältnisse wiedergeben. Das Atom verhält sich dabei scheinbar so, als ob es eine Bewegungskomponente senkrecht zur Elektronenbahn hätte." The transitions between "primed" and "unprimed terms," which correspond to the transitions with $\Delta k = 0$, also

the corresponding motion of the atom in a particular case, and not towards the physical core of the correspondence principle.

To summarize, Kronig's methods as well as his goals paralleled Sommerfeld and Hönl's work rather than Bohr and Heisenberg's: As we have seen, Bohr and Heisenberg had aimed for a physical explanation of the sum rules and had been content with showing their consistency with the core idea of the correspondence principle in general. This approach had left the actual intensity distribution aside as a desideratum for further work. As such, the arguments of Bohr, Kramers, and Heisenberg first of all safeguarded the correspondence principle against Sommerfeld's attack.

Kronig proceeded in the opposite direction. Just like Sommerfeld and Hönl, he established intensity formulas which described the actual intensity distribution instead of blackboxing it. To obtain these solutions, Kronig adapted the original correspondence formulas for multiplet and Zeeman intensities to the structure of the intensity schemes and thus reformulated the intensity problem along the lines of Sommerfeld's *Gesetzmäßigkeiten* approach.

Kronig's work had tremendous implications for the Copenhagen community. It showed that Heisenberg and Bohr's qualitative argument was not enough. It now presented little more than a proof of principle asserting the consistency of the correspondence principle and the sum rules, while the actual challenge had become to develop a quantitative formulation of the intensity problem. In light of Heisenberg's earlier insistence that the correspondence principle was a physical statement about the relation between the motion of the quantum system and its radiation, such a solution ultimately would have to be based on this core idea.

In this respect, Kronig's work presented a dilemma. His intensity formulas had not resulted from a correspondence argument in which the Fourier representation of the motion was associated with the respective transitions, expressing the resulting intensity ratios in terms of quantum numbers. Rather, Kronig's adaptation had taken up the original correspondence formulas and adapted them to the structure of the intensity schemes. This meant, on the one hand, that the new intensity formulas could no longer be interpreted directly in terms of a harmonic component in the original motion; on the other hand, there was still a connection to the original correspondence argument on a formal level. Without a direct connection to the core idea of the original correspondence principle, the question was how the new intensity formulas could be interpreted in terms of the underlying motion.

As we have seen, Kronig already attempted to answer this question. Going beyond his purely formal approach, he proposed an interpretation of the intensity formulas in terms of a physical model. This attempt did not find much appreciation from Heisenberg and Pauli. Commenting on Kronig's work on multiplets, for example, Heisenberg criticized the use of the term *Ersatzstrahler*:

---

played a role in Sommerfeld's work. As we saw in Chap. 4.4, they provided the counterexample to the *Ganzzahligkeitshypothese*.

> In the literature (Landé), the word "Ersatzstrahler" has taken the meaning: "the application of the correspondence principle which one cannot understand"; I ask you sincerely to eliminate this word, which, for me, evokes memories on *Kriegs-Ersatzmarmelade* etc.[30]

By the time he made this comment in May 1925, Heisenberg was already deeply involved in developing a different approach. In it, he no longer aimed to interpret specific features in multiplet spectra on the basis of a physical model. Abandoning the tradition of the *Zwischenbahn* or the *virtual oscillator*, he sought to establish a new quantum kinematics, i.e., a new conceptual framework for describing motion. Thereby he no longer adapted the operationalization of the connection between radiation and motion, but the concept of motion and the core idea of the correspondence principle itself.

7.3 Pauli, Heisenberg, and Kronig and the Search for a *Quantum Kinematics*

While Heisenberg was the one to develop this adaptation of the correspondence principle, establishing the point of departure for a new quantum mechanics in the process, the goal of constructing a new quantum kinematics had been formulated several months before. The first attempts in this direction can be traced back to a discussion between Pauli, Heisenberg, and Kronig that took place in Copenhagen in April 1925. As is reflected in their subsequent letters, the arguments and results of the discussions were tremendously important. Pauli and Heisenberg made frequent reference to them as they discussed the development of the new quantum kinematics.

As we do not have a more detailed description of these discussions, I will reconstruct their content by connecting Heisenberg's and Pauli's admittedly fragmentary and disparate comments. Nonetheless, this reconstruction shows that the discussions between Pauli, Heisenberg, and Kronig marked a pivotal point in the approach to the intensity problem in Copenhagen. It brought Kronig's formal approach together with the reflection on the core idea of the correspondence principle. This shifted the discussion of the intensity problem away from a specific phenomenon in the *patchwork of problems* towards the conceptual development of the correspondence principle itself. This reorientation was associated directly with the new goal of constructing a new quantum kinematics and, as we will see, affected the approach to the intensity problem, down to the very choice of physical systems deemed relevant for the discussion.

[30]Heisenberg to Kronig, 21 May 1925 (AHQP 16.6). "Das Wort 'Ersatzstrahler' hat in der Literatur (Landé) die Bedeutung angenommen: 'Eine unsaubere Anwendung des Korrespondenzprinzips, die man nicht verstehen kann'; ich bitte Sie dringend dieses Wort, das für mich Erinnerungen an Kriegs-Ersatzmarmelade u.s.w. hervorruft, zu eliminieren."

From Multiplets to a New Quantum Kinematics

Preparing a short stay for Pauli, who would come to Copenhagen during his Easter vacation, Heisenberg and Pauli continued their discussion of the intensity problem in February 1925. More specifically, they addressed the work on intensity formulas for multiplets and Kramers' as yet unpublished idea of extracting transition probabilities from the dispersion formula:

> In the meantime new possibilities to sharpen the calculation of the intensities of spectral lines have appeared. (In Copenhagen one says of course: "to sharpen the correspondence principle")—*id est*, the imperialism of the correspondence principle.[31]

For Pauli, the new intensity formulas for multiplets and the work on dispersion provided new ways of "sharpening the calculation of intensities." Valuable in this respect, he rejected the idea that they constituted a sharpening of the correspondence principle itself. This perspective on the new intensity formulas and the correspondence principle did not change fundamentally during Pauli's stay in Copenhagen, as we can see from his letter to Kronig on 21 May 1925. At the same time, Pauli's position on the significance of the intensity problem for specific phenomena and the correspondence principle itself changed considerably:

> I think that after all the mere extension of the formal zoology to ever more complicated cases is an unfruitful thing. I am still very interested, however, in the general formal problem of the calculation of the transition probabilities, in particular with respect to the remodeling and extension of Born's formalism that we talked about in Copenhagen. When I turn to this formalism in my article for Springer's handbook (which has already made progress in the meantime) I will consider it again.[32]

During the discussions in Copenhagen, Pauli's letter indicates, the work on multiplet intensities was contrasted with a new perspective on the intensity problem. While the latter might still have constituted a way of "sharpening the calculation of intensities," it revolved around specific, ultimately "unfruitful" questions of spectroscopic

[31]Pauli to Heisenberg, 28 February 1925 in Pauli (1979, 212). "Inzwischen sind ja da neue Möglichkeiten aufgetaucht die Berechnung der Intensitäten der Spektrallinien zu verschärfen. (in Kopenhagen sagt man natürlich: 'Das Korrespondenzprinzip zu verschärfen'—id est Imperialismus des Korrespondenzprinzips.)" To calculate the intensities of spectral lines in the Stark effect, Pauli planned to follow Kramers' strategy. Considering Kramers and Heisenberg's dispersion formula for incoherent scattering associated with the "so-called Smekal jumps" in the limit of vanishing frequencies, he wanted to know "what you know about the probabilities of these jumps" along with the "formulas of the classical theory."

[32]Pauli to Kronig, 21 May 1925 in (Pauli 1979, 215). "Das bloße Ausdehnen der formalen Zoologie auf immer kompliziertere Fälle halte ich im Grunde doch für eine unfruchtbare Sache. Dagegen interessiere ich mich noch sehr für das allgemeine formale Problem der Berechnung der Übergangswahrscheinlichkeiten, insbesondere für die Ummodelung und Erweiterung des Bornschen Formalismus, über die wir in Kopenhagen sprachen. Wenn ich in meinem Artikel für das Springersche Handbuch (der inzwischen schon Fortschritte gemacht hat) zu diesem Formalismus komme, will ich wieder darüber nachdenken".

"zoology." In addition, Pauli thought that it was more promising to consider "the general formal problem of calculating the transition probabilities."

In his letter to Kronig, Pauli connected this question with the "remodeling and extension of Born's formalism." Ultimately, however, this remodeling did not make it into his final article for Springer's *Handbuch der Physik*.[33] In it, Pauli nonetheless commented on the intensity problem and its significance for quantum theory as a whole. Discussing Hönl's and Kronig's work on the Zeeman effect, he made the following remarkable statement:

> [W]e want to consider the issue of the values for the transition probabilities in an axially symmetric field of force as a kinematic one [. . .]. Here one has a special case of the general quantum kinematics, as yet unknown to us, which will take the place of classical kinematics operating with definite electron orbits. The [intensity] formulas teach us, at least in a formal way, which operations of quantum kinematics will take the place of the decomposition of a harmonic oscillation into a linear component parallel to a fixed direction and two left- and right-handed circular components in the plane perpendicular to this direction in classical kinematics.[34]

Pauli's statement provide crucial insights for understanding the position that had emerged in Copenhagen. Dating this statement is difficult: Pauli's article did not appear until 1926, at which time it was outdated by the advent of matrix mechanics, and by Dirac's c and q-number formalism. At the same time, the text mainly reflects the situation before the emergence of the new theory, to which Pauli referred only in a general footnote in the beginning of his text. As such, Pauli's statement on the yet unknown, general quantum kinematics makes it highly plausible that he had indeed written the passage in question before getting to know Heisenberg's work on *Umdeutung*. In other words, it presents a summary of the position emerging during the discussions in Copenhagen, or even as the research agenda that influenced Heisenberg's subsequent work.[35]

In this respect, Pauli's statement indicates that Kronig, Heisenberg, and Pauli had discussed the physical interpretation of the intensity formulas found by Sommerfeld,

---

[33]Pauli (1925, 68).

[34]Pauli (1925, 68). "Wir wollen hier jedoch die Frage nach den Werten der Übergangswahrscheinlichkeiten in einem achsensymmetrischen Kraftfeld als eine kinematische ansehen, für die es nur darauf ankommt, daß der Bewegungstypus der säkularen Störung des Feldes der einer überlagerten gleichförmigen Drehung um die Feldachse ist. Daß diese Auffassung das Richtige trifft, scheint auch daraus hervorzugehen, daß die Energiewerte des Atoms im Magnetfeld in die Intensitätsformeln in keiner Weise eingehen. Man hat es hier mit einem Spezialfall der uns noch unbekannten allgemeinen Quantenkinematik zu tun, welche an die Stelle der mit eindeutig definierten Elektronenbahnen operierenden klassischen Kinematik treten wird. Die Formeln (101) lehren uns wenigstens in formaler Hinsicht, welche Operationen der Quantenkinematik an die Stelle des Zerlegens einer harmonischen Schwingung in eine lineare Komponente parallel zu einer festen Richtung und zwei links- und rechtszirkulare Komponenten in der Ebene senkrecht zu dieser Richtung in der klassischen Kinematik treten."

[35]Pauli first learned of Heisenberg's attempts to fabricate a "quantum mechanics" in Hamburg, when Heisenberg returned from Helgoland, where he had already found results for the anharmonic oscillator and the rotator. See Heisenberg to Pauli, 21 June 1925 in Pauli (1979, 219–221).

Hönl, and Kronig. This interpretation was no longer sought with respect to the motion of a particular system in a given classical frame of reference, as in Kronig's earlier work. Rather, Pauli, Kronig, and Heisenberg now connected it with the "general quantum kinematics," i.e., the framework in which a motion was described in quantum theory.

Further, one can see that Pauli, Heisenberg, and Kronig thought that the new quantum kinematics would introduce "operations that take the place of the decomposition of a harmonic oscillation" in classical mechanics. Heisenberg's work on *Umdeutung*, of course, aimed to do just that. What Pauli's statement does not show is how such a decomposition would look like or what this would imply for a new concept of motion in quantum theory. In the absence of such a conceptualization, he pointed towards the already existing "formal" solutions to the intensity problem as an essential resource. For him, these solutions "teach us at least in a formal way which operations of quantum kinematics will take the place of the decomposition of a harmonic oscillation."

To understand how Pauli, Kronig, and Heisenberg sought to learn lessons from the formal approach to the intensity problem, we need to turn from Pauli's *Handbuch* article to Heisenberg's work on the intensity problem, leading to *Umdeutung*. The chronology of the pathway to *Umdeutung* has been described by Heisenberg himself on various occasions. It can be tied to a set of letters from Heisenberg to Kronig and to Pauli and has not been revised in the literature: In May 1925, the story goes, Heisenberg studied the intensity of the hydrogen atom but soon dropped it as too difficult. Instead he turned to the simpler non-trivial case of the anharmonic oscillator, developing the central ideas of his later *Umdeutung* paper.[36]

In the analysis of his pathway, it has gone almost entirely unnoticed that Heisenberg made short but frequent references to the Copenhagen discussions in his letters. Intended as reminders of the results already obtained, these references are essential for the reconstruction of the Copenhagen discussions. Most importantly, they show that formulating the new goal of solving the "general formal problem of the calculation of the transition probabilities" changed the approach taken to the intensity problem. Heisenberg, Pauli, and Kronig did not just convince themselves that the problem of multiplet intensities and its extension to more complicated cases would not be sufficient. They also considered a different problem that would be more suited for achieving their new goal. This reflection is manifested in the choice of a new toy model, on which their discussions centered: the anharmonic oscillator.[37]

While the reasons for this choice were not made explicit by Heisenberg or Pauli in their letters, its implications are quite clear. Leaving multiplet spectra aside meant

[36]This course of events was presented by Heisenberg himself on various occasions. For example, see Interview of Werner Heisenberg by Thomas S. Kuhn on 1963 February 22, Niels Bohr Library & Archives, American Institute of Physics, College Park, MD USA, www.aip.org/history-programs/niels-bohr-library/oral-histories/4661-7. MacKinnon and Darrigol followed Heisenberg's description, corroborating it with Heisenberg's letters to Kronig and Pauli. See MacKinnon (1977) and (Darrigol 1992, 260–268).

[37]We know about this choice from a letter by Heisenberg to Kronig on 5 June 1925 (AHQP 16.6).

moving away from a class of phenomena that was empirically accessible and widely studied at the time. Focusing on the description of motion in quantum theory, Pauli, Kronig, and Heisenberg instead chose a system that was interesting primarily from this perspective. The anharmonic oscillator was important in this respect for two interlocking conceptual reasons. Multiplet atoms, as we have seen, present an example of simple periodic motions and are capable only of making transitions to adjacent states. The anharmonic oscillator is a system capable of transitions to all its states, and presents a more general type of periodic motion. Compared to multiplet atoms, the anharmonic oscillator is thus conceptually richer with respect to both the concept of motion in quantum theory and the state-transition model. As the simplest possible example embodying this richness, the anharmonic oscillator presented an epistemic vehicle which Heisenberg, Pauli, and Kronig identified as more suitable for arriving at their new goal.[38]

The Intensity Problem and the Anharmonic Oscillator

While we can thus reconstruct the shift in perspective on the intensity problem, the question remains as to how far Pauli, Kronig, and Heisenberg had actually come in achieving their new goal of constructing the new quantum kinematics. In this respect, Heisenberg's references to the Copenhagen discussions provide further information. In his letter to Kronig of 5 June 1925, in which he developed the core idea of his *Umdeutung*, Heisenberg also treated the anharmonic oscillator and established a "classical" and a "quantum theoretical" expression for the transition amplitudes as his central result:

$$a_\tau(n) = \lambda^{\tau-1}\kappa(\tau)\sqrt{n^\tau}$$
$$a_\tau(n) = \lambda^{\tau-1}\kappa(\tau)\sqrt{n(n-1)(n-2)\dots(n-\tau+1)}.$$

This result, he reminded Kronig, "was indeed the one discussed at the time (compare the discussions with Pauli)." The explicit formulas for the Fourier coefficients of the anharmonic oscillator provide the next central insight on the approach to the intensity problem developed during the Copenhagen discussions. According to Heisenberg, the first expression is "classical," while the second one is "quantum theoretical." Both determine the Fourier coefficient $a_\tau(n)$ of the overtone τ for the

[38]Against this reconstruction, one might argue that the anharmonic oscillator could have been a first attempt to develop a description of series spectra, particularly of the hydrogen spectrum. Choosing it would thus not constitute the adoption of a new epistemic vehicle for more conceptual considerations, but an attempt to extend the specific problem-solving techniques to another class of phenomena. This position is entirely plausible. It is in conflict, however, with Pauli's letter to Kronig and Pauli's statement in his *Handbuch* article quoted above. They clearly formulate a new perspective on the intensity problem in which specific problems are separated from more general conceptual considerations, and identify the "general formal problem of the calculation of the transition probabilities" as the object of the Copenhagen discussions.

oscillator in the state n as the product of the parameter λ introduced through an expansion of the motion, a factor $\kappa(\tau)$ independent of n and the quantum number n.

In light of the previous discussion, we are in a position to reconstruct how this result came to be established. From the perspective of the *vanishing at the edges* argument, Heisenberg's quantum theoretical formula is the sharpened version of the classical one. In it, the exponential expression n^τ is replaced by the product $n(n-1)(n-2)\dots(n-\tau+1)$, in analogy to the case of the Zeeman effect, where the square in the classical intensity was replaced by the product of two zeros $(j-m)(j-m+1)$.

Interpreted within the intensity schemes, the factors in this expression are determined by the edges of the scheme. Considering the case of spontaneous emission, transitions from a particular state n lead to states $n\dots 0$. Transitions to states lower than the ground state $(-1, -2, -3, \dots)$ do not occur. According to the principle of vanishing at the edges, the Fourier coefficients should be zero for these transitions. This situation is reflected in the sharpened correspondence formula: for each transition to a non-existing state, one factor in the polynomial is zero. For example, if the system is in the state $n = 4$, the transition to the state $n = -1$ $(\tau = 5)$ has zero intensity because the factor $n - \tau + 1$ is zero; for the transition to the state $n = -2$ $(\tau = 6)$ the factor $n - \tau + 2$ is zero, and so on.

These zeros can be read off the intensity scheme itself, which is given in Fig. 7.2. Following the previous work on the intensity problem, the basic assumption would be that the Fourier coefficients are given as a polynomial of n that vanishes beyond the edges of the scheme. To determine the zeros of this polynomial, one considers a given diagonal a_τ and considers only those transitions beyond the edges that correspond to an existing initial state. One can then simply read off the zeros by looking at the respective rows; the polynomial for Fourier coefficients a_3 has to have the zeros $(n-\tau+1)$,$(n-\tau+2)$,$(n-\tau+3)$. Multiplying these zeros yields the polynomial $n(n-1)(n-2)$, given that $\tau = 3$. In general, the transitions beyond the edges will run from -1 to $-\tau$ and the polynomial will thus be given by $n(n-1)(n-2)\dots(n-\tau+1)$.

-3	$a_3(0)=0$	...			
-2	$a_2(0)=0$	$a_3(1)=0$	...		
-1	$a_1(0)=0$	$a_2(1)=0$	$a_3(2)=0$	...	...
0	$a_0(0)$	$a_1(1)$	$a_2(2)$	$a_3(3)$	...
1		$a_0(1)$	$a_1(2)$	$a_2(3)$	...
2			$a_0(2)$	$a_1(3)$	...
3				$a_0(3)$	...
...					...
n''/n'	0	1	2	3	...

Fig. 7.2 Intensity scheme for the anharmonic oscillator

It is thus possible to extract a characteristic feature of the quantum theoretical expression, i.e., its product structure, from the intensity scheme itself. We do not know whether such considerations based on intensity schemes played a role in the Copenhagen discussions. Heisenberg did not mention them in his letters and characterized the *vanishing at the edges* argument in a different way. In the discussion of the intensity problem for the anharmonic oscillator, Kronig, Pauli, and Heisenberg had extended the approach taken in Kronig's work, in which the *vanishing at the edges* argument had been used to sharpen the original correspondence formulas. In this respect, the *vanishing at the edges* argument became a central tool for translating "classical" into "quantum theoretical" intensity formulas.[39]

This reinterpretation of the *vanishing at the edges* argument as a translation procedure meant an important shift. Pauli, Kronig, and Heisenberg no longer interpreted the products in the quantum intensity formulas as zero intensities at the edges of the scheme. Rather, they saw the characteristic product structure as a result of the translation procedure, in which power expressions in the classical formula were always turned into binomial coefficients in the quantum theoretical version.[40]

As is indicated by the formulas that Heisenberg identified as the result of the Copenhagen discussions in his letter to Kronig on 5 June 1925, Pauli, Kronig, and Heisenberg had applied the *vanishing at the edges* argument in this way as a translation procedure for turning classical into quantum theoretical expressions. These formulas do not only give the product structure obtainable from the intensity scheme. Rather, they come in the form of a classical and a quantum theoretical expression for the Fourier coefficient, featuring a square root expression of the quantum number n and the perturbation parameter λ to the power of $\tau-1$.

These features could not have been obtained from considerations on the intensity scheme alone. In fact, Heisenberg, Pauli, and Kronig needed to take three steps to implement the *vanishing at the edges* argument: first, they had to establish the Fourier coefficients for the anharmonic oscillator in classical mechanics to arrive at the power expressions $\lambda^{\tau-1}$ and a_1^{τ}. Second, they needed to quantize the Fourier coefficient a_1 to get the square root expression n. Finally, they translated the quantized classical Fourier coefficient via the *vanishing at the edges* argument, which they were able to do on the basis of the intensity schemes. While Heisenberg,

[39] The most explicit formulation of this interpretation was given by Heisenberg in a letter to Kronig, 8 May 1925 (AHQP 16.6). In it, Heisenberg discussed the intensities for hydrogen and stated that he used "the principle that the jumps at the edge ought to vanish" to translate a classical intensity formula into its quantum theoretical counterpart.

[40] See Heisenberg to Kronig, 8 May 1925 (AHQP 16.6). "[D].h. was neulich trivial war, an Stelle der Potenzen treten Binomialkoeffizienten." Heisenberg made this remark in the context of his work on hydrogen intensities, in which he also applied the replacement of powers by binomial coefficients. His reminder of the discussions in Copenhagen pointed either to the case of the Zeeman effect, in which simple quadratic expression like $(j - m)^2$ were replaced by products of the form $(j - m)(j - m + 1)$, or to the case of the anharmonic oscillator, in which the classical power expression n^{τ} was replaced by the product $n(n + 1) \ldots (n + \tau - 1)$.

Kronig, and Pauli might thus have considered a derivation on the basis of the intensity schemes, it is clear at the very least that in the course of their discussions they turned to considerations based on a mechanical description of the anharmonic oscillator.

Given the information in Heisenberg's letters, we can do more than establish the existence of such considerations. The specific features of their results suffice to reconstruct the main line of their argument in more detail. To take the first step in the *vanishing at the edges* argument, Heisenberg, Pauli, and Kronig did not resort to the purely geometric considerations on which the previous formulas for multiplets were based. Their result did not describe the relative intensities of a multiplet arising from some superimposed rotation. Instead, they gave the Fourier coefficients for the motion of the anharmonic oscillator associated with individual transitions. These Fourier coefficients were obtained from recursion relations among the Fourier coefficients, resulting from the solution of the equation of motion in classical mechanics.

Such a calculation was also performed by Heisenberg in his letter to Kronig and, as David Cassidy has observed, had been made by Heisenberg as early as 1922 in a letter to Pauli in a different context.[41] While it is unclear what kind of conventions were used by Pauli, Heisenberg, and Kronig,[42] they had to follow a similar line of argument in order to arrive at this result. In order to characterize the essential features of this calculation, I will thus analyze the argument in the earlier version developed by Heisenberg in 1922.

In this calculation, Heisenberg considered the equation of motion for the anharmonic oscillator:

$$\ddot{x} + \omega_0^2 x + \lambda x^2 = 0$$

and introduced the ansatz:

$$x = a_1 e^{i\omega t} + a_2 e^{2i\omega t} + a_3 e^{3i\omega t} \ldots$$

Making this ansatz,[43] one would write down the second derivative $\ddot{x}$:

$$\ddot{x} = -a_1\omega^2 e^{i\omega t} - 4a_2\omega^2 e^{i2\omega t} - 9a_3\omega^2 e^{i3\omega t} - 16\omega_4^2 e^{i4\omega t} \ldots$$

---

[41] See Cassidy (1976, 342–345) and Heisenberg to Pauli, 29 September 1922 in Pauli (1979, 67).

[42] It is doubtful that they followed Heisenberg's earlier calculation in all its detail. In his letter to Kronig in June 1925, for example, Heisenberg adopted a different ansatz for the coordinate x–choosing a real Fourier series with a constant term.

[43] This ansatz is not the most general one, as it does not have a constant term.

and the x^2-term:

$$
\begin{aligned}
x^2 &= (a_1 e^{i\omega t} + a_2 e^{i2\omega t} + a_3 e^{i3\omega t} \ldots)^2 \\
&= a_1^2 e^{i2\omega t} + 2a_1 a_2 e^{i3\omega t} + 2a_1 a_3 e^{i4\omega t} \ldots \\
&+ a_2^2 e^{i4\omega t} + 2a_2 a_3 e^{i5\omega t} + 2a_2 a_4 e^{i6\omega t} \ldots \\
&+ \ldots
\end{aligned}
$$

Introducing these equations into the equation of motion, the central step in the calculation is to collect all terms with the factors $e^{i\omega t}$, $e^{2i\omega t}$, $e^{3i\omega t}$, ... These terms need to vanish individually to fulfill the equation of motion:

$$
\begin{aligned}
a_1(\omega_0^2 - \omega^2)\ e^{i\omega t} &= 0 \\
(a_2(\omega_0^2 - 4\omega^2) + \lambda a_1^2)\ e^{2i\omega t} &= 0 \\
(a_3(\omega_0^2 - 9\omega^2) + 2\lambda a_1 a_2)\ e^{3i\omega t} &= 0 \\
(a_4(\omega_0^2 - 16\omega^2) + \lambda(2a_1 a_3 + a_2^2))\ e^{4i\omega t} &= 0 \\
&\ldots
\end{aligned}
$$

These equations are then resolved one after the other. Taking $\omega_0^2 = \omega^2$ as shown by the first equation, the Fourier coefficient a_2 becomes:

$$
a_2 = \frac{1}{3\omega^2}\lambda a_1^2.
$$

The expression for a_2 is then introduced into the equation for the Fourier coefficient a_3:

$$
\begin{aligned}
a_3 &= \tfrac{1}{8\omega^2} 2\lambda\ a_1 a_2 \\
&= \tfrac{1}{12\omega^4}\lambda^2\ a_1^3.
\end{aligned}
$$

Using the results for a_2 and a_3, another iteration of this procedure yields a_4:

$$
\begin{aligned}
a_4 &= \tfrac{1}{15\omega^2}\lambda\ (2a_1 a_3 + a_2^2) \\
&= \tfrac{1}{54\omega^6}\lambda^3\ a_1^4.
\end{aligned}
$$

In each iteration of this procedure, the Fourier coefficient a_τ is thus given by the τth power of the Fourier coefficient a_1 multiplied by the parameter $\lambda^{\tau-1}$, leading to the general expression for a Fourier coefficient a_τ:

$$
a_\tau \propto \lambda^{\tau-1} a_1^\tau.
$$

As suggested by Heisenberg's letter to Kronig in June 1925, Pauli, Heisenberg, and Kronig had made a calculation of this type as the first step in their *vanishing at the edges* argument.

In the second, far less laborious step, they needed to introduce the quantum number n into this classical expression. To do so, they used the simple harmonic oscillator with the frequency ω and Fourier coefficient a_1 as an approximation for the first term in the series expansion of the anharmonic oscillator. This made it possible to quantize the Fourier coefficient in a straightforward way by simply considering the energy of the harmonic oscillator:

$$E = nh\omega = 2\pi^2 m a_1^2 \omega^2$$
$$a_1 = const\sqrt{n}.$$

Introducing this result into the recursion relation, Pauli, Kronig, and Heisenberg thus obtained what Heisenberg called the "classical" expression for the Fourier coefficients:

$$a_\tau = \kappa(\tau)\lambda^{\tau-1}\sqrt{n^\tau},$$

where κ is a factor independent of n. Having come to this position, Pauli, Heisenberg, and Kronig were able to apply the *vanishing at the edges* argument as a translation of classical into quantum formulas, and replace the power expression by the product $n(n-1)(n-2)\ldots(n-\tau+1)$ to obtain the "quantum theoretical" Fourier coefficient:

$$a_\tau(n) = \lambda^{\tau-1}\kappa(\tau)\sqrt{n(n-1)(n-2)\ldots(n-\tau+1)}.$$

Having reconstructed the approach to the intensity problem in the Copenhagen discussions, we are in a position to return to the question as to how far Pauli, Kronig, and Heisenberg had come in formulating the envisaged new quantum kinematics. As we have seen, the Copenhagen discussions essentially used the same techniques of making correspondence arguments as Kronig had done. Applying the *vanishing at the edges* argument for translating a classical into a quantum expression, they discovered new aspects of this procedure, like the replacement of powers by binomials.[44]

Moreover, Pauli, Kronig, and Heisenberg developed the *vanishing at the edges* argument in a new direction. In the earlier version of arguments for multiplets and their Zeeman effects, the classical intensity formulas had been obtained from purely geometric arguments involving coordinate transformations between different frames of reference. For the anharmonic oscillator, this was no longer the case. Instead, the

[44] At the same time other aspects, like the idea that the sum of the transition probabilities was constant, ceased to play a role.

expression for the Fourier coefficient a_τ was now established by considering the equation of motion. Within the formal approach to the intensity problem, however, this shift was of little importance. After all, the translation from a classical into a quantum theoretical expression occurred on the level of formal solutions. Whether the classical solutions were obtained from geometric relations or dynamic laws did not play a role.

Overall, this extension of the formal translation procedure to the new epistemic vehicle of the anharmonic oscillator and the associated reinterpretation of these techniques did not allow Pauli, Heisenberg, and Kronig to formulate a new quantum kinematics. On the basis of their formal procedures, however, they arrived at quantum theoretical expressions which such a physical approach had to reproduce. As such, they saw the approach as a point of departure for learning "which operations of quantum kinematics will take the place of the decomposition of a harmonic oscillation [...] in classical kinematics."[45]

Interim Conclusion

Heisenberg's and Pauli's comments on the Copenhagen discussions yield the following picture: Pauli, Heisenberg, and Kronig had focused on the problem of determining transition probabilities. Their approach was based on the "principle of vanishing at the edges" and thus extended the formal approach to the intensity problem developed by Sommerfeld, Hönl, and Kronig. At the same time, the goal was to go beyond this purely formal approach and, ultimately, to solve the intensity problem on the basis of the physical core of the correspondence principle. Such a solution, Pauli and Heisenberg agreed, would require a new "quantum kinematics," i.e., a fundamental reconceptualization of the concept of motion underlying the correspondence principle.

From a broader perspective, the discussions between Pauli, Heisenberg, and Kronig played a pivotal role in the Copenhagen reaction to the intensity problem. After Bohr's and Heisenberg's qualitative physical arguments based on the original correspondence principle and Kronig's formal solution, the intensity problem was no longer seen as a challenging problem within the research field of multiplet spectroscopy. Instead, it was associated with the search for a new description of motion and thus with the development of quantum theory as a conceptual framework.

This new perspective had major implications for the approach to the intensity problem on the level of argumentation: Kronig, Pauli, and Heisenberg came to the conclusion that an extension of the zoology of multiplets to more complex cases would be unfruitful. Aspiring to establish a new quantum kinematics, they searched for a different test case and found it in the anharmonic oscillator. This

---

[45]Pauli (1925, 68). See Footnote 34 for the full quote.

system was important, first of all, with respect to conceptual questions concerning the description of motion in quantum theory. It presented a system capable of transitions to all its states and a more general type of periodic motion. From this perspective, the anharmonic oscillator was conceptually richer with respect to both the concept of motion in quantum theory and the state-transition model. As the simplest possible example embodying these two features, the anharmonic oscillator presented an epistemic vehicle that Heisenberg, Pauli, and Kronig identified as more suitable for arriving at their new goal.

Making this choice and applying the techniques developed in the context of multiplet spectroscopy was not enough, however, to develop a new quantum kinematics. As we will see in the next section, the situation changed only when Heisenberg abandoned the purely formal approach and sought to adapt the description of motion underlying the correspondence principle to the emerging formal structure. In his attempt, the choice made in the Copenhagen discussions and the considerations developed in them played a central role.

7.4 The Intensity Problem and Heisenberg's *Umdeutung*

The previous reconstruction—if correct—has important implications for understanding Heisenberg's work on *Umdeutung*. It allows us to reevaluate the development of Heisenberg's work, which has often been described and analyzed without mentioning the previous discussion of the intensity problem[46]: After he left Copenhagen in April or May 1925, the standard description of the course of events tells us, Heisenberg attempted to work out the intensity for the hydrogen atom and famously failed.

This attempt, we can now see, was part of the general approach to the intensity problem. It presented an extension of the approach developed in Copenhagen, in which Heisenberg considered a classical expression for the intensity of the hydrogen lines and tried to translate it using the "principle that the intensity vanishes at the edges." Failing in this attempt, Heisenberg returned to the anharmonic oscillator and developed the central idea of his *Umdeutung*.

In it, Heisenberg formulated his version of the new quantum kinematics envisaged in the Copenhagen discussions. Like in Pauli's statement in the *Handbuch* article, he replaced the classical Fourier representation of the motion with a quantum theoretical description. In this description, the motion of a quantum system was given by the transition frequencies $\omega(n, n-\tau)$ and the transition amplitudes $A(n, n-\tau)$. This new kinematic description provided a new mathematical formulation of the physical core idea of the correspondence principle. As such, it took the

[46]See Jammer (1966), Mehra and Rechenberg (1982a), and Darrigol (1992). For a discussion of how Heisenberg's work on the intensities of hydrogen has been interpreted in the secondary literature, see Blum et al. (2017).

place of the former operationalizations of the correspondence principle in the form of the *Zwischenbahn* or the *virtual oscillator*. It did so, however, not by introducing a new physical model representing the correspondence relation. Instead, Heisenberg adapted the concept of motion itself and thus adapted not just the operationalization of the principle's physical core, but the physical core itself.

It was this adaptation which Olivier Darrigol has argued resulted from "an attempt to symbolically translate classical mechanics into a form expressed in terms of genuine quantum-theoretical concepts." As he admitted himself, however, such a "symbolic translation:"

> finds its roots in the general context of this principle, namely, the idea that a formal analogy exists between the laws of quantum theory and those of classical theory. The precise expression of this analogy as formulated by Heisenberg must be traced back to a more specific aspect of the same principle: the correspondence between quantum-theoretical spectrum [sic!] and the harmonics of a classical motion.[47]

Indeed as we have seen, the formal analogy between classical and quantum theory played a marginal role in the discussions of the intensity problem in Copenhagen in 1924 and 1925, at best framing the discussions. Rather, what was at stake was the connection between radiation and motion which Heisenberg, Pauli, and Kronig understood to be the principle's core idea in prolongation of Bohr's original formulation.

The earliest known exposition of Heisenberg's adaptation stems from a letter to Kronig on 5 June 1925. Following up on the reconstruction presented so far, Alexander Blum, Christoph Lehner, Jürgen Renn and myself have revisited this central document and analyzed the conceptual development underlying Heisenberg's work and the subsequent emergence of matrix mechanics.[48] Within the present chapter, I will summarize our reconstruction and discuss Heisenberg's work as part of the Copenhagen reaction to the intensity problem in 1924 and 1925. This discussion is the basis for a reflection on the relation of the correspondence principle as a tool for problem solving and its role in the conceptual development of quantum theory emerging from Heisenberg's work.

Heisenberg's Quantum Kinematics

In developing his quantum kinematics, Heisenberg followed two different approaches. In the first, he formulated the idea that "knowledge of the Fourier series of the motion is sufficient in the classical theory to calculate *everything*" and extrapolated this idea to quantum theory.[49] While this point of departure can indeed

---

[47]Darrigol (1992, 265–267).

[48]Blum et al. (2017). For reconstructions of Heisenberg's letter to Kronig in the secondary literature, see MacKinnon (1977, 164–171) and Darrigol (1992, 265–267).

[49]Heisenberg to Kronig, 5 June 1925 (AHQP 16.6). "In der klassischen Theorie gent die Kenntnis der Fourierreihe der Bewegung, um alles auszurechnen [...]".

be interpreted as motivated in terms of a formal analogy, Heisenberg's central aim was to develop his new quantum kinematics, and to show how every physical quantity could be expressed once the motion of the quantum system was known. In the second approach, Heisenberg used his new quantum kinematics to determine the transition amplitudes that describe the motion of the quantum system.

For the present discussion of the Copenhagen reaction to the intensity problem, the second approach is central. To understand its role in Heisenberg's work, however, we need to consider the rationale of his first approach. Developing the idea to "calculate everything" from a new kinematic description of the quantum system, Heisenberg focused on a specific physical problem: the force exerted on a point P by an anharmonic oscillator. As he had done in his discussions with Kronig and Pauli, he first solved the problem within classical mechanics and then translated this solution into quantum theory.[50]

As we have shown in our detailed reconstruction, this calculation led Heisenberg to his new multiplication rule. To express the force in terms of the motion of the anharmonic oscillator in the classical case, Heisenberg developed the force into a power series for small displacements of the coordinate x. Introducing the Fourier series of the motion of the anharmonic oscillator into this expansion, he determined the Fourier coefficients of the force b in terms of the Fourier coefficients a_τ of the motion:

$$b_0 = 1 - \frac{a_0^2 + \frac{a_1}{2} + \ldots}{a^2}$$

$$b_1 = -\frac{2(a_0 a_1 + \frac{a_1 a_2}{2}) + \ldots}{a^2}$$

$$\ldots,$$

where a is the distance of the oscillator from the point of origin.[51]

Heisenberg's calculation was straightforward in classical mechanics. Introducing the classical expression for x into the power series, he simply calculated its higher powers like x^2, x^3, x^4. To determine the overtone b_τ, he then collected the terms associated with a particular overtone τ and identified them with the respective Fourier coefficient of the force. This identification was similar to the previous calculations for the anharmonic oscillator in Copenhagen insofar as it also involved collecting terms associated with a particular frequency arising from expressions for x, its higher power x^2 and its second derivative $\ddot{x}$.

As he turned to the translation of these expressions into quantum theory, the situation became challenging. As Heisenberg realized, higher powers of x could not be calculated by means of simple multiplication as they were in the classical calculation. Within the classical kinematic description, the product of two Fourier series:

[50] Heisenberg to Kronig, 5 June 1925 (AHQP 16.6).

[51] Heisenberg to Kronig, 5 June 1925 (AHQP 16.6).

$$\sum_{\tau_1,\tau_2} a_{\tau_1} e^{(i\tau_1\omega t)} a_{\tau_2} e^{(i\tau_2\omega t)} = \sum_{\tau_1,\tau_2} a_{\tau_1} a_{\tau_2} e^{i(\tau_1\omega+\tau_2\omega)t}$$

had a straightforward interpretation. As both Fourier series have the same fundamental frequency, their product corresponds to an overtone with the frequency $(\tau_1 + \tau_2)\omega$. Within Heisenberg's new quantum kinematics, on the other hand, two frequencies $\omega(n, n - \tau_1)$ and $\omega(n, n - \tau_2)$ are not multiples of such a fundamental frequency (except in the special case of the harmonic oscillator). Their sum thus does not correspond to a frequency in the motion.[52]

For Heisenberg, this implied that a new rule for constructing products of x was needed; otherwise, the physical quantity could not be expressed in terms of transition frequencies and transition probabilities. The solution to this problem came in the form of Ritz's combination principle. Following it, Heisenberg argued in his letter to Kronig:

> One will thus try to reinterpret equation (1) [the expression for the Fourier coefficients of the force] quantum theoretically [...] the essential thing in this reinterpretation, it seems to me, is that the arguments of the quantum theoretical amplitudes have to be picked in such a way that they conform to the relation of the frequencies.[53]

Heisenberg explained how he "picked" the right transition amplitudes for the product terms in the form of an example. He considered the classical product:

$$b_2 e^{i\omega t} = (a_1 e^{i\omega t})^2$$

and gave the quantum expression

$$b(n, n-2)e^{i\omega(n,n-2)t} = a(n, n-1)a(n-1, n-2)e^{it(\omega(n,n-1)+\omega(n-1,n-2))}.$$

Writing down this expression, he stressed that the combination $\omega(n, n-1) + \omega(n-1, n-2)$ on the right-hand side was identical with the frequency $\omega(n, n-2)$ on the left-hand side. At first glance, this identification appears to be Heisenberg's way of emphasizing that both Fourier series actually have the same frequency. However, it also indicates how he translated classical products into quantum theoretical counterparts. It appears that he first considered the frequency $\omega(n, n-2)$ and the ways in which this frequency could be constructed in terms of frequencies $\omega(n, n-1)$ and $\omega(n-1, n-2)$. Fixing the frequencies on both sides of the equation, he then "picked" the transition amplitudes accordingly.

---

[52] Blum et al. (2017).

[53] Heisenberg to Kronig, 5 June 1925 (AHQP 16.6). "Man wird dann versuchen, die Gleichungen (1) quantentheoretisch umzudeuten, [...] das wesentliche an dieser Umdeutung scheint mir, dass die Argumente der quantentheoretischen Amplituden so gewählt werden müssen, wie es dem Zusammenhang der Frequenzen entspricht."

Following this prescription, the Fourier coefficients of the force in the classical calculation would be translated by writing down the quantum theoretical expression $b(n, n-\tau)$. The products of different a_τ in the classical expression are then turned into products $a(n, n-\alpha)a(n-\alpha, n-\beta)\dots a(n-\gamma, n-\tau)$, where the first factor has the same initial state n and the last factor has the same final state $n-\tau$, in keeping with $b(n, n-\tau)$.

Having thus found a way to translate his classical expressions into quantum theoretical ones, Heisenberg's attempt to calculate a physical quantity from this approach remained inconclusive. The main reason was not that Heisenberg's new scheme was unsatisfactory. Rather, it was unclear how a force exerted on a point P had to be interpreted within the state-transition model or what it meant to express it in terms of transition frequencies and transition amplitudes.[54]

Heisenberg did not provide such an interpretation or pursue his initial attempt any further at this point. Instead, he developed a second approach towards the new quantum kinematics in which he determined the transition probabilities for the anharmonic oscillator by solving the equation of motion. This brought him back to a problem with a much clearer interpretation in terms of the state-transition model, and to the calculation developed by Pauli, Kronig, and himself a few months earlier.

These calculations now became the blueprint for Heisenberg's argument. Again, he first developed a classical solution for the problem by introducing a classical Fourier representation into the equation of motion. This calculation was conceptually equivalent to the Copenhagen discussions and led to recursion relations among the Fourier coefficients:[55]

$$
\begin{aligned}
a_2(\omega_0^2 - 4\omega^2) + \tfrac{a_1^2}{2} &= 0 \\
a_3(\omega_0^2 - 9\omega^2) + \tfrac{a_1 a_2}{2} &= 0 \\
&\dots
\end{aligned}
$$

His next step was to translate the classical recursion relations into a quantum theoretical version. It was here that his new approach departed from the earlier argument. Using his new kinematic description and its multiplication rule, Heisenberg replaced the classical Fourier coefficients with their quantum theoretical counterparts and obtained new quantum theoretical recursion relations. He did so by first considering the Fourier coefficients $a_1, a_2, \dots$ and replacing them with $a_1(n, n-1), a_2(n, n-2), \dots$. This translation determined that transitions would

[54]For a discussion of the conceivable possibilities, see Blum et al. (2017).

[55]Note that Heisenberg's recursion relations are different. Due to the specific ansatz, there is a constant term a_0 and the parameter λ does not occur in the recursion relations. For a discussion of these points see Blum et al. (2017).

begin at n and end at $n - \tau$. The quantum theoretical versions of the product terms in the classical expression would also begin and end with these states and range through all states in between. Heisenberg thus translated a product in the classical expression into a quantum theoretical version by writing down all possible ways to get from n to $n - \tau$, e.g. the classical product $a_1 a_2$ would be given by $a_1(n, n-1)a_2(n-1, n-3) + a_2(n, n-2)a_1(n-2, n-3)$:

$$\begin{aligned} a_2(n, n-2)(\omega_0^2 - 4\omega^2) &= -\frac{1}{2}a_1(n, n-1)a_1(n-1, n-2) \\ a_3(n, n-3)(\omega_0^2 - 9\omega^2) &= -\frac{1}{2}(a_1(n, n-1)a_2(n-1, n-3) \\ &\quad + a_2(n, n-2)a_1(n-2, n-3)) \\ &\quad \ldots . \end{aligned}$$

After this translation, Heisenberg solved the new recursion relations as he had done in the classical case: he expressed $a_2(n, n-2)$ in terms of products of a_1, introduced this result into the expression for $a_3(n, n-3)$ and so on:

$$\begin{aligned} a_2(n, n-2) = &\quad \frac{1}{6\omega_0^2}a_1(n, n-1)a_1(n-1, n-2) \\ a_3(n, n-3) = &\quad \frac{1}{16\omega_0^2}(a_1(n, n-1)a_2(n-1, n-3) \\ &\quad + a_2(n, n-2)a_1(n-2, n-3)) \\ = &\ \frac{1}{48\omega_0^4}a_1(n, n-1)a_1(n-1, n-2)a_1(n-2, n-3) \\ \ldots . \end{aligned}$$

Each iteration of this procedure introduced another factor to the expression, so that the transition amplitude $a_\tau(n, n-\tau)$ was given by:

$$a_\tau(n, n-\tau) \propto \kappa(\tau)a_1(n, n-1)a_1(n-1, n-2)\ldots a_1(n-\tau+1, n-\tau).$$

This expression, Heisenberg understood, closely resembled the result that he had found together with Pauli and Kronig in April on the basis of the *vanishing at the edges* procedure. The only difference was that the transition amplitude a_1 was not yet quantized because he had not incorporated a quantum condition into his new kinematic description. To arrive at such a description, as he put it, one had to "take over from the theory of the harmonic oscillator that concerning the dependence on n, a_1 is given by $\sqrt{n}$" into his new scheme.[56] This meant nothing other than resorting to the solution already employed during the Copenhagen discussions and

[56] Heisenberg to Kronig, 5 June 1925 (AHQP 16.6). "Entnimmt man aus der Theorie des harmonischen Oscillators, dass, was die Abhängigkeit von n betrifft, a_1 gegeben ist durch $\sqrt{n}$ (auch in der Quantentheorie), [...]".

connecting the new expression to the result already obtained. Heisenberg already believed that this approach was little more than a patch-up. A satisfactory solution for "the determination of the constant," i.e., the quantization of his scheme, he concluded, was still a "chapter of its own."[57]

Based on this reconstruction of Heisenberg's pivotal letter to Kronig, we are in a position to discuss his work on *Umdeutung* as it relates to the work on the intensity problem in Copenhagen in 1925. Following the reconstruction, we can see that central elements of Heisenberg's work on *Umdeutung* built on key elements of the Copenhagen discussions. He followed up on the goal of constructing a new quantum kinematics formulated in Copenhagen, and developed his considerations for the model of the anharmonic oscillator, which had been identified as the epistemic vehicle to achieve this goal.

In these considerations, Heisenberg dropped the formal approach to the intensity problem underlying the Copenhagen discussions. Instead he proposed a new kinematic description based on transition frequencies and transition amplitudes, and thus gave a new operationalization of the core idea of the correspondence principle. Developing this new description, Heisenberg first considered the calculation of a force in quantum theory, whose physical interpretation was unclear, and then immediately turned to the intensity problem.

In both approaches, his argument extended the line of argument of the Copenhagen discussions. Instead of developing his consideration from scratch using his new quantum kinematic description, he still translated a calculation based on a classical kinematic description into a quantum theoretical version. In particular, Heisenberg's second approach was modeled after the calculation developed by Pauli, Kronig, and himself two months earlier. There he had also solved the problem of the anharmonic oscillator classically and determined the Fourier coefficients by means of recursion relations before translating this expression into quantum theory. This approach, Heisenberg thought, would be applicable in general:

> This formula is now indeed the one discussed at the time (cf. the discussions with Pauli), and I dare say that one really thereby has a general law for calculating the intensities: From the equations of motion one obtains simple relations between the a_τ that determine the a_τ (for f degrees of freedom up to f independent constants). After the quantum theoretical transformation, take over these relations directly into quantum theory and one has the intensities (again up to f independent constants). The determination of the constants is a chapter of its own and I do not want to write about it today. But one can show, for example, that your intensity formulas for multiplets and Zeeman effects appear to follow from the above developed scheme.[58]

[57] Heisenberg to Kronig, 5 June 1925 (AHQP 16.6). See Footnote 58 for the full quote.

[58] Heisenberg to Kronig, 5 June 1925 (AHQP 16.6). "Diese Formel ist nun in der Tat die seinerzeit schon besprochene (vgl. die Diskussionen mit Pauli), und ich könnte mir denken, dass man damit wirklich ein allgemeines Gesetz zur Berechnung der Intensitäten hat: Aus den Bewegungsgleichungen ergeben sich einfache Beziehungen zwischen den a_τ, die (bei f Freiheitsgraden bis auf f unabhängige Konstante) die a_τ bestimmen. Diese Beziehungen übernehme man, nach quantentheoretischer Verwandlung, direkt in die Quantentheorie und hat (wieder bis auf f unabhängige Konstante) die Intensitäten. Die Bestimmung der Konstante ist noch ein

Heisenberg's summary thus linked his new approach directly to the previous discussions of the intensity problem. It did so in a dual sense. First, it showed that his new quantum kinematics led to results already obtained, and that the new scheme thus provided a reasonable candidate for the new quantum kinematics. Second, Heisenberg's summary formulated a prescription for getting results, which applied both to Heisenberg's new physical approach as well as to the earlier formal argument.

As such, Heisenberg pointed to the genetic relationship between his *Umdeutung* and the Copenhagen discussions on the level of argumentation. In this respect, the change was minor. He merely translated classical calculations into quantum theory at a slightly earlier stage. In the Copenhagen discussions, as we have seen, the solution to the recursion relation had been translated into quantum theory. Using his new kinematic description, Heisenberg now translated the recursion relations themselves.

However minute the change in the calculation might appear, Heisenberg's new approach marked a crucial step. On the one hand, his new kinematic description provided the desired solution to the intensity problem in terms of the core idea of the correspondence principle. On the other hand, Heisenberg's return to the intensity problem and the previous calculation was important for the conceptual development of his scheme. It led to the introduction of the equation of motion into his scheme and thus provided the point of departure for the transformation of his new kinematic description into a mechanical framework.

Given the classical interpretation of the equation of motion, it at first appears that his purely kinematic description was now transformed into a full-blown dynamic description analogous to classical mechanics. For Heisenberg, this was not the case at the time. As he understood it, the equation of motion no longer played the same role as in classical mechanics. Rather it provided "relations among the Fourier coefficients." In this respect, they were no different from the relations obtained from geometric considerations in the case of multiplets, and the physical significance of the equation of motion remained unclear for Heisenberg for some time. As he confessed to Pauli in a letter a month later: "Also, I would like to understand what the equations of motion actually mean if one considers them as relations among transition probabilities."[59]

Kapitel für sich u. ich will darüber heut nichts schreiben. Aber man kann z.B. zeigen, dass Ihre Intensitätsformeln der Multipletts u. Zeemaneffekte auch aus dem eben ausgeführten Schema zu folgen scheinen."

[59]Heisenberg to Pauli, 24 June 1925 in Pauli (1979, 228). "Auch würd' ich gern verstehen, was eigentlich die Bewegungsgleichungen bedeuten, wenn man sie als Relation zwischen [den] Übergangswahrscheinlichkeiten auffasst."

The Completion of Heisenberg's Scheme: The Quantum Condition and the F-sum Rule

These uncertainties about the interpretation of the new scheme played a minor role in Heisenberg's work at the time. For him, it was important first of all to complete his new computational scheme, and to further extend its range of application. From this perspective, the main problem was the "determination of the constants," which Heisenberg had identified as the missing element of his new scheme in his letter to Kronig. As is well known, Heisenberg solved this problem about a month later by introducing a new quantum condition. In a letter to Pauli,[60] he argued that one should translate the quantum condition of the old quantum theory:

$$J = \oint pdq = nh$$

into a quantum theoretical version using the translation techniques of dispersion theory. He showed how this is done for a multiply periodic system with the classical coordinate $q = \sum a_\tau e^{i\omega t\tau}$ and momentum $p = m\dot{q} = \sum mia_\tau(\tau\omega)e^{i\omega t\tau}$. Introducing these expressions into the quantum condition:

$$J = 2\pi m \sum a_\tau^2(\tau\omega)\tau,$$

he calculated the derivative with respect to J:

$$1 = 2\pi m \sum \tau \frac{d}{dJ} a_\tau^2(\tau\omega).$$

Following the scheme of dispersion theory, the operator $\tau d/dJ$, which acts on the classical Fourier coefficient and the overtone $\tau\omega$, is replaced by the difference between the corresponding transition amplitude and the frequency divided by h:

$$1 = \frac{2\pi m}{h} \sum_\tau (a^2(n+\tau, n)\omega(n+\tau, n) - a^2(n, n-\tau)\omega(n, n-\tau)).$$

This relation, Heisenberg realized, allowed him to quantize simple harmonic systems like the harmonic oscillator, which can only make transitions to adjacent

[60]Heisenberg to Pauli, 24 June 1925 in Pauli (1979, 227). There is no direct evidence indicating how Heisenberg actually arrived at his new quantum condition. In his presentation of the derivation to Pauli, Heisenberg did not explain why the old quantum condition underlying the solution in the Kronig letter was insufficient, or why his new quantum condition was more satisfactory. Presenting the quantum condition in essentially the same way in the final publication, Heisenberg still made a spurious argument. The old quantum condition, he argued, only fit into a mechanical scheme in an arbitrary way because the action was only determined up to a constant. For a discussion of a possible conceptual motivation for the new quantum condition, see Blum et al. (2017).

states. In this case, there is only one difference to evaluate. Considering the situation in the ground state, Heisenberg assumed that there are no transitions to lower states and thus obtained the transition amplitude $a(0, 1)$:

$$a(0, 1) = \sqrt{\frac{h}{2\pi m\omega_0}}.$$

He could then determine the transition probabilities for higher transitions recursively: He assumed that transition probabilities for inverse transitions were identical, i.e., $a(0, 1) = a(1, 0)$, so that his first result determines one summand in the next difference:

$$1 = \frac{2\pi m}{h} \sum_{\tau} (a^2(2, 1)\omega_0 - a^2(1, 0)\omega_0)$$

$$a(2, 1) = \sqrt{\frac{2h}{2\pi m\omega_0}}.$$

In general, this procedure gave the transition probability:

$$a(n, n + \tau) = \sqrt{\frac{h}{2\pi m\omega_0} n}.$$

Heisenberg's argument is a familiar one. We already discussed it in Chap. 6 in the context of the work on the f-sum rule developed by Reiche and Thomas and their attempts to determine transition probabilities for harmonic systems. They, too, used the translation techniques of dispersion theory to translate a relation for the Fourier coefficients of a multiply periodic system into a relation among transition probabilities. Arriving at the same statement as Heisenberg, they used this quantum expression to determine the transition probabilities for simple periodic systems on the assumption that there are no transitions below the ground state.

As was argued in Chap. 6, Reiche and Thomas arrived at this result on their own after Kramers and Bohr had claimed priority on the subject for the Copenhagen community. A further indication of the independence of the two approaches is that Heisenberg made a connection between his new quantum condition, which he had formulated after leaving Copenhagen, with the work of Reiche and Thomas and the dispersion theory only in the published version of his *Umdeutung*. The independence of these two pathways has important implications. It allows us to compare the trajectory of Heisenberg's work with that of Reiche and Thomas and to identify the role of the f-sum rule / quantum condition in their respective approaches.

In their work on the f-sum rule, we have seen, Reiche and Thomas ran into a dead end. As they realized, the new relation allowed them to determine the transition probability only for simple harmonic systems capable of transitions to adjacent states. For systems not restricted by selection rules, the f-sum rule was no longer sufficient. There is evidence that Reiche worked on this problem in the fall of 1924 as he tackled the hydrogen atom. However, he did not get very far, as is reflected in Reiche and Thomas's paper on the f-sum rule, which stuck to simple periodic systems. While the f-sum rule was a constraint for the transition probabilities of every quantum system, its actual range of application was limited to a single class of systems.

For Heisenberg, the situation was different. For him, the search for a new quantum condition was not at the focus of his research as the f-sum rule had been for Reiche from 1923 to 1925. Instead, as we have seen, he had been working on a new kinematic description of the quantum system to follow up on the intensity problem for multiplets. Having developed such a new kinematic description within the state-transition model, he developed the new quantum condition to fill an argumentative gap in his approach. This did not change the fact that only the transition probabilities of simple harmonic systems could be determined directly from the f-sum rule / quantum condition. Within Heisenberg's approach, however, the quantum condition was not just a singular equation. It was embedded in a more general calculational scheme. This allowed Heisenberg to use it to tackle other systems like the anharmonic oscillator by considering the case of the simple harmonic oscillator as an approximation in the anharmonic case. Of course, finding a way to establish such an approximation was difficult, and Heisenberg admitted that it was so far only possible for systems like the anharmonic oscillator. In principle, however, the new kinematic description could apply to systems capable of transitions between all of its states and provided a framework into which the f-sum rule could function as a quantum condition.

With the incorporation of the quantum condition, Heisenberg solved the intensity problem for the anharmonic oscillator completely within his new quantum kinematical scheme. As such, Heisenberg's treatment of the anharmonic oscillator marks a point of closure for the discussion of the intensity problem in Copenhagen. Concluding my analysis, I will discuss the Copenhagen reaction with respect to the relation between the *patchwork of problems* and the conceptual development of quantum theory. In this discussion, I will look at the trajectory of the work done in Copenhagen and contrast it with the case studies presented in the previous chapters.

7.5 Conclusion

In this chapter, I have followed the discussion of the intensity problem in Copenhagen from 1924 to 1925 and presented a detailed analysis of the shifting perspective on it and its role in the development of Heisenberg's *Umdeutung*. This analysis departs considerably from the reconstruction of Heisenberg's pathway in the secondary literature, in which dispersion theory or the virtual oscillators of BKS theory were seen as the direct precursors to Heisenberg's work. Against this view, I have argued for a new contextualization: Heisenberg's work emerged from an extensive discussion of the multiplet intensity problem and its relation to the correspondence principle.

This discussion, as I have shown, was a reaction of the Copenhagen community to the work of Sommerfeld on multiplet intensities and his "attack on the correspondence principle." Initially, it was a defensive attempt to save the correspondence principle, in which the Copenhagen community acknowledged only gradually that the intensity problem presented a challenge to the principle. In reaction to this challenge, Bohr and Heisenberg argued that the core idea of the correspondence principle was not affected by it, and that it merely meant adopting the *virtual oscillator* model as a new physical model representing this idea. This argument, as we have seen, suggested that the Utrecht sum rules were consistent with the core idea of the correspondence principle. At the same time, Bohr, Kramers, and Heisenberg blackboxed the quantitative description of multiplet intensities.

The Copenhagen community went beyond this position in January 1925, when Ralph Kronig developed quantitative intensity formulas. Kronig's methods as well as his goals paralleled Sommerfeld and Hönl's work rather than Bohr and Heisenberg's: he established intensity formulas that described the actual intensity distribution instead of blackboxing it. To obtain these solutions, Kronig adapted the original correspondence formulas for multiplet and Zeeman intensities to the structure of the intensity schemes. This reformulation of the intensity problem along the lines of Sommerfeld's *Gesetzmäßigkeiten* approach had tremendous implications for the Copenhagen community. It showed that Heisenberg and Bohr's argument presented little more than a qualitative consistency argument, while the actual challenge was to develop a quantitative formulation of the intensity problem.

More importantly, Kronig's formal solution to the intensity problem was no longer connected directly to the kinematic description of atomic motion. It was thus detached from the core idea of the correspondence principle. In light of the commitment to the correspondence principle as a statement about the relation between radiation and motion, a solution to the problem would have to be based on this core idea.

In the spring of 1925, Pauli, Heisenberg, and Kronig began to work on a kinematic interpretation of the new intensity formulas. Their discussions, as we have seen, played a pivotal role in the Copenhagen reaction to the intensity problem and proved important for Heisenberg's subsequent work on *Umdeutung* in many ways. First of all, they abandoned the idea of establishing a new physical model for the correspondence principle. Instead, they formulated the goal of establishing a new quantum kinematics. To arrive at this goal, Pauli, Heisenberg, and Kronig moved away from multiplet intensities and identified the anharmonic oscillator as the central epistemic vehicle. Working on this new test case, the Copenhagen discussions led to a solution of the intensity problem for the anharmonic oscillator that became a blueprint for Heisenberg's calculations based on his new quantum kinematic description. As such, Heisenberg's *Umdeutung* emerged in large part from an attempt to solve the intensity problem on the basis of the core idea of the correspondence principle.

The present reconstruction of the work on the intensity problem in Copenhagen touches on several of the main themes of this book, which I will discuss in the following. First of all, by discussing the reaction to Sommerfeld's work on multiplet intensities, it provides a case study of how the applications of the correspondence principle were received in Copenhagen. In this respect, it is important for understanding the dissemination and adaptation of the correspondence principle. On the one hand, it captures the relation between problem solving in the *patchwork of problems* and the conceptual development of quantum theory. On the other hand, it offers the possibility of comparing the approach to the correspondence principle taken in Copenhagen in 1924/1925 with the approaches developed in Munich and Breslau discussed in the previous chapters. Last but not least, the present reconstruction offers a new understanding of the transformation of the concept of motion that took place in Heisenberg's *Umdeutung*.

Conceptual Development Within the Patchwork of Problems

The discussion of the intensity problem in Copenhagen was different from the applications of the correspondence principle in Munich, Breslau or Göttingen in one major respect. As we have seen in the preceding chapters, Sommerfeld, Reiche, and Franck applied the correspondence principle within their work on particular research problems, focusing on the determination of transition probabilities. By contrast, physicists in the Copenhagen community focused on the correspondence principle itself, considering it to be an essential part of quantum theory. For them, the intensity problem was connected, first of all, with its core idea of a connection between radiation and motion. The intensity of multiplets or the number of dispersion electrons were manifestations of this problem, possibly providing *Fingerzeige* for developing the principle.

These different perspectives translated directly into the adaptations of the correspondence principle emerging from them. Integrating the principle into their work on specific problems, Sommerfeld, Heisenberg and Hönl, Reiche and Thomas, and Franck and Hund used the correspondence principle as a tool. Their adaptive reformulations changed either the physical core of the correspondence principle qualitatively or the computational recipes of the correspondence arguments. In both cases, physicists hardly problematized their adaptations, be it by reflecting on the significance for the formulation of the principle in general or by developing consequences of their reformulation for other phenomena.

These two types of adaptation, as we have seen, were also present in the Copenhagen reaction: In the fall of 1924, Bohr, Kramers, and Heisenberg invoked a qualitative adaptation of the principle when they associated the transition probabilities with the motion in the initial state rather than with the *Zwischenbahn*. For them, the intensity problem was equivalent to the explanation of the Utrecht sum rules. These rules presented a qualitative challenge for the operationalization of the principle in terms of a physical model, and hence called for a qualitative adaptation of this operationalization.

In the spring of 1925, Kronig, Pauli, and Heisenberg adapted the correspondence arguments for the intensities of multiplets or the anharmonic oscillator in a formal way and arrived at a quantitative formulation of the problem. For them, the intensity problem had become to determine the intensities of multiplet lines explicitly. In this reformulation, the intensity problem presented the quantitative challenge of finding a computational scheme for such a determination, and the qualitative challenge of interpreting the resulting intensity formulas in terms of the physical core of the principle. These two challenges led to an adaptive reformulation on both the computational and the conceptual level. Initially, the Copenhagen community developed and explored a formal approach to the intensity problem based on intensity schemes and adapted the correspondence argument on a computational level. In a next step, Heisenberg adapted the concept of a corresponding motion, prolonging and reshaping the approach to the intensity problem on a computational level and providing a physical interpretation of the intensity problem.

The attempt to develop such an adaptation was rooted ultimately in the commitment to the physical core of the correspondence principle and the belief in its importance for quantum theory. This commitment, however, was not sufficient for Heisenberg's quantitative formulation of the correspondence principle. As we have seen, it emerged only when the Copenhagen community went beyond general consistency arguments and took the intensity problem seriously in its specific quantitative formulation. A solution to the problem, Pauli, Heisenberg, and Kronig believed, would require an adaptation of the correspondence principle in the form of a new quantum kinematics.

This insight came with a shift in perspective on the intensity problem. With it, the determination of transition probabilities ceased to be associated primarily with a specific physical problem in multiplet spectroscopy or dispersion theory. In other words, it was removed in part from the work on the *patchwork of problems*. Instead the problem was seen as a way to develop quantum theory as a conceptual framework. This shift, as we have seen, affected the approach to the intensity problem, down to the decision of crucial exemplary cases. In the search for a new quantum kinematics, the models used to represent various phenomena in the *patchwork of problems* were considered insufficient: The precessional motions in the case of multiplets, and the harmonic oscillator in the case of dispersion, belonged to the class of harmonic motions and thus represented systems capable of making transitions to adjacent states. While they were suited to describe the phenomena in question and also led to epistemic conflicts in these contexts, Heisenberg, Pauli, and Kronig saw that these simple harmonic systems were not rich enough from a conceptual point of view. To develop a new concept of motion adapted to the formal structures of the state-transition model, they thought one needed a representation capable of describing transitions between all possible states. As such, they opted for the anharmonic oscillator as a new epistemic vehicle embodying this feature.

Heisenberg's Umdeutung *Between Spectroscopic Classification and Quantum Mechanics*

The above reconstruction provides a new understanding of the transformation embodied in Heisenberg's *Umdeutung*. As we saw in this chapter and in Chap. 4, Heisenberg's main experiential resource—the intensity problem for multiplets—was situated in atomic spectroscopy. Ultimately, it emerged from practices of classifying and systematizing experimental data. In the 1910s, these practices had encoded spectroscopic knowledge into general formal descriptions like Ritz's combination principle, and had been amalgamated with the state-transition model. In the first half of the 1920s, these spectroscopic classifications were further developed by physicists like Alfred Landé and Arnold Sommerfeld, who championed the analysis of spectroscopic data based on diagrammatic term schemes.

The developing spectroscopic classifications were largely independent of concepts of classical physics. If anything, the notion of a spectroscopic term and a combination of two such terms was interpreted in terms of the state-transition model. While this interpretation encompassed the idea that spectroscopic terms corresponded to states of different energies, the state-transition model was important, first of all, for labeling spectroscopic terms with quantum numbers and formulating empirical regularities like selection rules.

With the emergence of new photometric techniques, as we have seen, spectroscopic classifications were enriched by incorporating intensities and formulating the regularities governing them. Establishing a new representation of spectroscopic classifications, Sommerfeld and others encoded these regularities into intensity schemes. Using them, they went beyond the mere classification of spectral data and instead aimed to predict intensities without recourse to observation. In the attempt to develop such predictions, as we have seen, they turned to the correspondence principle and the quantum theory of the atom. This allowed them to complete their formal descriptions by adapting the results of the correspondence approach to the structures of the intensity schemes.

As discussed above, it was this successful description of spectral intensities that provided the central experiential resource for Heisenberg's work on *Umdeutung*. In this respect, the formal solutions to the intensity problem developed for multiplets, and that for the anharmonic oscillator developed by Kronig, Heisenberg, and Pauli played a central role: they provided a formulation of the intensity problem within the state-transition model and explored how it was related to the calculation based on the correspondence principle and a classical description of atomic motion.

This formulation of the intensity problem provided the first half of Heisenberg's adaptation. The second half was rooted in the idea of a connection between the motion of a quantum system and the radiation emitted by it, which presented the stable core of the correspondence principle throughout the 1920s. Emerging in large part as a reflection on the new description of the intensity problem within the state-transition model, Heisenberg's *Umdeutung* thus presents an integration of the spectroscopic classification embodied in the intensity schemes into the description of motion in classical mechanics. This integration resulted in an adaptation of the concept of motion, in which transition probabilities—as a central property of the state-transition model—became transition amplitudes and were used to construct kinematic and dynamic variables. This meant a transformation of the state-transition model into a quantum mechanical description, and thus a physicalization of spectroscopic classifications. At the same time, it became the starting point for a reinterpretation of classical mechanics into quantum mechanics. It operated with fundamentally different physical quantities, whose interpretation was anything but clear when Heisenberg developed his new quantum kinematics, and became the starting point of the interpretational debate of quantum mechanics.

Chapter 8
Conclusion

The history of the correspondence principle presented in this book has followed a theoretical tool and its use in many strands of research embedded in the *patchwork of problems* of quantum physics during the 1920s. In these highly local contexts, the implementation of the correspondence principle led to transformations of both the problems and the tool. The developments arising from it led in various directions, which the historical actors did not perceive as pointing toward a new theory of quantum mechanics. The emergence of such a theory in the form of matrix mechanics appeared as a reflection on the applications of the correspondence principle on a higher theoretical level, rather than as the achievement of a long-sought research goal.

Concluding this study, I will return to the discussion of its more general themes. This discussion comes in two parts. First, I will look at the dissemination of the correspondence principle in the quantum network and its use as a tool in the work on the *patchwork of problems* from a bird's eye perspective. This discussion leads to a reflection on the implications for the analysis of problem solving and conceptual development when writing a history of theoretical tools.

Second, I will analyze how individual correspondence arguments were made. Given the different empirical situations, the variety of theoretical approaches, and institutional settings, it is difficult to capture this process within a single narrative. Nonetheless, I believe it makes sense to characterize the features of this process in more detail by contrasting the individual cases and identifying similarities and differences among them. This makes it possible to characterize the fine structure of the conceptual development of the correspondence principle resulting from its *transformation through implementation* and reflect on the historiographic implications for studying the *practice of theory* and the analysis of conceptual development.

M. Jähnert, *Practicing the Correspondence Principle in the Old Quantum Theory*, Archimedes 56, https://doi.org/10.1007/978-3-030-13300-9_8

8.1 The Correspondence Principle and the *Patchwork of Problems*

Developed by Bohr and Kramers in their work on the quantum theory of multiply periodic systems, the correspondence principle was understood as a kinematic relation. Its core idea was a connection between the motion of a quantum system and its radiation. It governed the possibility of transitions in the *state-transition model* and linked the radiative transitions to the motion of the states.

In this capacity, the correspondence principle initially served as an explanation of the selection rules governing atomic and molecular spectra. As we saw in Chap. 2, this explanation was central for Bohr as a means to integrate various representations of atoms and molecules like the harmonic oscillator, the rotator and Bohr's planetary model of the atom into his quantum theory of multiply periodic systems. At the same time, Bohr gave only a general and qualitative formulation of his correspondence idea. This formulation remained ambiguous, as it left major conceptual issues like the initial-final-state problem untouched and did not lend itself to straightforward operationalization or even mathematization. Taking a step in this direction, as we have seen, Kramers developed a model representation for the correspondence relation that became known as the *Zwischenbahn*. It presented a tentative solution to the initial-final-state problem and yielded a first quantitative formulation of the correspondence relation, allowing for the determination of transition probabilities from the corresponding motion.

The conception of the correspondence principle developed by Bohr and Kramers was studied by physicists outside of Copenhagen mostly from published papers. From these sources, as the overview on the dissemination of the correspondence principle in Chap. 3 and the case studies in Chaps. 4 through 6 have shown, physicists outside of Copenhagen understood the core idea of the correspondence principle and its initial range of application in a homogeneous way. Without a mathematically precise formulation, the principle's core was explicated in a general way; nonetheless, it was clearly formulated: the correspondence principle established a connection between the motion of a quantum system and its radiative transitions.

This understanding, however, did not lead to the adoption of the correspondence principle from 1918 onwards. Rather, the year 1922/1923 became the watershed for the dissemination of the correspondence principle within the communities in Europe and the U.S., as the previous reception turned into the active use of the principle. Suddenly, it was applied in research on multiplet and molecular spectroscopy, dispersion theory and electron scattering, becoming significant beyond Bohr's quantum theory of multiply periodic systems in the process.

Bringing the correspondence principle into contact with existing theoretical notions, techniques and models, as will be discussed in more detail below, the implementation of the correspondence principle provided the basis for the emergence of tensions and, consequently, for the adaptive reformulation of the correspondence principle. What is important here is that the integration of the correspondence principle into different research fields in the *patchwork of problems* presents the key source for the diversification of the principle. As physicists adopted the correspondence principle, they bridged the gaps between their new tool and the existing frameworks in different ways and developed their own correspondence arguments. This integration was far more decisive for the diversification of correspondence arguments than the more subtle variations in the interpretation of the principle or personal connections to Copenhagen.

This can be seen most clearly in the case of Franck and Hund. Their work was shaped more than any other by communication with Copenhagen. Franck's visit motivated a transfer of the correspondence argument from one field to another and transformed the perspective on the problem of scattering in Göttingen. Still, Franck and Hund integrated the principle into their work and developed their own quantum theoretical conception of scattering, leading to a correspondence argument that was strikingly different from the one developed in Copenhagen. In this sense, Franck and Hund were no different from Reiche, who developed and reshaped his correspondence arguments within his changing view on the problem of absorption and dispersion in quantum theory, or Sommerfeld, who first integrated the principle into his model of multiplet atoms based on space quantization and then adapted it within his *Gesetzmäßigkeiten* approach.

The Dissemination of the Correspondence Principle: A Self-Stabilizing Process

The late onset and rapid increase of correspondence arguments calls for an explanation, and I suggested two candidates in Chap. 3: on the one hand, the growing importance of transition probabilities in quantum physics; on the other hand, the rise of Copenhagen as a research center within the quantum network. Both of these candidates could count as central forces driving the adoption of the correspondence principle as a tool: As we saw in the case studies, the determination of transition probabilities was central in either way and emerged as a problem in the majority of cases before the correspondence principle was adopted. Likewise, personal communication with Copenhagen played a central role.

While this is the case, I believe that neither candidate qualifies as an explanation for the adoption of the correspondence principle, at least not in a strong causal sense. By 1922, none of the leading theoretical figures like Bohr, Sommerfeld or Born was actively promoting the determination of transition probabilities as a research goal on a larger scale. Likewise, they did not call for improvement of the description of the radiation process within the state-transition model.[1] Similarly, there was no single experimental innovation that put the determination of transition probabilities on the

[1]In general, there were almost no programmatic statements that might have led in this direction. Chap. 4 discussed one prominent example: the idea of the young Heisenberg to characterize atomic

agenda. Even the Utrecht intensity measurements, which boosted work on transition probabilities in multiplet spectroscopy, by no means affected all research fields in the *patchwork of problems* in the same way.

Instead of trying to identify a single theoretical or experimental driving force, we need to accept that the motivations for making correspondence arguments were extremely diverse: Sommerfeld saw the opportunity to account for the *Gesetzmäßigkeiten* of multiplet intensities, Heisenberg sought to extract a model interpretation from empirical data, Franck saw the possibility to explain the Ramsauer effect, Reiche wanted to complement the comparison of empirical values for transition probabilities with theoretical predictions. In yet another case, Edwin Kemble turned to the correspondence principle to refute a theoretical proposal independent of the nature of the transition process (i.e. half-integral quantization introduced by Adolf Kratzer).

In a similar way, the institutional settings and personal situations in which physicists took up the correspondence principle differed so widely from each other that they provide no clear causal explanation for the emergence of correspondence arguments. Unlike the dispersion of Feynman diagrams, the primary users of the correspondence principle were not young postdoc, who had been trained to do so. Rather, the principle was used by leaders of scientific schools like Sommerfeld or Born, by established professors like Ehrenfest, Franck, Reiche, Kemble, or Richard Tolman, and by postdocs and Ph.D. students like Thomas, Heisenberg, Hönl, Fermi, Kronig, or Hund.

For these physicists, Bohr and his institute did not play a role that was similar to the one played by Freeman Dyson and the *Institute for Advanced Study* for the dispersion of Feynman diagrams in the 1950s. For the most part, physicists became acquainted with the correspondence principle and developed their first correspondence arguments on their own without contact to Copenhagen. This is illustrated by Sommerfeld's case or by the large group of physicists in the U.S. like Edwin Kemble or Richard Tolman, who had little chance of meeting Bohr in person and did not engage in private discussions with Copenhagen as they developed their correspondence arguments.

In some cases, like those of Franck, Reiche, or Kronig and Heisenberg, interaction with Copenhagen was central. With the exception of Franck, however, these physicists had already developed their correspondence arguments. Rather than being introduced to a new tool in Copenhagen, they received additional input from Bohr or Kramers and developed new perspectives on the principle by themselves. As such, they actively incorporated the correspondence principle into their work and then developed their correspondence arguments further in a dialog with the Copenhagen community.[2]

---

orbits through the transitions leading to and from it. Tellingly, this idea was presented only in a letter to Alfred Landé, rejected as nonsense, and remained unpublished.

[2]For the importance of dialogs in the development of quantum physics, see Hendry (1984) as well as Beller (1999).

Given the diversity of motivations and situations involved in the making of new correspondence arguments, we can thus hardly attribute their emergence to a single cause. At the same time, it is all the more remarkable that the applications of the correspondence principle took off around 1922. Unless we want to accept that various actors working independently on different problems miraculously redirected their research in similar ways, we need to strike a balance between pure historical contingency and a monocausal explanation. To do so, I suggest to stop focusing primarily on the emergence of new correspondence arguments and thus to stop looking for primary causes. Instead, the dissemination of the correspondence principle is better understood as a process that stabilized itself under certain conditions in analogy with a crystallization process.

At first sight, it might appear that this perspective does not resolve the problem discussed above. In a way, it does not. The interpretation I am proposing here does not explain in a strong causal sense why the applications of the correspondence principle emerged in 1922 and not 2 years before or after. Instead, I suggest a perspective from which we can understand how the applications of the principle became a relatively stable, communal phenomenon between 1922 and 1926, despite being triggered in multiple ways on the level of individual historical actors.

In this respect, thinking about the applications of the correspondence principle as a self-stabilizing process allows us to identify the determination of transition probabilities as a nucleus for new correspondence arguments. These arguments relied on two main theoretical resources: the state-transition model, on the one hand, and the idea of a corresponding motion, on the other.

This description alone, however, was not enough; other stabilizing factors were needed. Where these factors were not in place, historical actors moved on to other problems rather than prolonging their applications of the correspondence principle. This was the case in the work of Kramers and Sponer on atomic and molecular spectroscopy in 1919, or in the work of Franck and Hund on the Ramsauer effect in 1922. Sommerfeld and Heisenberg's work could have ended in a similar way. Initially, they considered the qualitative description of multiplet intensities to be all that could be hoped for. Similarly, Reiche's implementation appeared to merely confirm Ladenburg's argument in general without opening up new theoretical avenues.

Fortunately, we can clearly identify why the latter applications led to continued discussions. New experimental results put Sommerfeld and Heisenberg's research on a new foundation and turned the study of transition probabilities into a sizable field of interest. In Reiche's case, the study of dispersion received important input from Copenhagen, leading to a revision of his argument. In other words, the interaction with Copenhagen, developing theoretical approaches, and new experimental results all stabilized the applications of the correspondence principle. They provided communal, theoretical and empirical resources available to physicists making correspondence arguments.

Some of these resources were available prior to 1922 and were perceived as such. What was needed, however, were combinations of these resources that provided something like a state of supersaturation, to remain within the metaphor of a crystallization process. As indicated by the preceding case studies, such combinations were in place around 1922. By this point, several highly context-specific motivations were able to set the development of new correspondence arguments in motion, which then stabilized themselves rather than dropping out of sight. This perspective provides a way to understand the stability of the applications of the correspondence principle without a central driving force.[3]

The description of the applications of the correspondence principle as a crystallization process is similar to the idea of an *autocatalytic process* proposed by Alexander Blum, Roberto Lalli and Jürgen Renn.[4] They used it to explain the emergence of a field of "general relativity and gravitation" 40 years after the genesis of general relativity, and argued that neither the emergence of new empirical phenomena in astrophysics nor the availability of new funding opportunities was enough. Rather, several interlocking factors—conceptual, experimental, and environmental—needed to come together so that a new community of general relativists could be formed at key conferences, eventually leading to the establishment of a new research field.

While the renaissance of general relativity and the *autocatalytic process* bear some resemblance to the applications of the correspondence principle and the idea of a self-stabilizing crystallization process, there are key differences. First, in the case of general relativity, it is possible to identify a single event—the Berne conference celebrating the fiftieth anniversary of general relativity—triggering the new development. Such an event is missing in the case of the correspondence principle. Second, the two phenomena differ greatly in their temporal stability. The renaissance of general relativity was closely connected with the building of a new community and triggered the emergence of a new research field. Thereby its development is one of considerable growth and expansion. The applications of the correspondence principle, by contrast, were part of work on the *patchwork of problems* in quantum theory, which presents a much more fragmented communal and epistemic landscape. In the latter case, instead of continued growth, we find immense expansion followed by a fading out.

These two differences are reflected in the different metaphors. The autocatalytic process describes a process that perpetuates and strengthens itself, set in motion

[3]Moreover, we can use this picture to think about how the correspondence principle ceased to be a tool. As has been observed by Helge Kragh, correspondence arguments did not just disappear with the advent of new theories like matrix or wave mechanics, or even quantum electrodynamics. Rather, physicists developed correspondence arguments in parallel with the new theories. At the same time, the applications of the correspondence principle after 1926 were in no way as ubiquitous as the prior ones and eventually faded out altogether. See Kragh (2012, 218–220). A way to think about this is that the newly arriving theories had an impact on the conditions under which correspondence arguments were made. The principle no longer presented the only tool for determining transition probabilities. Instead, it was confronted with several competitors claiming to be more fundamental, to be more soundly formulated and to offer new theoretical avenues. Belonging to what was now called the old quantum theory, the principle was weakened, but still remained available as a tool.

[4]Blum et al. (2015).

by a central driving force. In the analogy of the crystallization process, the focus is on the stability and fragility of the process, while the actual events setting new developments in motion are accidental.

The *Patchwork of Problems* and the Emergence of Quantum Mechanics

For the majority of physicists, the application of the correspondence principle first of all provided a solution for concrete problems. With this focus, correspondence arguments led them to new perspectives on their research problems. They did not follow a goal of developing the correspondence principle into a sound, quantitative principle, much less of building a new theory of quantum mechanics upon it. Rather, the applications of the correspondence principle demonstrate most pointedly that physicists were driven by their work on the *patchwork of problems*.

From this perspective, the implications for the formulation of the principle were of secondary importance, and the historical actors did not draw far-reaching conclusions from their adaptations of the correspondence principle. Nonetheless, physicists extended and developed the initial formulation of the principle, arriving at different formulations.

Reflections on the formulation of the correspondence principle remained confined to the Copenhagen community. Bohr, Kramers and others reacted to the applications of the principle and came to work on new topics such as dispersion theory or the intensity of multiplets. From 1924 onwards physicists around Bohr focused on the correspondence principle itself and aimed to "sharpen" it. This implied to move away from solving problems in one particular field and to develop a formulation of the correspondence principle that could integrate different applications. This reaction to the applications of the correspondence principle paved the way for its development from the qualitative idea Bohr had formulated in 1918 into its first mathematically self-consistent formulation in Heisenberg's *Umdeutung* in 1925.

This perspective on the old quantum theory resonates with Suman Seth's notion of a "physics of problems" and his account of the transition from the old quantum theory to quantum mechanics. As Seth has put it, "for the physicist of problems [...] there were not anomalies, crises and revolutions, but problems and their methods of solution."[5] Likewise, I do not see that the users of the correspondence principle identified their problems as anomalies in a Kuhnian sense. Rather, they applied and adapted a tool to solve specific problems.

While sharing Seth's assessment, I believe that setting up an opposition between "physicists of problems" like Sommerfeld, and "physicists of principles" like Planck, Einstein or Bohr, is a personification that is both unnecessary and hard

[5] Seth (2010, 267).

to justify.[6] Just like any other physicist within the quantum network, I have shown, Bohr and Sommerfeld used the correspondence principle to address specific problems. At the same time, they discussed it as part of a larger conceptual framework and the role it should play in a future quantum theory.

If we thus want to use and develop the central idea underlying notions like "patchwork of problems" or "physics of problems"—and I think this is essential for understanding quantum physics in the 1920s—we should not use it as a label to subdivide physicists into camps of pragmatist problem-solvers and foundationalist physicist-philosophers. As I see it, it is far more fruitful to consider two interlocking modes of practice, one in which theoretical resources are put to work while larger theoretical issues are bracketed, and one that reflects upon this work, sometimes touching on larger theoretical issues.

The central difference between the approach of Bohr, Kramers or Heisenberg in 1925 and Sommerfeld, Reiche or Franck, as I see it, was that they drew different conclusions from these reflections. These conclusions depended on the individual pathways of the historical actors, on their approach to quantum physics and the specific problems they were working on. As a rule, however, physicists outside of Copenhagen focused on the "problems and their methods of solution" and drew conclusions in this respect: Sommerfeld and Hönl ended up reflecting on their new formal description of multiplet intensities and came up with the notion of multiplicative terms; Reiche came to focus on transitions associated with a particular state and interpreted his f-sum rule as a permanence relation for the transition probabilities in question; Franck and Hund formulated a new concept of scattering. These reflections paid little attention to the implications for the formulation of the correspondence principle itself, yet they are reflections on problem solving. In Copenhagen, as we have seen, the commitment to the correspondence principle as a "law of quantum theory" was considerably stronger. Here, applications of the principle were considered with respect to the formulation of the principle itself and became part of attempts to develop a new theory of quantum mechanics.

Developing correspondence arguments thus meant very different things for Sommerfeld, Franck, and Reiche than for Heisenberg, Pauli, and Kronig. The former aimed to account for complex physical problems in particular research fields, while the latter searched for a new concept of motion. Following these approaches, they opted for different physical models, suited either to describe the phenomenon in question or to embody the central feature deemed important for the new concept. This division, however, does not imply that conceptual development came into play only when physicists searched for a new overarching theory, nor that problem-solving was always part of the work on specific phenomena. As I have shown, the implementation of the correspondence principle in a specific case was essential for both activities, bearing the potential for transformation on the level of isolated physical problems and on the level of larger conceptual frameworks.

[6] For a similar position, see Eckert (2015, 21). Moreover, see James and Joas (2015, 669) for a discussion of Max Born as a physicist of principle working on specific problems.

8.2 *Transformation Through Implementation*: The Conceptual Development of the Correspondence Principle

After the discussion of the applications of the correspondence principle from a perspective on the *patchwork of problems*, I will turn to the discussion of the principle's *transformation through implementation* and thereby to the fine structure of its conceptual development. As has been shown in Chaps. 4 through 6, the history of the correspondence principle and its applications is a prime example for elaborating a description of this process. The correspondence principle did not present an immutable, ready-made tool: Without a mathematically precise formulation, the original correspondence principle remained a general, if not vague idea. At the same time, it was nonetheless clearly formulated in its core assumption and remained stable with respect to this core throughout the period under study. Implementing the correspondence principle in a specific case, physicists needed to deal with this vagueness in one way or the other and to appropriate it to address their problems.

Making correspondence arguments, physicists took up Bohr's formulation of the correspondence principle and Kramers' *Zwischenbahn* and integrated the principle into their work. Through this integration, they developed new correspondence arguments that introduced new conceptual elements. This did not leave the initial correspondence idea untouched; indeed, physicists encountered conflicts between their approaches and their new tool. In almost every case, the application of the principle in a new field meant that the correspondence principle was not applied in Bohr's or Kramers' way. Rather, the principle had to be adapted in order to provide solutions to specific problems at hand, changing both the correspondence principle and the perspective on the problems.

The reconstruction of the principle's development presented here differs considerably from the account by Olivier Darrigol. He focused on Bohr's thinking about the correspondence principle and identified four separate uses of the correspondence principle. First, the asymptotic relation between quantum theory and classical electrodynamics; second, the formal use of classical mechanics, in which atomic motion is identical with classical motion and governed by classical dynamical laws; third, the virtual use of space-time pictures in BKS theory, which still operates with a classical concept of motion while eschewing a dynamical description; finally, a symbolic translation of classical mechanics into a quantum mechanics, in which the classical space-time description of motion and classical dynamics are abandoned.[7]

The perspective on the development of the correspondence principle emerging from Darrigol's account is thus one of abstraction and symbolization, in which classical mechanics is translated into quantum mechanics while the classical concept of motion is abandoned and replaced by a quantum theoretical counterpart. In Dar-

[7]Darrigol (1997, especially 559).

rigol's account, these two steps are separated from each other. The transformation of mechanics arises from a formal analogy between two conceptual frameworks, while the classical concept of motion is first abandoned and then redefined in quantum theory after the new quantum mechanics is established. In this way, the new concept of motion becomes purely "symbolic" insofar as the new kinematic description no longer refers to the description of stationary states.[8]

At first glance, this description is plausible, especially when looking primarily at the work of Bohr and physicists working with him in Copenhagen, as Darrigol did. However, as I have argued in Chap. 7, this interpretation is problematic, even if we limit ourselves to the perspective of the Copenhagen community. Darrigol's reconstruction tacitly introduces a subtle shift in the interpretation of the correspondence principle: in his discussion of Bohr's work from 1918, he introduces the principle as a statement about radiation and motion in quantum theory, arguing explicitly that it is not a metatheoretical statement about classical and quantum theory. In his reconstruction of the emergence of quantum mechanics, which he sees as the end point of the principle's conceptual development, he argues that precisely this metatheoretical analogy played the decisive role, while the concept of motion is redefined in the aftermath.[9]

In contrast to this tacit assumption, the present study has argued that no such shift occurred. The metatheoretical interpretation of the correspondence principle may be found in Bohr's work and gained some importance for Heisenberg in 1925; nonetheless, the connection between radiation and motion remained the conceptual core of the correspondence principle from 1918 until 1926. This conceptual continuity was particularly strong in Copenhagen. Due to the commitment to the correspondence principle as a statement about the motion of a quantum system, the work of Bohr, Heisenberg and others naturally focused on the concept of motion and not on the formal analogy between classical and quantum mechanics.

Focusing on the connection between radiation and motion as the principle's conceptual core, its transformation appears in a different light. In short, the conceptual development of the correspondence principle can be summarized in three steps: first, the conceptual adaptation of the principle as a statement about the motion of the quantum system and its radiation on a qualitative level; second, the formal adaptation of correspondence arguments to descriptions based on the state-transition model; and, third, the conceptual adaptation of the correspondence principle as a statement about the motion of a quantum system on a quantitative level, consistent with the formal description of the state-transition model.

I have pointed to some more general features of this process throughout this book. Building on these discussions, I will characterize the triad of the implementation of the correspondence principle, the recognition of problems, and the adaptive reformulation from an overarching perspective, and discuss how these processes transformed the correspondence principle itself.

[8] Darrigol (1997, 558–559).

[9] Darrigol (1997, 550 and 559).

Implementation as an Epistemic Motor

Implementing the correspondence principle, as we have seen, physicists relied on the state-transition model and the classical description of motion. These two resources were essential, as is illustrated by the work of Sommerfeld and Heisenberg, who could only make their argument by adopting a kinematic description of the multiplet atom and thus going beyond the classification of multiplet spectra on the basis of the state-transition model. Likewise, Franck and Hund needed to describe scattering in terms of the state-transition model by identifying the initial and final states before they could make a correspondence argument.

Where physicists could not identify a corresponding motion, their correspondence arguments failed, or at least suffered from severe limitations. In Kramers' work on electron scattering and X-ray spectra, for example, a corresponding motion for a transition between free and bound states was lacking. In order to make the argument, he needed to consider a process allowing for such an identification. In Reiche's case, the determination of transition probabilities was limited to a few exceedingly simple cases without the corresponding motion given by the *Zwischenbahn*.

On an operational and a conceptual level, the state-transition model and the classical conception of motion were thus more than available resources. Rather, they were the conceptual pillars of the correspondence principle, such that the possibility of describing a problem in their terms was a precondition for making correspondence arguments.

Building on these two preconditions, correspondence arguments were made by setting up a Fourier series and associating its harmonic components with the respective transitions. Operationalizing the correspondence principle took different forms: In Bohr's formulation, correspondence arguments were made by writing down a Fourier series for the motion and mapping a Fourier representation qualitatively onto the radiation spectrum without calculating transition probabilities from it. This was different for correspondence arguments based on Kramers' *Zwischenbahn* model. Here, the correspondence relation was expressed in the form of a mathematical equation that linked this representation directly to the transition probabilities.[10]

Whether formulated as a qualitative mapping or as a quantitive determination of transition probabilities, the operationalization of the correspondence principle was a central step in making correspondence arguments. As I have argued, identifying which component in the motion corresponded to a particular transition in a specific situation forced the actors to fill in the blanks that were left open in the general formulation of the correspondence principle and to adopt tentative solutions to conceptual problems.

[10]In this form, the principle was easier to handle and found application in the majority of correspondence arguments around 1922, becoming closer to what Ursula Klein calls a paper tool: a representation on paper that was "exterior to mental processes, visual and maneuverable." (Klein 2001, 293)

It is this general emphasis on the operationalization of the principle that connects my work to other studies on the *practice of theory*, like Ursula Klein's and David Kaiser's work on paper tools. Their studies have put much emphasis on manipulating formalisms on paper. This focus was certainly powerful and useful in moving away from an analysis of theories as purely logical structures; at the same time, I believe it is too narrow. To incorporate other kinds of theoretical practices like the ones described in this book, the *practice of theory* needs to be conceptualized more broadly, including different ways of manipulating and appropriating ideas that range from finding solutions to mathematical equations, to drawing diagrams, and to making inferences on the basis of qualitative verbal statements. Such a conception of theoretical practices does not imply a return to the old perspective on theories as logical artifices detached from the actual work done with them. Rather, it becomes a study of work with conceptual resources and the way in which this work reshapes these resources.

For this process, operationalization was the central starting point. Filling in the gaps in the formulation of the correspondence principle, as we have seen, necessarily involved bringing the correspondence principle into contact with existing theoretical notions, techniques and models. This became the source for the emergence of tensions and, consequently, for the adaptive reformulation of the correspondence principle. The integration of the correspondence principle into different research fields in the *patchwork of problems* was thus not only responsible for the diversification of correspondence arguments, it was also the epistemic motor for the conceptual development of the correspondence principle.

Recognizing Problems: The Emergence of Epistemic Challenges

Following up on the implementation of the correspondence principle, we turn to the emergence of challenges within correspondence arguments. Within the pattern of *transformation through implementation*, this step appears to follow logically and inevitably from the integration of the correspondence principle into different frameworks. For the historical actors, as we have seen, this was not the case. As a rule, tensions between the correspondence principle and concepts already in place were often unforeseen and emerged only gradually: Franck and Hund, Sommerfeld and Heisenberg, and Reiche initially thought that the correspondence principle fit into their work in a straightforward way. Arriving at predictions that did not match the experimental results, they pointed to uncertainties in the averaging procedure of Kramers' *Zwischenbahn* or the approximate character of the theoretical calculations rather than to deeper conceptual conflicts. Moreover, tensions disappeared again, as in Sommerfeld and Heisenberg's case in which a contradiction pointing to a "flaw" in the correspondence approach was interpreted as an unexpected consequence on the basis of contemporary experimental evidence.

Where pointing to future developments or reevaluations in light of experimental results was not a viable option, discrepancies turned into epistemic challenges.[11] These challenges emerged both on the level of physical models and basic assumptions, and on the level of computational schemes. Franck and Hund's case is a particularly clear example for the former scenario. Here, the conceptualization of scattering in terms of the state-transition model clashed qualitatively with the idea of mapping the Fourier representation of the corresponding motion onto a continuous radiation spectrum. In a similar way, Sommerfeld and Heisenberg found that the *Zwischenbahn* model was not consistent with the physical expectation regarding the total intensity of a multiplet in the state-transition model.

Sommerfeld and Hönl's and Reiche's works show how epistemic challenges emerged on the level of formal descriptions. In these cases, the clash emerged on the level of quantitative results obtained on the basis of the correspondence principle on the one hand, and independent, equally quantitative representations, such as Sommerfeld's intensity schemes, on the other. Similar to the qualitative inconsistencies, these clashes were related to more general physical expectations, like the asymptotic relation between quantum theory and classical radiation theory, or the expectation on the total intensity of a multiplet. Unlike qualitative arguments and their rough estimates, however, the two representations were given in a digital form. These digital representations, as we have seen, were interpreted by the historical actors as noise-free, and their clash thus implied contradictions that could not be explained away by pointing to approximations.

Adaptive Reformulation as a Mode of Development

The two types of epistemic challenges affected the correspondence principle on different levels and played different roles in its development. Qualitative challenges, as we have seen, led to adaptations of physical models or the physical core of the correspondence principle. In this case, physicists kept the core idea of a connection between radiation and motion intact and adapted the principle's reference system or its model representation. For example, Hund understood that he needed to identify that part of the classical trajectory associated with the deflection as the corresponding motion in order to be able to apply the principle. Likewise, Heisenberg saw in the fall of 1924 that the Utrecht sum rules implied that the transition probabilities had to be determined by the motion of the system in the initial state. This requirement, he thought, made it necessary to opt for a new operationalization of the principle's core idea in terms of *virtual oscillators*, replacing the *Zwischenbahn* model.

Usually, these adaptations were developed by incorporating the implications of the specific epistemic challenge into the correspondence principle, and in this sense they present prime examples of ad-hoc adaptations. This was important to

[11] Steinle (2009).

conceptualize the problem in the first place, or to show that a particular phenomenon was consistent with the correspondence principle in general. As we have seen, these adaptations rarely affected the actual calculations and did not lead physicists to problematize the formulation of the correspondence principle. As such, they did not play a central role in the development of the correspondence principle, neither as a problem-solving tool nor as a foundational idea of quantum theory.

This was different for the challenges on the level of formal descriptions. Reacting to them, physicists abandoned the *Zwischenbahn* model and sidelined the core idea of the correspondence principle along with it. Instead, they integrated their correspondence arguments into the formal description of the problem based on the state-transition model. For this integration, it was central that the problem could be formulated independently on the basis of the state-transition model: for example, Reiche's adaptation only emerged in light of his treatment of absorption in the state-transition model. Likewise, Sommerfeld and Hönl returned to the correspondence principle to complete their intensity schemes.

Integrating the correspondence formulas into these descriptions led to the adaptation of the computational scheme underlying the correspondence arguments. For example, the work on multiplet intensities led to the development of the *vanishing at the edges* argument as a translation procedure for turning classical into quantum formulas. Likewise, Reiche adopted the translation procedure of Kramers' dispersion theory to connect his new formal description of transitions associated with a particular state to the correspondence principle. In this respect, the adaptations resulting from challenges on the level of formal descriptions lost contact with the physical content of the correspondence principle. At the same time, they led to the elaboration of a set of formalisms that provided successful, albeit partial, descriptions for various phenomena on the basis of the state-transition model.

Formalization and Conceptual Development

This formalization of correspondence arguments, as we have seen, became the focal point for the adaptation of the correspondence principle as a statement about the motion of a quantum system. In other words, the adaptation on the computational level translated into the adaptation of the conceptual core of the principle. To understand how this transformation came about, it is helpful to think about epistemic conflicts as clashes between two independent theoretical descriptions: the state-transition model, on the one hand, and the concept of motion, on the other.[12]

[12]These clashes might be interpreted as a borderline problem. Unlike boundary problems, which arise between elaborate conceptual frameworks like classical mechanics, electrodynamics and thermodynamics or quantum field theory and general relativity, here we are considering clashes between two highly incomplete descriptions. For the concept of a borderline problem and its application in the case of the development of special and general relativity, see Renn (2007, especially 29–32).

As we have seen, these two descriptions were the conceptual pillars of the correspondence principle. For most of the historical actors, however, there was no conceptual boundary between the state-transition model and the concept of motion in quantum theory. Moreover, the users of the correspondence principle seldom reflected on the corresponding motion as classical or quantum theoretical when they first implemented the principle.[13] This unquestioned union of a classical and a quantum theoretical concept of motion was problematized and destabilized through the implementations of the correspondence principle: As we have discussed above, the formalization of the correspondence arguments led to the description of different phenomena on the basis of the state-transition model, while cutting the ties between this description and the motion of a quantum system.

Calling for the reestablishment of the physical interpretation of the correspondence relation, the formalization of correspondence arguments had major implications for the interpretation of the original formulation of the correspondence principle. After their adaptations, Sommerfeld and Hönl, Reiche, and Franck and Hund no longer identified the corresponding motion with the actual motion of the quantum system. Rather, they used the corresponding motion as an auxiliary to make their argument and, in some cases, even identified it as a placeholder for a yet unknown, entirely different quantum theoretical description of motion. Pauli, Heisenberg, and Kronig extended this line of thought when they formulated the goal of developing a quantum theoretical concept of motion to replace its classical counterpart.

With the emergence of this separation, the state-transition model and the classical concept of motion came into open conflict. As in the case of other borderline problems, this allowed for the mediation between the two theoretical descriptions involved. This mediation relied crucially on the formalization of the correspondence arguments, which, as we have seen, emerged primarily from the applications of the correspondence principle within the *patchwork of problems*. It had provided a description of the multiplet intensity problem or of dispersion in terms of the state-transition model. Moreover, it led to the formulation of a calculational scheme that was essential for the development of the new quantum kinematics.

Thus formulating the first half of a borderline problem, as we have seen in Chap. 7, the formalization of the correspondence arguments allowed for a transformation of the classical concept of motion into a quantum theoretical version by adapting it to the formal structure underlying spectroscopic classifications. In the resulting quantum kinematics, developed in Heisenberg's *Umdeutung*, transition probabilities became transition amplitudes and were used to construct kinematic and dynamic variables. This can be seen as both a transformation of the state-transition model into a (quantum) mechanical description and as a reinterpretation of classical mechanics on the basis of the state-transition model.

---

[13]This might appear baffling from today's perspective, as the concept of motion underlying correspondence arguments (or more generally the old quantum theory) is obviously classical. However, Sommerfeld is a notable exception to this description. He clearly identified the concept of motion as a "foreign element" that was conceptually independent, and incompatible with quantum theory.

as we have seen, these two descriptions were the conceptual pillars of the correspondence principle for most of our historical actors. [illegible] there was no conceptual boundary between the semi-classical model and [illegible] In quantum theory, however, the terms of the correspondence principle [illegible] reflection on the [illegible] [illegible] these [illegible] implementing the principle [illegible] This [illegible] a classical and a quantum theoretical concept of motion was [illegible] the implementations of the correspondence principle. As we have discussed above, the formulation of the correspondence arguments led to [illegible] on the basis of [illegible] between the [illegible]

Appendix A
Applications of the Correspondence Principle, 1918–1928

Table A.1 List of papers resulting from work done outside of Copenhagen in the context of atomic spectroscopy, 1918–1921

Place	Publications	Main problem addressed through correspondence arguments
Munich	Sommerfeld and Kossel (1919); Sommerfeld (1920d)	Selection rules
Tübingen	Landé (1921a,b)	Selection rules, Intensity distribution for the Zeeman effect

Table A.2 List of papers resulting from work done outside of Copenhagen in the context of molecular spectroscopy, 1918–1922

Place	Publications	Main problem addressed through correspondence arguments
Munich	Kratzer (1920); Lenz (1920); Sommerfeld (1920c)	Selection rules
Lund	Heurlinger (1920)	Selection rules
Berlin	Reiche (1920)	Selection rules
Göttingen	Sponer (1920, 1921)	Selection rules, Intensity distribution for the infrared bands

M. Jähnert, *Practicing the Correspondence Principle in the Old Quantum Theory*, Archimedes 56, https://doi.org/10.1007/978-3-030-13300-9

Table A.3 List of papers resulting from work done outside of Copenhagen in the context of atomic spectroscopy, 1922–1928

Place	Publications	Main problem addressed through correspondence arguments
Munich	Sommerfeld and Heisenberg (1922a,b), Hönl (1924), Sommerfeld (1925), Sommerfeld and Hönl (1925), Hönl (1926)	Intensity distribution for multiplets and their Zeeman splittings
Leiden	Fermi (1924a, 1925), Goudsmit and Kronig (1925a,b),	Intensity distribution for multiplets and their Zeeman splittings
Varia	Landé (1921b, 1922b), Thomas (1924), Tolman (1924), Bartels (1925, 1926)	Intensity distribution for Zeeman splittings, Intensity in series spectra

Table A.4 List of papers resulting from work done outside of Copenhagen in the context of molecular spectroscopy, 1922–1928

Place	Publications	Main problem addressed through correspondence arguments
Pasadena	Birge (1922, 1926), Tolman (1924, 1925), Tolman and Badger (1926), Condon (1928)	Intensity distribution in band spectra
Harvard	Kemble (1923, 1924, 1925a,b), Kemble and Bourgin (1926), Bourgin (1927)	Intensity distribution in band spectra
Munich	Kratzer (1922, 1923), Pauli (1922), Hönl and London (1925a,b)	Intensity distribution in band spectra, Stability of stationary states
Varia	Lenz (1924), Dieke (1925a,b), Tamm (1925), Franck (1926)	Intensity distribution in band spectra, Coupling of vibrational and rotational degrees of freedom

Table A.5 List of papers resulting from work done outside of Copenhagen in the context of dispersion theory, 1922–1928

Place	Publications	Main problem addressed through correspondence arguments
Breslau	Ladenburg and Reiche (1923, 1924), Thomas (1925), Reiche and Thomas (1925), Reiche (1926b)	Determination of transition probabilities
Göttingen	Born (1924), Nordheim (1925), Placinteanu (1926)	Quantum formula for dispersion
Minnesota	Breit (1924, 1925a,b), Van Vleck (1924a,b)	Quantum formula for absorption and dispersion
Munich	Wentzel (1924a)	Quantum formula for dispersion

Table A.6 List of papers resulting from work done outside of Copenhagen in the context of the study of collision processes 1922–1928

Place	Publications	Main problem addressed through correspondence arguments
Göttingen	Hund (1922, 1923), Born and Jordan (1925), Born et al. (1925), Jordan (1925), Nordheim (1926)	Ramsauer effect, Black-body radiation, Atom-atom collisions
Varia	Wentzel (1924b), Fermi (1924b,c) Halpern (1924), Mie (1925), Smekal (1925), Breit (1926)	Intensity distribution of the continuous X-ray spectrum and the Compton effect

Table A.7 List of papers written by physicists in Copenhagen, 1922–1928

Research Field	Publications	Main problem addressed through correspondence arguments
Atomic spectroscopy	Heisenberg (1925b,a), Hoyt (1923a,b, 1925a,b, 1926), Kronig (1925a,b), Sugiura (1927)	Intensity of series spectra, Intensity distribution for multiplets and their Zeeman splittings
Molecular spectroscopy	Fowler (1925a,b), Dennison (1928)	Intensity distribution in band spectra
Dispersion theory	Kramers (1924a,b), Kramers and Heisenberg (1925), Kuhn (1925)	Quantum dispersion formula
Electron scattering	Kramers (1922, 1923)	Intensity distribution of the continuous X-Ray spectrum
Quantum theory	Bohr (1923b,c), Bohr et al. (1924a,b), Heisenberg (1925a)	

Archives

AHQP	Archive for the History of Quantum Physics, American Philosophical Society, Philadelphia.
BSC	Bohr Scientific Correspondence, Niels Bohr Archive, Copenhagen
BSM	Bohr Scientific Manuscripts, Niels Bohr Archive, Copenhagen
Franck Papers	Franck, James. Papers, Special Collections Research Center, University of Chicago Library
Hönl Papers	Nachlass Helmut Hönl, Nachlass-Deposita, Universitätsarchiv Albert-Ludwigs-Universität Freiburg
Hund Papers	Nachlass Friedrich Hund, Handschriften und Nachlässe, Niedersächsische Staats- und Universitätsbibliothek Göttingen
Schwarzschild Papers	Nachlass Karl Schwarzschild, Handschriften und Nachlässe, Niedersächsische Staats- und Universitätsbibliothek Göttingen
Sommerfeld Papers	Nachlass Arnold Sommerfeld (NL 89), Deutsches Museum, Archiv, Munich
UAG	Universitätsarchiv Göttingen, Niedersächsische Staats- und Universitätsbibliothek Göttingen

M. Jähnert, *Practicing the Correspondence Principle in the Old Quantum Theory*, Archimedes 56, https://doi.org/10.1007/978-3-030-13300-9

References

Arabatzis, Theodore. 1992. The Discovery of the Zeeman Effect: A Case Study of the Interplay between Theory and Experiment. *Studies in History and Philosophy of Science* 23: 365–388.

Assmus, Alexi J. 1990. *Molecular Structure and the Genesis of the American Quantum Physics Community, 1916–1926*. Ph.D. thesis, Harvard University, Cambridge, MA.

Assmus, Alexi J. 1992a. The Americanization of Molecular Physics. *Historical Studies in the Physical and Biological Sciences* 23: 1–34.

Assmus, Alexi J. 1992b. The Molecular Tradition in Early Quantum Theory. *Historical Studies in the Physical and Biological Sciences* 22: 209–231.

Assmus, Alexi J. 1993. The Creation of Postdoctoral Fellowships and the Siting of American Scientific Research. *Minerva* 31: 151–183.

Bartels, Hans. 1925. Zur Verteilung der Übergangswahrscheinlichkeiten in den Alkaliatomen. *Zeitschrift für Physik* 32: 415–438.

Bartels, Hans. 1926. Das Problem der spektralen Intensitätsverteilung und die Kaskadensprünge im Bohrschen Atommodell. *Zeitschrift für Physik* 37: 35–59.

Bederson, Benjamin. 2005. Fritz Reiche and the Emergency Committee in Aid of Displaced Foreign Scholars. *Physics in Perspective* 7: 453–472.

Beller, Mara. 1999. *Quantum Dialogue: The Making of a Revolution*. Chicago: University of Chicago Press.

Birge, Raymond. 1922. The Quantum Theory of Band Spectra and Its Application to the Determination of Temperature. *The Astrophysical Journal* 55: 273–290.

Birge, Raymond. 1926. Electronic Bands. In *Molecular Spectra in Gases*. Bulletin of the National Research Council 11, Part 3, 69–259. National Research Council.

Blum, Alexander, Roberto Lalli, and Jürgen Renn. 2015. The Reinvention of General Relativity: A Historiographical Framework for Assessing One Hundred Years of Curved Space-time. *Isis* 106: 598–620.

Blum, Alexander S. 2015. QED and the Man Who Didn't Make It: Sidney Dancoff and the Infrared Divergence. *Studies in History and Philosophy of Modern Physics* 50: 70–94.

Blum, Alexander S., Martin Jähnert, Christoph Lehner, and Jürgen Renn. 2017. Translation as Heuristics. Heisenberg's Turn to Matrix Mechanics. *Studies in History and Philosophy of Modern Physics* 60: 3–22.

Bohr, Niels. 1913a. On the Constitution of Atoms and Molecules. *Philosophical Magazine Series 5* 26: 1–25, 476–502, 857–875.

Bohr, Niels. 1913b. On the Effect of Electric and Magnetic Fields on Spectral Lines. *Philosophical Magazine Series 5* 26: 506–524.

M. Jähnert, *Practicing the Correspondence Principle in the Old Quantum Theory*, Archimedes 56, https://doi.org/10.1007/978-3-030-13300-9

Bohr, Niels. 1918a. On the Quantum Theory of Line Spectra, Part I: On the General Theory. *Det Kongelige Danske Videnskabernes Selskab. Skrifter. Naturvidenskabelig og Matematisk Afdeling* 4: 1–36.
Bohr, Niels. 1918b. On the Quantum Theory of Line Spectra, Part II: On the Hydrogen Spectrum. *Det Kongelige Danske Videnskabernes Selskab. Skrifter. Naturvidenskabelig og Matematisk Afdeling* 4: 36–100.
Bohr, Niels. 1920. Über die Serienspektra der Elemente. *Zeitschrift für Physik* 2: 423–469.
Bohr, Niels. 1921. Zur Frage der Polarisation der Strahlung in der Quantentheorie. *Zeitschrift für Physik* 6: 1–9.
Bohr, Niels. 1922. The Effect of Electric and Magnetic Fields on Spectral Lines. *Proceedings of the Physical Society of London* 35: 275–302.
Bohr, Niels. 1923a. L'application de la théorie des quanta aux problèmes atomiques. In *Atomes et électrons. Rapports et discussions du Conseil de Physique tenu à Bruxelles du 1er au 6e avril 1921 sous les auspices de l'Institut International de Physique Solvay*, 228–248. Paris: Gauthier-Villars.
Bohr, Niels. 1923b. Über die Anwendung der Quantentheorie auf den Atombau, I. Die Grundpostulate der Quantentheorie. *Zeitschrift für Physik* 13: 117–165.
Bohr, Niels. 1923c. *Über die Quantentheorie der Linienspektren*. Braunschweig: Vieweg.
Bohr, Niels. 1924. Zur Polarisation des Fluorescenzlichtes. *Naturwissenschaften* 12: 1115–1117.
Bohr, Niels. 1976. *Bohr's Collected Works: The Correspondence Principle (1918–1923)*, vol. 3. Amsterdam: North-Holland.
Bohr, Niels. 1977. *Bohr's Collected Works: The Periodic System (1920–1923)*, vol. 4. Amsterdam: North-Holland.
Bohr, Niels. 1981. *Bohr's Collected Works: Work on Atomic Physics: (1912–1917)*, vol. 2. Amsterdam: North-Holland.
Bohr, Niels. 1987. *Bohr's Collected Works: The Penetration of Charged Particles through Matter: (1912–1954)*, vol. 8. Amsterdam: North-Holland.
Bohr, Niels, and Dirk Coster. 1923. Röntgenspektren und periodisches System der Elemente. *Zeitschrift für Physik* 12: 342–374.
Bohr, Niels, Hendrik A. Kramers, and John Slater. 1924a. The Quantum Theory of Radiation. *Philosophical Magazine Series 6* 47: 785–822.
Bohr, Niels, Hendrik A. Kramers, and John C. Slater. 1924b. Über die Quantentheorie der Strahlung. *Zeitschrift für Physik* 24: 69–87.
Bokulich, Alisa. 2009. Three Puzzles about Bohr's Correspondence Principle, 1–24. http://philsci-archive.pitt.edu/archive/00004826/01/3Puzzles-BohrCP-Bokulich.pdf
Born, Max. 1924. Über Quantenmechanik. *Zeitschrift für Physik* 26: 379–395.
Born, Max. 1925. *Vorlesungen über Atommechanik*. Berlin: Springer.
Born, Max. 1926a. *Probleme der Atomdynamik*. Berlin: Springer.
Born, Max. 1926b. *Problems of Atomic Dynamics*. Cambridge, MA: Massachusetts Institute of Technology.
Born, Max, Hedwig Born, and Albert Einstein. 1969. *Briefwechsel 1916–1955*. Munich: Nymphenburger Verlagshandlung.
Born, Max, Werner Heisenberg, and Pascual Jordan. 1926. Zur Quantenmechanik II. *Zeitschrift für Physik* 35: 557–615.
Born, Max, and Pascual Jordan. 1925. Zur Quantentheorie aperiodischer Vorgänge. *Zeitschrift für Physik* 33: 479–505.
Born, Max, Pascual Jordan, and Lothar Nordheim. 1925. Zur Theorie der Anregung von Atomen durch Stöße. *Die Naturwissenschaften* 13: 969–970.
Borrelli, Arianna. 2009. The Emergence of Selection Rules and their Encounter with Group Theory: 1913–1927. *Studies in History and Philosophy of Modern Physics* 40: 327–337.
Borrelli, Arianna. 2011. Angular Momentum between Physics and Mathematics. In *Mathematics Meets Physics*, ed. Karl-Heinz Schlote and Martina Schneider, 395–440. Frankfurt am Main: Verlag Harri Deutsch.

Bourgin, David G. 1927. Line Intensities in the Hydrogen Chloride Fundamental Band. *Physical Review* 29: 794–816.

Breit, Gregory. 1924. The Quantum Theory of Dispersion. *Nature* 114: 310.

Breit, Gregory. 1925a. Note on the Theory of Optical Dispersion. *Journal of the Washington Academy of Sciences* 15: 34–39.

Breit, Gregory. 1925b. Some Remarks on Two-Coupled Multiply Periodic Systems, the Statistics of Quantum Theory, and the Theory of Dispersion. *Journal of the Washington Academy of Sciences* 15: 269–278.

Breit, Gregory. 1926. A Correspondence Principle in the Compton Effect. *Physical Review* 27: 362–372.

Brillouin, Léon. 1922. *La théorie des quanta et l'atome de Bohr*. Recueil des conférences-rapports de documentation sur la physique, vol. 2. Paris: Blanchard.

Buchwald, Eberhard. 1923. *Das Korrespondenzprinzip*. Sammlung Vieweg: Tagesfragen aus den Gebieten der Naturwissenschaften und der Technik, vol. 67. Braunschweig: Vieweg.

Burger, Hans C., and Hendrik B. Dorgelo. 1924. Beziehung zwischen inneren Quantenzahlen und Intensitäten von Mehrfachlinien. *Zeitschrift für Physik* 23: 258–266.

Burgers, Johann M. 1917. The Spectrum of a Rotating Molecule According to the Theory of Quanta. *Koninklijke Akademie van Wetenschappen te Amsterdam. Section of Science. Proceedings* 20: 170–177.

Cassidy, David C. 1976. *Werner Heisenberg and the Crisis in Quantum Theory, 1920–1925*. Ph.D. thesis, Purdue University.

Cassidy, David C. 1979. Heisenberg's First Core Model of the Atom. *Historical Studies in the Physical Sciences* 10: 187–224.

Castagnetti, Guiseppe, and Jürgen Renn. forthcoming. *Quantum Networks: Research Strategies and Research Politics in the Emergence of Quantum Physics*. Max Planck Research Library for the History and Development of Knowledge Studies. Berlin: Edition Open Access.

Condon, Edward U. 1928. Nuclear Motions Associated with Electron Transitions in Diatomic Molecules. *Physical Review* 32: 858–872.

Darrigol, Olivier. 1992. *From "C"-Numbers to "Q"-Numbers: The Classical Analogy in the History of Quantum Theory*. Berkeley: University of California Press.

Darrigol, Olivier. 1997. Classical Concepts in Bohr's Atomic Theory. *Physis: Rivista internazionale di storia della scienza* 34: 545–567.

Darrigol, Olivier. 2010. Quantum Theory and Atomic Structure, 1900–1927. In *The Modern Physical and Mathematical Sciences*. The Cambridge History of Science, ed. Marie Joe Nye, vol. 5, 331–349. Cambridge: Cambridge University Press.

Davis, Bergen. 1915. Wave-length Energy Distribution in the Continuous Spectrum. *Physical Review* 9: 64–77.

Debye, Peter. 1916a. Quantenhypothese und Zeeman-Effekt. *Physikalische Zeitschrift* 17: 507–512.

Debye, Peter. 1916b. Quantenhypothese und Zeeman-Effekt. *Nachrichten von der Königlichen Gesellschaft der Wissenschaften zu Göttingen. Mathematisch-physikalische Klasse*, 142–153.

Dennison, David M. 1926. The Rotation of Molecules. *Physical Review* 28: 318–333.

Dennison, David M. 1928. The Shape and Intensities of Infrared Absorption Lines. *Physical Review* 31: 503–519.

Dieke, Gerhard H. 1925a. Intensities in Band Spectra. *Nature* 115: 875.

Dieke, Gerhard H. 1925b. Über die Intensitäten in den Bandenspektren. *Zeitschrift für Physik* 33: 161–168.

Dorgelo, Hendrik B. 1923. Die Intensität der Mehrfachlinien. *Zeitschrift für Physik* 13: 206–211.

Dorgelo, Hendrik B. 1924. Die Intensität mehrfacher Spektrallinien. *Zeitschrift für Physik* 22: 170–177.

Dresden, Max. 1987. *H.A. Kramers Between Tradition and Revolution*. Berlin: Springer.

Duane, William, and Franklin L. Hunt. 1915. On X-ray Wave-Lengths. *Physical Review* 6: 166–172.

Duncan, Anthony, and Michel Janssen. 2007a. On the Verge of *Umdeutung* in Minnesota: Van Vleck and the Correspondence Principle. Part I. *Archive for History of Exact Sciences* 61: 553–624.

Duncan, Anthony, and Michel Janssen. 2007b. On the Verge of *Umdeutung* in Minnesota: Van Vleck and the Correspondence Principle. Part II. *Archive for History of Exact Sciences* 61: 625–671.

Duncan, Anthony, and Michel Janssen. 2014. The Trouble with Orbits: The Stark Effect in the Old and the New Quantum Theory. *Studies in History and Philosophy of Modern Physics* 48: 68–83.

Eckert, Michael. 2001. The Emergence of Quantum Schools: Munich, Göttingen and Copenhagen as New Centers of Atomic Theory. *Annalen der Physik* 10: 151–162.

Eckert, Michael. 2013a. *Arnold Sommerfeld. Atomphysiker und Kulturbote*. Göttingen: Wallstein.

Eckert, Michael. 2013b. *Die Bohr-Sommerfeldsche Atomtheorie: Sommerfelds Erweiterung des Bohrschen Atommodells 1915/1916. Klassische Texte der Wissenschaft*, vol. 39. Berlin: Springer.

Eckert, Michael. 2014. How Sommerfeld Extended Bohr's Model of the Atom: 1913–1916. *European Physical Journal History* 39: 141—156.

Eckert, Michael. 2015. From Aether Impulse to QED: Sommerfeld and the *Bremsstrahlen* Theory. *Studies in History and Philosophy of Modern Physics* 51: 9–22.

Eckert, Michael, and Karl Märker. 2004. Atombau und Spektrallinien. In *Arnold Sommerfeld: Wissenschaftlicher Briefwechsel*, ed. Michael Eckert and Karl Märker, vol. 2: 1919–1951, 13–43. Munich: Verlag für Geschichte der Naturwissenschaft und Technik.

Ehrenfest, Paul, and Gregory Breit. 1922. Ein bemerkenswerter Fall von Quantisierung. *Zeitschrift für Physik* 9: 207–211.

Einstein, Albert. 1916. Strahlungs-Emission und -Absorption nach der Quantentheorie. *Deutsche Physikalische Gesellschaft, Verhandlungen* 18: 318–323.

Epstein, Paul S. 916a. Zur Quantentheorie. *Annalen der Physik* 50: 168–188.

Epstein, Paul S. 1916b. Zur Theorie des Starkeffektes. *Annalen der Physik* 50: 489–521.

Epstein, Paul S. 1916c. Zur Theorie des Starkeffektes. *Physikalische Zeitschrift* 17: 148–150.

Eucken, Arnold. 1914. *Die Theorie der Strahlung und der Quanten: Verhandlungen auf einer von E. Solvay einberufenen Zusammenkunft (30. Oktober bis 3. November 1911)*. Abhandlungen der Deutschen Bunsen-Gesellschaft für angewandte physikalische Chemie. Halle: Knapp.

Ewald, Paul Peter. 1920. Review of "Arnold Sommerfeld: Atombau und Spectrallinien". *Jahrbuch der Radioaktivität und Elektronik* 16: 301–305.

Fermi, Enrico. 1924a. Berekeningen over de intensiteiten van spectrallijnen. *Physica* 4: 340–343.

Fermi, Enrico. 1924b. Sulla teoria dell'utro tra atomi e corpuscoli elettrici. *Atti della Reale Accademia Nazionale dei Lincei Rendiconti* 33: 243–245.

Fermi, Enrico. 1924c. Über die Theorie des Stoßes zwischen Atomen und elektrisch geladenen Teilchen. *Zeitschrift für Physik* 24: 315–327.

Fermi, Enrico. 1925. Sopra l'intensità delle righe multiple. *Atti della Reale Accademia Nazionale dei Lincei Rendiconti* 1: 120–124.

Forman, Paul. 1970. Alfred Landé and the *Anomalous Zeeman Effect*, 1919–1921. *Historical Studies in the Physical Sciences* 2: 153–261.

Fowler, Ralph. 1925a. A Note on the Summation Rules for the Intensities of Spectral Lines. *Philosophical Magazine Series 6* 50: 1079–1083.

Fowler, Ralph. 1925b. Applications of the Correspondence Principle to the Theory of Line-Intensities in Band-Spectra. *Philosophical Magazine Series 6* 49: 1272–1288.

Franck, James. 1923. Neuere Erfahrungen über quantenhaften Energieaustausch bei Zusammenstößen von Atomen und Molekülen. In *Ergebnisse der exakten Naturwissenschaften*, vol. 2, 106–123. Berlin: Springer.

Franck, James. 1926. Elementary Processes of Photochemical Reactions. *Transactions of the Faraday Society* 21: 536–542.

Franck, James, and Fritz Reiche. 1920. Über Helium und Parhelium. *Zeitschrift für Physik* 1: 154–160.

Fues, Erwin. 1922a. Die Berechnung wasserstoffunähnlicher Spektren aus Zentralbewegungen der Elektronen I. *Zeitschrift für Physik* 11: 364–378.
Fues, Erwin. 1922b. Die Berechnung wasserstoffunähnlicher Spektren aus Zentralbewegungen der Elektronen II. *Zeitschrift für Physik* 12: 1–12.
Gearhart, Clayton A. 2010. "Astonishing Successes" and "Bitter Disappointment": The Specific Heat of Hydrogen in Quantum Theory. *Archive for History of Exact Sciences* 64: 113–202.
Gearhart, Clayton A. 2012. Fritz Reiche's 1921 Quantum Theory Textbook. In *Research and Pedagogy: A History of Quantum Physics through Its Textbooks*. Max Planck Research Library for the History and Development of Knowledge Studies, ed. Massimiliano Badino and Jaume Navarro, vol. 2, 101–116. Berlin: Edition Open Access.
Gearhart, Clayton A. 2014. The Franck-Hertz Experiments, 1911–1914: Experimentalists in Search of a Theory, with an Appendix, 'On the History of our Experiments on the Energy Exchange between Slow Electrons and Atoms' by Gustav Hertz. *Physics in Perspective* 16: 293–343.
Goudsmit, Samuel, and Ralph de Laer Kronig. 1925a. Die Intensität der Zeemankomponenten. *Koninklijke Akademie van Wetenschappen te Amsterdam. Section of Science. Proceedings* 28: 418–422.
Goudsmit, Samuel, and Ralph de Laer Kronig. 1925b. Die Intensität der Zeemankomponenten. *Die Naturwissenschaften* 13: 90.
Gyeong Soon Im. 1995. The Formation and Development of the Ramsauer Effect. *Historical Studies in the Physical and Biological Sciences* 25: 269–290.
Gyeong Soon Im. 1996. Experimental Constraints on Formal Quantum Mechanics: The Emergence of Born's Quantum Theory of Collision Processes in Göttingen, 1924–1927. *Archive for History of Exact Sciences* 50: 73–101.
Halpern, Otto. 1924. Zur Theorie der Röntgenstrahlstreuung. *Zeitschrift für Physik* 30: 153–172.
Heilbron, John L. 1964. *A History of the Problem of Atomic Structure from the Discovery of the Electron to the Beginning of Quantum Mechanics*. Ph.D. thesis, University of California, Berkeley.
Heilbron, John L., and Thomas S. Kuhn. 1969. The Genesis of the Bohr Atom. *Historical Studies in the Physical Sciences* 1: 211–290.
Heisenberg, Werner. 1922. Zur Quantentheorie der Linienstruktur und der anomalen Zeemaneffekte. *Zeitschrift für Physik* 8: 273–297.
Heisenberg, Werner. 1925a. Über die quantentheoretische Umdeutung kinematischer und mechanischer Beziehungen. *Zeitschrift für Physik* 33: 879–893.
Heisenberg, Werner. 1925b. Über eine Anwendung des Korrespondenzprinzips auf die Frage nach der Polarisation des Fluoreszenzlichtes. *Zeitschrift für Physik* 31: 617–626.
Heisenberg, Werner. 1960. Erinnerungen an die Zeit der Entwicklung der Quantenmechanik. In *Theoretical Physics in the Twentieth Century. A Memorial Volume to Wolfgang Pauli*, ed. Markus Fierz and Viktor Weisskopf, 40–47. New York: Interscience Publishers.
Heisenberg, Werner, and Pascual Jordan. 1926. Anwendung der Quantenmechanik auf das Problem der anomalen Zeemaneffekte. *Zeitschrift für Physik* 37: 263–277.
Hendry, John. 1981. Bohr-Kramers-Slater: A Virtual Theory of Virtual Oscillators. *Centaurus* 25: 189–221.
Hendry, John. 1984. *The Creation of Quantum Mechanics and the Bohr-Pauli Dialogue*. Studies in the History of Modern Science, vol. 14. Dordrecht: Reidel.
Hentschel, Klaus. 2002. *Mapping the Spectrum: Techniques of Visual Representation in Research and Teaching*. Oxford: Oxford University Press.
Hertz, Gustav. 1922a. On the Mean Free Path of Slow Electrons in Neon and Argon. *Koninklijke Akademie van Wetenschappen te Amsterdam. Section of Science. Proceedings* 25: 90–98.
Hertz, Gustav. 1922b. Ueber die mittlere freie Weglänge von langsamen Elektronen in Neon und Argon. *Physica* 2: 87–92.
Heurlinger, Torsten. 1920. Über Atomschwingung und Molekülspektra. *Zeitschrift für Physik* 1: 82–91.
Hönl, Helmut. 1924. Die Intensitäten der Zeemankomponenten. *Zeitschrift für Physik* 31: 340–354.

Hönl, Helmut. 1926. Zum Intensitätsproblem der Spektrallinien. *Annalen der Physik* 76: 273–323.
Hönl, Helmut, and Fritz London. 1925a. Über die Intensität der Bandenlinien. *Zeitschrift für Physik* 33: 803–809.
Hönl, Helmut, and Fritz London. 1925b. Über die Intensität der Bandenlinien. *Die Naturwissenschaften* 13: 756.
Hoyt, Frank C. 1923a. Relative Probabilities of the Transitions Involved in the Balmer Series Lines of Hydrogen. *Philosophical Magazine Series 6* 47: 826–831.
Hoyt, Frank C. 1923b. The Relative Intensity of X-Ray Lines. *Philosophical Magazine Series 6* 46: 135–145.
Hoyt, Frank C. 1925a. Application of the Correspondence Principle to Relative Intensities in Series Spectra. *Physical Review* 26: 749–760.
Hoyt, Frank C. 1925b. Harmonic Analysis of Electron Orbits. *Physical Review* 25: 174–186.
Hoyt, Frank C. 1926. Transition Probabilities and Principle Quantum Numbers. *Proceeding National Academy of Sciences* 12: 227–231.
Hund, Friedrich. 1922. *Versuch einer Deutung der grossen Durchlässigkeit einiger Edelgase für sehr langsame Elektronen*. Ph.D. thesis, Georg-August-Universität Göttingen, Göttingen.
Hund, Friedrich. 1923. Theoretische Betrachtungen über die Ablenkung von freien langsamen Elektronen in Atomen. *Zeitschrift für Physik* 13: 241–263.
Jähnert, Martin. 2013. The Correspondence Idea in the Early Bohr Atom 1913-1915. *Annalen der Physik* 525: A155–A158.
Jähnert, Martin. 2015. Practising the correspondence principle in the Old Quantum Theory: Franck, Hund and the Ramsauer Effect. In *One Hundred Years of the Bohr Atom: Proceedings from a Conference*, ed. Aaserud Finn and Helge Kragh, 198–214. Copenhagen: Royal Danish Academy of Sciences and Letters.
James, Jeremiah and Christian Joas. 2015. Subsequent and Subsidiary? Rethinking the Role of Applications in Establishing Quantum Mechanics. *Historical Studies in the Natural Sciences* 45: 641–702.
Jammer, Max. 1966. *The Conceptual Development of Quantum Mechanics*. New York: McGraw-Hill.
Jordan, Pascual. 1925. Zur Quantentheorie aperiodischer Vorgänge. II. Bemerkung über die Integration der Störungsrechnung. *Zeitschrift für Physik* 33: 506–508.
Jordi Taltavull, Marta. 2013. Challenging the Boundaries between Classical and Quantum Physics: The Case of Optical Dispersion. In *Traditions and Transformations in the History of Quantum Physics HQ–3: Third International Conference on the History of Quantum Physics*, Berlin, 28 June–2 July 2010. Max Planck Research Library for the History and Development of Knowledge Proceedings, ed. Shaul Katzir, Christoph Lehner, and Jürgen Renn, vol. 5, 29–59. Berlin: Edition Open Access.
Kaiser, David. 2005. *Drawing Theories apart: The Dispersion of Feynman Diagrams in Postwar Physics*. Chicago: University of Chicago Press.
Kayser, Heinrich. 1900–1912. *Handbuch der Spectroscopie*, 6 vols. Leipzig: Verlag von S. Hirzel.
Kemble, Edwin C. 1923. The Relative Intensities of Band Components in the Infra-Red Spectra of Diatomic Gases. *Physical Review* 21: 715.
Kemble, Edwin C. 1924. Quantization in Space and the Relative Intensities of the Components of Infra-Red Absorption Bands. *Proceeding National Academy of Sciences* 10: 274–279.
Kemble, Edwin C. 1925a. The Application of the Correspondence Principle to Degenerate Systems and the Relative Intensities of Band Lines. *Physical Review* 25: 1–22.
Kemble, Edwin C. 1925b. Über die Intensität der Bandenlinien. *Zeitschrift für Physik* 35: 286–292.
Kemble, Edwin C., and David G. Bourgin. 1926. Relative Intensities of Band Lines in the Infra-red Spectrum of a Diatomic Gas. *Nature* 117: 789.
Klein, Ursula. 1998. Paving a Way through the Jungle of Organic Chemistry. Experimenting with Changing Systems of Order. In *Experimental Essays—Versuche zum Experiment*, ed. Michael Heidelberger and Friedrich Steinle, 251–271. Baden-Baden: Nomos Verlagsgesellschaft.

Klein, Ursula. 1999. Techniques of Modelling and Paper Tools in Classical Chemistry. In *Models as Mediators: Perspectives on Natural and Social Sciences*, ed. Mary Morgan and Margaret Morrison, 146–167. Cambridge: Cambridge University Press.

Klein, Ursula. 2001. Paper Tools in Experimental Cultures. *Studies in History and Philosophy of Science Part A* 32: 265–302.

Klein, Ursula. 2003. *Experiments, Models, Paper Tools*. Stanford: Stanford University Press.

Kojevnikov, Alexei. forthcoming. Niels Bohr, the Construction of the Copenhagen Network, and the Interpretation of Quantum Mechanics. In *Quantum Networks: Research Strategies and Research Politics in the Emergence of Quantum Physics*. Max Planck Research Library for the History and Development of Knowledge Studies, ed. Guiseppe Castagnetti and Jürgen Renn. Berlin: Edition Open Access.

Konno, Hiroyuki. 1993. Kramers' Negative Dispersion, the Virtual Oscillator Model, and the Correspondence Principle. *Centaurus* 36: 117–166.

Kox, Anne J. 1997. The Discovery of the Electron: II. The Zeeman Effect. *European Journal of Physics* 18: 139–144.

Kragh, Helge. 2002. *Quantum Generations: A History of Physics in the Twentieth Century*, 2nd ed. Princeton: Princeton University Press.

Kragh, Helge. 2012. *Niels Bohr and the Quantum Atom: The Bohr Model of Atomic Structure 1913–1925*. Oxford: Oxford University Press.

Kramers, Hendrik A. 1919. The Intensities of Spectral Lines. *Det Kongelige Danske Videnskabernes Selskab. Skrifter. Naturvidenskabelig og Matematisk Afdeling* 8: 285–386.

Kramers, Hendrik A. 1922. Om Absorption af Røntgenstraaler. *Fysisk Tidsskrift* 20: 130–132.

Kramers, Hendrik A. 1923. On the Theory of X-Ray Absorption and of the Continuous X-Ray Spectrum. *Philosophical Magazine Series 6* 46: 836–871.

Kramers, Hendrik A. 1924a. The Law of Dispersion and Bohr's Theory of Spectra. *Nature* 133: 673–676.

Kramers, Hendrik A. 1924b. The Quantum Theory of Dispersion. *Nature* 114: 310–311.

Kramers, Hendrik A. 1956. *Collected scientific papers*. Amsterdam: North-Holland.

Kramers, Hendrik A., and Werner Heisenberg. 1925. Über die Streuung von Strahlung durch Atome. *Zeitschrift für Physik* 31: 681–708.

Kratzer, Adolf. 1920. Die ultraroten Rotationsspektren der Halogenwasserstoffe. *Zeitschrift für Physik* 3: 289–307.

Kratzer, Adolf. 1922. Die Gesetzmäßigkeiten der Bandensysteme. *Annalen der Physik* 67: 127–153.

Kratzer, Adolf. 1923. Die Feinstruktur einer Klasse von Bandenspektren. *Annalen der Physik* 71: 72–103.

Kronig, Ralph de Laer. 1925a. Über die Intensität der Mehrfachlinien und ihrer Zeemankomponenten. *Zeitschrift für Physik* 31: 885–897.

Kronig, Ralph de Laer. 1925b. Über die Intensität der Mehrfachlinien und ihrer Zeemankomponenten II. *Zeitschrift für Physik* 33: 261–272.

Kronig, Ralph de Laer. 1960. The Turning Point. In *Theoretical Physics in the Twentieth Century. A Memorial Volume to Wolfgang Pauli*, ed. Markus Fierz and Viktor Weisskopf, 5–39. New York: Interscience Publishers.

Kuhn, Werner. 1925. Über die Gesamtstärke der von einem Zustande ausgehenden Absorptionslinien. *Zeitschrift für Physik* 31: 408–412.

Ladenburg, Rudolf. 1921. Die quantentheoretische Deutung der Zahl der Dispersionselektronen. *Zeitschrift für Physik* 4: 451–468.

Ladenburg, Rudolf, and Fritz Reiche. 1923. Absorption, Zerstreuung und Dispersion in der Bohrschen Atomtheorie. *Die Naturwissenschaften* 11: 584–598.

Ladenburg, Rudolf, and Fritz Reiche. 1924. Dispersionsgesetz und Bohrsche Atomtheorie. *Die Naturwissenschaften* 12: 672–673.

Landé, Alfred. 1921a. Über den anomalen Zeemaneffekt (Teil I). *Zeitschrift für Physik* 5: 231–241.

Landé, Alfred. 1921b. Über den anomalen Zeemaneffekt (Teil II). *Zeitschrift für Physik* 7: 231–241.

Landé, Alfred. 1922a. *Fortschritte der Quantentheorie*. Dresden, Leipzig: Theodor Steinkopff.

Landé, Alfred. 1922b. Zur Theorie der anomalen Zeeman- und magneto-mechanischen Effekte. *Zeitschrift für Physik* 11: 353–363.

Landé, Alfred. 1926. *Die Neuere Entwicklung der Quantentheorie*, 2nd ed. Wissenschaftliche Forschungsberichte. Dresden, Leipzig: Theodor Steinkopff.

Landé, Alfred, and Ernst Back. 1925. *Zeemaneffekt und Multiplettstruktur der Spektrallinien*. Berlin: Springer.

Lenz, Wilhelm. 1920. Über einige speziellere Fragen aus der Theorie der Bandenspektren. *Physikalische Zeitschrift* 21: 691–695.

Lenz, Wilhelm. 1924. Einige korrespondenzmäßige Betrachtungen. *Zeitschrift für Physik* 25: 299–311.

MacKinnon, Edward. 1977. Heisenberg, Models, and the Rise of Matrix Mechanics. *Historical Studies in the Physical Sciences* 8: 137–188.

MacKinnon, Edward. 1982. *Scientific Explanation and Atomic Physics*. Cambridge: Cambridge University Press.

Mehra, Jagdish, and Helmut Rechenberg. 1982a. *The Historical Development of Quantum Theory: The Discovery of Quantum Mechanics 1925*, vol. 2. New York: Springer.

Mehra, Jagdish, and Helmut Rechenberg. 1982b. *The Historical Development of Quantum Theory: The Quantum Theory of Planck, Einstein and Sommerfeld: Its Foundation and The Rise of Its Difficulties, 1900–1925*, vol. 1, Parts I and II. New York: Springer.

Mie, Gustav. 1925. Bremsstrahlung und Comptonsche Streustrahlung. *Zeitschrift für Physik* 33: 33–41.

Nielsen, Ruth. 1976. Introduction. In *Bohr's Collected Works: The Correspondence Principle (1918–1923)*, ed. Ruth Nielsen, vol. 3, 3–46. Amsterdam: North-Holland.

Nordheim, Lothar. 1925. Zur Frage der Polarisation des Streu- und Fluoreszenzlichtes. *Zeitschrift für Physik* 33: 729–740.

Nordheim, Lothar. 1926. Zur Theorie der Anregung von Atomen durch Stöße. *Zeitschrift für Physik* 36: 496–539.

Ornstein, Leon S., and Hans C. Burger. 1924. Intensitäten der Komponenten im Zeemaneffekt. *Zeitschrift für Physik* 28: 135–142.

Pauli, Wolfgang. 1922. Über das Modell des Wasserstoffmolekülions. *Annalen der Physik* 68: 177–240.

Pauli, Wolfgang. 1925. Quantentheorie. In *Handbuch der Physik*, ed. Karl Scheel, 1–279. Berlin: Springer.

Pauli, Wolfgang. 1979. *Wissenschaftlicher Briefwechsel mit Bohr, Einstein, Heisenberg u.a. (1919-1929)*. New York: Springer.

Placinteanu, Ioan. 1926. Über die Wechselwirkung zwischen Strahlung und Quadrupolatomen. *Zeitschrift für Physik* 39: 276–298.

Planck, Max. 1921. *Vorlesungen über die Theorie der Wärmestrahlung*, 4th ed. Leipzig: Johann Ambrosius Barth.

Radder, Hans. 1988. *The Material Realization of Science*. Assen: Van Gorcum.

Radder, Hans. 1991. Heuristics and the Generalized Correspondence Principle. *British Journal for the Philosophy of Science* 42: 195–226.

Rademacher, Hans, and Fritz Reiche. 1927. Die Quantelung des symmetrischen Kreisels nach Schrödingers Undulationsmechanik. Intensitätsfragen. *Zeitschrift für Physik* 41: 453–492.

Ramsauer, Carl. 1921a. Über den Wirkungsquerschnitt der Gasmoleküle gegenüber langsamen Elektronen. *Physikalische Zeitschrift* 22: 613–615.

Ramsauer, Carl. 1921b. Über den Wirkungsquerschnitt der Gasmoleküle gegenüber langsamen Elektronen. *Annalen der Physik* 64: 513–540.

Ramsauer, Carl. 1921c. Über den Wirkungsquerschnitt der Gasmoleküle gegenüber langsamen Elektronen. I. Fortsetzung. *Annalen der Physik* 66: 546–558.

Ramsauer, Carl. 1923. Über den Wirkungsquerschnitt der Gasmoleküle gegenüber langsamen Elektronen II. Fortsetzung und Schluss. *Annalen der Physik* 72: 345–352.

Reiche, Fritz. 1920. Zur Theorie der Rotationsspektren. *Zeitschrift für Physik* 1: 283–293.

Reiche, Fritz. 1921. *Die Quantentheorie: Ihr Ursprung und ihre Entwicklung*. Berlin: Springer.
Reiche, Fritz. 1926a. Die Quantelung des symmetrischen Kreisels nach Schrödingers Undulationsmechanik. *Zeitschrift für Physik* 39: 444–464.
Reiche, Fritz. 1926b. Über Beziehungen zwischen den Übergangswahrscheinlichkeiten beim Zeemaneffekt (magnetischer f-Summensatz). *Die Naturwissenschaften* 14: 275–276.
Reiche, Fritz. 1929. Zur quantenmechanischen Dispersionsformel des atomaren Wasserstoffs im Grundzustand. *Zeitschrift für Physik* 53: 168–191.
Reiche, Fritz, and Willy Thomas. 1925. Über die Zahl der Dispersionselektronen, die einem stationären Zustand zugeordnet sind. *Zeitschrift für Physik* 31: 510–525.
Renn, Jürgen. 2007. Classical Physics in Disarray. In *The Genesis of General Relativity*. Einstein's Zurich Notebook: Introduction and Source, ed. Michel Janssen, John D. Norton, Jürgen Renn, Tilman Sauer, and John Stachel, vol. 1, 21–80. New York and Berlin: Springer.
Renn, Jürgen. 2013. Schrödinger and the Genesis of Wave Mechanics. In *Erwin Schrödinger–50 Years After*, ed. Wolfgang L. Reiter and Jakob Yngvason, 9–36. Zurich: European Mathematical Society.
Robertson, Peter. 1979. *The Early Years-The Niels Bohr Institute 1921–1930*. Copenhagen: Akademisk Forlag Universitetsforlaget i Københaven.
Schrödinger, Erwin. 2011a. *Eine Entdeckung von ganz außerordentlicher Tragweite: Schrödingers Briefwechsel zur Wellenmechanik und zum Katzenparadoxon*, vol. 1. Heidelberg: Springer.
Schrödinger, Erwin. 2011b. *Eine Entdeckung von ganz außerordentlicher Tragweite: Schrödingers Briefwechsel zur Wellenmechanik und zum Katzenparadoxon*, vol. 2. Heidelberg: Springer.
Schwarzschild, Karl. 1916. Zur Quantenhypothese. *Sitzungsberichte der Preussischen Akadademie der Wissenschaften* 1916: 548–568.
Schweber, Silvan S. 1986. The Empiricist Temper Regnant: Theoretical Physics in the United States 1920–1950. *Historical Studies in the Physical and Biological Sciences* 11: 55–98.
Serwer, Daniel. 1977. Unmechanischer Zwang: Pauli, Heisenberg and the Rejection of the Mechanical Atom. *Historical Studies in the Physical Sciences* 8: 189–256.
Seth, Suman. 2007. Crisis and the Construction of Modern Theoretical Physics. *British Journal for History of Science* 40: 25–51.
Seth, Suman. 2010. *Crafting the Quantum: Arnold Sommerfeld and the Practice of Theory, 1890–1926*. Cambridge, Mass.: MIT Press.
Seth, Suman. 2013. Quantum Physics. In *The Oxford Handbook of the History of Physics*, ed. Jed Z. Buchwald and Robert Fox, 814–859. Oxford: Oxford University Press.
Slater, John C. 1975. *Solid State and Molecular Theory: A Scientific Biography*. New York: Wiley.
Smekal, Adolf. 1925. Zur Quantentheorie der Streuung und Dispersion. *Zeitschrift für Physik* 32: 241–244.
Sommerfeld, Arnold. 1916a. Zur Quantentheorie der Spektrallinien. *Annalen der Physik* 51: 1–94, 125–167.
Sommerfeld, Arnold. 1916b. Zur Quantentheorie des Zeemaneffektes der Wasserstofflinien, mit einem Anhang über den Stark-Effekt. *Physikalische Zeitschrift* 17: 491—-507.
Sommerfeld, Arnold. 1917. Zur Quantentheorie der Spektrallinien. Intensitätsfragen. *Sitzungsberichte der Mathematisch-physikalischen Klasse der Königlich-Bayerischen Akadademie der Wissenschaften zu München*, 83–109.
Sommerfeld, Arnold. 1919. *Atombau und Spektrallinien*, 1st ed. Braunschweig: Vieweg.
Sommerfeld, Arnold. 1920a. Allgemeine spektroskopische Gesetze, insbesondere ein magnetooptischer Zerlegungssatz. *Annalen der Physik* 63: 221–263.
Sommerfeld, Arnold. 1920b. Bemerkungen zur Feinstruktur der Röntgenspektren. *Zeitschrift für Physik* 1: 135–146.
Sommerfeld, Arnold. 1920c. Eine einheitliche Auffassung des Balmer'schen und Deslandres'schen Termes. *Arkiv för Mathematikm Astronomi och Fysik* 15: 1–5.
Sommerfeld, Arnold. 1920d. Zur Kritik der Bohrschen Theorie der Lichtemission. *Jahrbuch der Radioaktivität und Elektronik* 17: 417–429.
Sommerfeld, Arnold. 1921. *Atombau und Spektrallinien*, 2nd ed. Braunschweig: Vieweg.
Sommerfeld, Arnold. 1922a. *Atombau und Spektrallinien*, 3rd ed. Braunschweig: Vieweg.

Sommerfeld, Arnold. 1922b. Über die Deutung verwickelter Spektren (Mangan, Chrom usw.) nach der Methode der inneren Quantenzahl. *Annalen der Physik* 70: 32–62.

Sommerfeld, Arnold. 1923. *Atomic Structure and Spectral Lines*, 3rd ed. London: Methuen & Co.

Sommerfeld, Arnold. 1924a. *Atombau und Spektrallinien*, 4th ed. Braunschweig: Vieweg.

Sommerfeld, Arnold. 1924b. Grundlagen der Quantentheorie und des Bohrschen Atommodelles. *Die Naturwissenschaften* 12: 1047–1049.

Sommerfeld, Arnold. 1925. Über die Intensität der Spektrallinien. *Zeitschrift für technische Physik* 6: 2–11.

Sommerfeld, Arnold. 1968a. *Arnold Sommerfeld: Gesammelte Schriften*, 4 vols. Braunschweig: Vieweg.

Sommerfeld, Arnold. 1968b. Autobiographische Skizze. In *Arnold Sommerfeld: Gesammelte Schriften*, ed. Fritz Sauter, vol. 4, 673–679. Braunschweig: Vieweg.

Sommerfeld, Arnold. 2000. *Arnold Sommerfeld: Wissenschaftlicher Briefwechsel*, vol. 1: 1892–1918. Munich: Verlag für Geschichte der Naturwissenschaft und Technik.

Sommerfeld, Arnold. 2004. *Arnold Sommerfeld: Wissenschaftlicher Briefwechsel*, vol. 2: 1919–1951. Munich: Verlag für Geschichte der Naturwissenschaft und Technik.

Sommerfeld, Arnold, and Werner Heisenberg. 1922a. Die Intensität der Mehrfachlinien und ihrer Zeemankomponenten. *Zeitschrift für Physik* 11: 131–154.

Sommerfeld, Arnold, and Werner Heisenberg. 1922b. Eine Bemerkung über relativistische Röntgendubletts und Linienschärfe. *Zeitschrift für Physik* 10: 393–398.

Sommerfeld, Arnold, and Helmut Hönl. 1925. Über die Intensität der Multiplett-Linien. *Sitzungsberichte der preussischen Akademie der Wissenschaften. Sitzung der physikalisch-mathematischen Klasse vom 12. März* 9: 141–161.

Sommerfeld, Arnold, and Walter Kossel. 1919. Auswahlprinzip und Verschiebungssatz bei Serienspektren. *Verhandlungen der Deutschen Physikalischen Gesellschaft* 21: 240–259.

Sponer, Hertha. 1920. *Über ultrarote Absorption zweiatomiger Gase*. Ph.D. thesis, Georg-August-Universität Göttingen, Göttingen.

Sponer, Hertha. 1921. Über ultrarote Absorption zweiatomiger Gase. *Jahrbuch der Philosophischen Fakultaet der Georg August-Universitaet zu Goettingen. Teil II. Auszüge aus den Dissertationen der mathematische-naturwissenschaftlichen Abteilung* 34: 153–160.

Stark, Johannes. 1914. Beobachtungen über den Effekt des elektrischen Feldes auf Spektrallinien. I. Quereffekt. *Annalen der Physik* 43: 965–982.

Stark, Johannes. 1916. Beobachtungen über den zeitlichen Verlauf der Lichtemission in Spektralserien. *Annalen der Physik* 49: 731–768.

Stark, Johannes, and Heinrich Kirschbaum. 1914a. Beobachtungen über den Effekt des elektrischen Feldes auf Spektrallinien. III. Abhängigkeit von der Feldstärke. *Annalen der Physik* 43: 991–1016.

Stark, Johannes, and Heinrich Kirschbaum. 1914b. Beobachtungen über den Effekt des elektrischen Feldes auf Spektrallinien. IV. Linienarten, Verbreiterung. *Annalen der Physik* 43: 1017–1047.

Stark, Johannes, and Georg Wendt. 1914. Beobachtungen über den Effekt des elektrischen Feldes auf Spektrallinien. II. Längseffekt. *Annalen der Physik* 43: 983–990.

Steinle, Friedrich. 2009. Scientific Change and Empirical Concepts. *Centaurus* 51: 305–313.

Stern, Otto, and Max Volmer. 1919. Über die Abklingungszeit der Fluoreszenz. *Physikalische Zeitschrift* 20: 183–188.

Stuewer, Roger H. 1975. *The Compton Effect: Turning Point in Physics*. New York: Science History Publications.

Sugiura, Yoshkazu. 1927. Application of Schrödinger's Wave Functions to the Calculation of Transition Probabilities for the Principal Series of Sodium. *Philosophical Magazine Series 7* 4: 495–504.

Tamm, Igor. 1925. Versuch einer quantitativen Fassung des Korrespondenzprinzips und die Berechnung der Intensitäten der Spektrallinien. I. *Zeitschrift für Physik* 34: 59–80.

Tanona, Scott D. 2002. *From Correspondence to Complementarity: The Emergence of Bohr's Copenhagen Interpretation of Quantum Mechanics*. Ph.D. thesis, Indiana University.

Thomas, Willy. 1924. Näherungsweise Berechnung der Bahnen und Übergangswahrscheinlichkeiten des Serienelektrons im Natriumatom. *Zeitschrift für Physik* 24: 169–196.
Thomas, Willy. 1925. Über die Zahl der Dispersionselektronen, die einem stationären Zustand zugeordnet sind. Vorläufige Mitteilung. *Die Naturwissenschaften* 12: 627.
Tolman, Richard C. 1924. Duration of Molecules in Upper Quantum States. *Physical Review* 23: 693–709.
Tolman, Richard C. 1925. On the Estimation of Maximum Coefficients of Absorption. *Physical Review* 26: 431–432.
Tolman, Richard C., and Richard M. Badger. 1926. The New Kind of Test of the Correspondence Principle Based on the Prediction of the Absolute Intensities of Spectral Linies. *Physical Review* 27: 383–396.
van der Waerden, Bartel Leendert. 1968. Introduction Part I. Towards Quantum Mechanics. In *Sources of Quantum Mechanics*, ed. Bartel Leendert van der Waerden, 1–18. New York: Dover.
Van Vleck, John H. 1924a. The Absorption of Radiation by Multiply Periodic Orbits, and Its Relation to the Correspondence Principle and the Rayleigh-Jeans Law: Part I. Some Extensions of the Correspondence Principle. *Physical Review* 24: 330–346.
Van Vleck, John H. 1924b. The Absorption of Radiation by Multiply Periodic Orbits, and Its Relation to the Correspondence Principle and the Rayleigh-Jeans Law: Part II. Calculation of Absorption of Multiply Periodic Orbits. *Physical Review* 24: 347–365.
Van Vleck, John H. 1926. *Quantum Principles and Line Spectra*. Bulletin of the National Research Council 10, Part 4. Washington, DC: National Research Council.
Warwick, Andrew. 1989. *The Electrodynamics of Moving Bodies and the Principles of Relativity in British Physics 1894–1919*. Ph.D. thesis, Cambridge University, Cambridge.
Warwick, Andrew. 2003. *Masters of Theory: Cambridge and the Rise of Mathematical Physics*. Chicago: The University of Chicago Press.
Wehefritz, Valentin. 2002. *Verwehte Spuren. Prof. Dr. phil Fritz Reiche*. Universität im Exil, vol. 5. Dortmund: Universitätsbibliothek Dortmund.
Wentzel, Gregor. 1924a. Dispersion und Korrespondenzprinzip. *Zeitschrift für Physik* 29: 306–310.
Wentzel, Gregor. 1924b. Zur Quantentheorie des Röntgenbremsspektrums. *Zeitschrift für Physik* 27: 257–284.
Wheaton, Bruce R. 1983. *The Tiger and the Shark: Empirical Roots of Wave-Particle Dualism*. Cambridge: Cambridge University Press.
Wien, Wilhelm. 1919. Über Messungen der Leuchtdauer der Atome und der Dämpfung der Spektrallinien. I. *Annalen der Physik* 60: 597–639.
Wien, Wilhelm. 1921. Über Messungen der Leuchtdauer der Atome und der Dämpfung der Spektrallinien. II. *Annalen der Physik* 66: 229–236.
Zeeman, Pieter. 1897a. Doublets and Triplets in the Spectrum Produced by External Magnetic Forces. *Philosophical Magazine Series 5* 44: 55–60.
Zeeman, Pieter. 1897b. On the Influence of Magnetism on the Nature of the Light Emitted by a Substance. *Philosophical Magazine Series 5* 43(262): 226–239.

Index

M. Jähnert, *Practicing the Correspondence Principle in the Old Quantum Theory*, Archimedes 56, https://doi.org/10.1007/978-3-030-13300-9